Eco-Engineered Bioreactors

Advanced Natural Wastewater Treatment

Eco-Engineered Bioreactors

Advanced Natural Wastewater Treatment

James Higgins
Al Mattes
William Stiebel
Brent Wootton

CRC Press is an imprint of the
Taylor & Francis Group, an informa business

Cover photo: Copyright ©2008 James Cavanaugh

CRC Press
Taylor & Francis Group
6000 Broken Sound Parkway NW, Suite 300
Boca Raton, FL 33487-2742

First issued in paperback 2020

ISBN 13: 978-0-367-57257-0 (pbk)
ISBN 13: 978-1-138-05446-2 (hbk)

Library of Congress Cataloging-in-Publication Data

Names: Higgins, James, author. | Mattes, Allan, author. | Stiebel, William, author. | Wootton, Brent, author.
Title: Eco-engineered bioreactors : advanced natural wastewater treatment / James Higgins, Allan Mattes, William Stiebel, and Brent Wootton.
Description: Boca Raton : Taylor & Francis, CRC Press, 2018. | Includes bibliographical references.
Identifiers: LCCN 2017018296 | ISBN 9781138054462 (hardback : alk. paper)
Subjects: LCSH: Sewage--Purification--Biological treatment.
Classification: LCC TD755 .H5385 2018 | DDC 628.3--dc23
LC record available at https://lccn.loc.gov/2017018296

Visit the Taylor & Francis Web site at
http://www.taylorandfrancis.com

and the CRC Press Web site at
http://www.crcpress.com

Contents

Preface

The purpose of this book, *Eco-Engineered Bioreactors: Advanced Natural Wastewater Treatment* (*EEB Ecotechnology*), is to provide a detailed understanding of a highly innovative method of natural wastewater treatment using in-ground, fixed film bioreactors now known as Eco-Engineered Bioreactors (EEBs). *EEB Ecotechnology* traces the evolution of EEBs, beginning with the development of the earliest aerated surface flow (SSF), gravel-bed varieties* that are now known as *BREW Bioreactors* [BBRs]† all the way to today's wide slate of them.

EEB Ecotechnology reviews the results of R&D, bench-, pilot-, and demonstration-scale testing, full-scale projects and other work involving EEBs designed and managed by the Lead Author, James Higgins, first for the ecological landscape contracting company, *Soil Enrichment Systems Inc.* (SESI),‡ then later with the environmental consulting company, *Jacques Whitford Environment Limited* (Jacques Whitford), a legacy firm of a major engineering consulting company, *Stantec Consulting Inc.* (Stantec), and currently through his ecological engineering consulting firm, *Environmental Technologies Development Corporation* (ETDC).

The Lead Author was ably assisted by Co-Author, Al Mattes, of *Nature Works Remediation, Corp.* (Nature Works), the developer of the most advanced type of *Biochemical Reactor* (BCR),§ a kind of anaerobic EEB, and by the following other Co-Authors who provided editing and other assistance in preparing the drafts of this book, and contributed text for some of the sections in it: William Stiebel of Jacques Whitford & Stantec, and Brent Wootton of the Centre for Alternative Wastewater Treatment (CAWT).

As was mentioned, there are now many kinds of EEBs and *EEB Ecotechnology* presents information on all of their types; on their morphologies; on the

* As described in Chapter 1, EEBs evolved from Constructed Treatment Wetlands (CWs). Background relevant to EEBs on both Natural Wetlands and CWs is provided in Appendices A and B, respectively, while information on other technologies and processes germane to EEB development (e.g., Aerated Lagoons and phytoremediation) is found in Appendix C. Since EEBs for airports are an important market, Appendix D provides the needed background on wastewaters (WWs) generated at them. Appendix E provides background on the problems and potential from treating WWs contaminated with mercury. The Glossary presents definitions for many of the technical terms used in this book.

† BBRs are called "engineered wetlands (EWs)" in older references and EW is still sometimes used to describe them at airports, but BBR is preferred as most cannot any longer be described as "wetlands" of any sort and their gravel surfaces may be vegetated with terrestrial plants or not vegetated at all.

‡ Further information on the companies, organizations, and individuals that contributed to the development of the EEB ecotechnology is found in Appendix F (Acknowledgments), References are found after the Glossary.

§ Biochemical reactors remove metals and metalloids from WWs largely (but not exclusively) using sulfate reduction (see Chapters 4 and 5).

testing methods used to develop and define them; on their current (and still evolving and improving) design and engineering; on how they operate so efficiently and on their microbiology. It updates and expands on preliminary information already published on-line by the Lead Author in a preview, *Engineered Bioreactor Systems*, which forms a volume of the *Encyclopedia of Environmental Management* (Higgins 2014). This is now greatly expanded and supplemented by new material from the Lead Author and the other Co-Authors as well as some of their associates, and from articles published in trade and peer-reviewed journals, and presented at various scientific and industry conferences, symposia, workshops, and courses. *EEB Ecotechnology* is written from the perspective of ecological engineers designing EEBs.

EEB Ecotechnology will be valuable to engineers and scientists seeking information on a sustainable and effective method to treat wastewaters of all kinds. It will be of particular interest to airport operating authorities, regulators, municipal and industrial wastewater treatment engineers, environmental officers for mining companies and waste management firms, operators of landfills, and others wishing to consider having WWT systems involving "passive" and sustainable long-term treatment for wastewaters at their facilities. *EEB Ecotechnology* also will be used as a textbook for university and college courses.

Some people deserve special thanks as they were the "Champions" who realized the potential of the evolving EEB Ecotechnology early on and encouraged and promoted it, often in the face of skepticism in their own organizations. These include Mike MacLean of the Edmonton Regional Airports Authority (ERAA)*; Andre Bachand of the Town of Alexandria, Township of North Glengarry ON (Alexandria); Todd Pepper of the Essex-Windsor Solid Waste Authority (EWSWA); Scott Wallace and Mark Liner of North American Wetlands Engineering (NAWE); Kim Minkel of the Niagara Frontier Transportation Authority (NFTA); Bill Duncan of Teck Resources Limited (Teck Cominco); Rubin Wallin of Rosebel Gold Mines NV (RGM); Professor Emeritus Les Evans of the School of Environmental Sciences (SES) of the University of Guelph (UoG); Dr. Doug Gould of Natural Science Canada's Mining and Mineral Sciences Laboratory (CANMET-MMSL); Martin Hildebrand of Nexom (formerly Nelson Environmental Limited) and one of the Co-Authors, Bill Stiebel. Stantec and its major legacy firm, Jacques Whitford, also deserve special mention as it was with their participation and encouragement that the ecotechnology was developed.

EEB Ecotechnology is dedicated to the two pioneers whose work on wetlands engineering opened the way for the development of EEBs: Dr. Robert Kadlec of Wetlands Management Systems and the late Sherwood (Woody) Reed of Environmental Engineering Consultants.

* The companies and organizations listed in this paragraph were those at which the cited individuals were involved at the time of their major contributions. More information is provided in Appendix F (Acknowledgments).

Authors

James Higgins, PhD, PEng is the Lead Author of this book and prepared most of its text. Until his retirement in 2013, he was a senior engineer and executive consultant at Stantec and earlier at Jacques Whitford. He has held positions as president, CEO, chairman and director in several consulting, environmental and ecological engineering companies. Higgins has been an adjunct professor and lecturer at the Universities of Guelph, Ottawa and Toronto, and currently works closely with CAWT where there are EEB treatability test facilities. He is the developer of the Eco-Engineered Bioreactor Ecotechnology, and initiates, organizes, and participates in EEB pilot- and demonstration-scale test projects and R&D at test facilities such as those at CAWT, and at client sites.

Dr. Higgins is a professional engineer with background and expertise in areas such as conventional and natural wastewater treatment, constructed wetlands, engineered bioreactors, site remediation, phytoremediation, risk assessment, and the clean-up of soils and waters contaminated with metals, metalloids and hydrocarbons such as aliphatics and aromatic organics and chlorinated materials.

Al Mattes, BSc, BFA, PhD is an ETDC associate and a principal of Nature Works. His activities have been mainly focused on the management of metals in contaminated soils and water. Working with other Nature Works principals, Dr. Mattes designed a BCR-based demonstration-scale EEB system (the Teck Demo) that efficiently removes high concentrations of dissolved metals and metalloids (collectively: metal[loid]s) from landfill leachates at the site of Teck Cominco's major lead zinc smelter in Trail BC, Canada. The Teck Demo, which involved two BCR in-ground basins (cells) and three downstream polishing CW cells, was designed to operate year-round and to treat up to 25 m^3/day of contaminated leachates from old landfills. The leachate treated in the Teck Demo had zinc concentrations of approximately 500 mg/L (peaks to 3800 mg/L), cadmium concentrations of approximately 25 mg/L (peaks to 100 ppm) and arsenic concentrations of >100 mg/L (peaks to 3300 mg/L). The Teck Demo operated successfully for a number of years removing >97%

of the zinc, 99.8% of the Cd, and 99.7% of the arsenic. It was able to withstand high shock loads and has demonstrated year-round operational capability. Dr. Mattes earned a PhD in geomicrobiology at the UoG, with special emphasis on the chemistry and biochemistry of arsenic. Mattes' activities have been mainly focused on the management of metals in contaminated soils and water.

William H. Stiebel, MSc, PGeo., FGC, Senior Principal, Natural Resources & Environmental Management for Stantec, holds a BSc (Hon., geology) from Queen's University and an MSc (earth sciences, hydrogeology) from the University of Waterloo (UoW). Stiebel has over 25 years of Canadian and overseas experience in environmental consulting. His expertise includes development of strategic environmental management options, groundwater resource management for irrigation, industrial and municipal water supplies, contaminant hydrogeology, hazardous/toxic waste management, industrial site decommissioning and clean-up, environmental site assessments, compliance auditing, environmental impact assessments, and the development of environmental information management systems. In addition, he has specialized expertise in the areas of aquifer analysis and contaminant transport evaluation, site clean-up criteria development, drainage, dewatering, and groundwater resource management.

Brent Wootton, MES, PhD has a Master's degree in the ecology and management of wetlands from the UoW and a PhD in aquatic ecology from Trent University. His research interests include the design and function of treatment wetlands and tundra treatment wetlands. He has been a member of several government boards and commissions, and has worked extensively with government, industry and community groups on environmental projects. Wootton led CAWT at Fleming College from 2005 to 2016. Under his leadership, CAWT quickly grew to become an internationally recognized research institute committed to excellence in water and wastewater research and development. Dr. Wootton now serves as the associate vice-president of Business Development, Applied Research, Government and Partner Relations at Fleming College; with institutional responsibility for all college-wide applied research, including oversight of the CAWT.

1

Introduction

1.1 Overview

1.1.1 Introducing Eco-Engineered Bioreactors

Eco-engineered bioreactors (EEBs) are kinds of attached growth, in-ground, natural wastewater treatment (WWT) methods that evolved from subsurface flow (SSF) Constructed Treatment Wetlands (CWs), but now differ greatly from them. With EEBs, morphology, substrates, operating methods, microbiology, flows and/or other process conditions can be manipulated and controlled to allow superior performance in comparison to that of CWs and many kinds of active (mechanical) wastewater treatment plants (WWTPs).

Treatment in EEBs involves passing wastewaters (WWs) through large excavated basins (cells*) containing permeable substrate media and, in these cells, contacting the WWs with microbes in fixed films on the media (Higgins 1997; Wallace 2001a,b). EEBs provide superior removals of many kinds of pollutants and other contaminants-of-concern (CoCs) over wide temperature ranges and flow rates. As is outlined herein, EEBs have been demonstrated and proven to provide superior treatment at locations from the jungles of the Amazon to the high Arctic.

Also, unlike the situation in the biological parts of active WWTPs (e.g., activated sludge units) where an "aliquot" of water being treated may have a "nominal residence time" as short as hours, residence times in EEBs are usually several days,† allowing higher levels of CoC transformations.

As was mentioned, the first kinds of EEBs were aerated gravel bed ones referred to as engineered wetlands (EWs). Now EEBs come in both aerobic and anaerobic types and there are several varieties of each.

Aerobic EEBs are usually configured as beds of aerated substrates. The most common kinds are BREW bioreactors (BBRs) and these involve the SSF flow of WW through/in granular media such as gravel into which

---

* In addition to describing the macroscopic basins of EEBs and associated facilities, the word "cell" may be used in this book to describe an individual microbe. The context of its use will make clear which meaning is involved.

† See Equation 8.2 in Chapter 8 for a definition of nominal residence time.

air from a blower is injected below, resulting in high removal levels for oxidizable contaminants.*

Figure 1.1 (source: Naturally Wallace) illustrates a gravel-bed horizontal sub-surface flow (HSSF) EW cell of the kind used early in the development of the EEB ecotechnology (late 1990s to early 2000s). As may be seen, wastewater flow through it was more or less horizontal and such cells were vegetated with emergent wetland plants.†

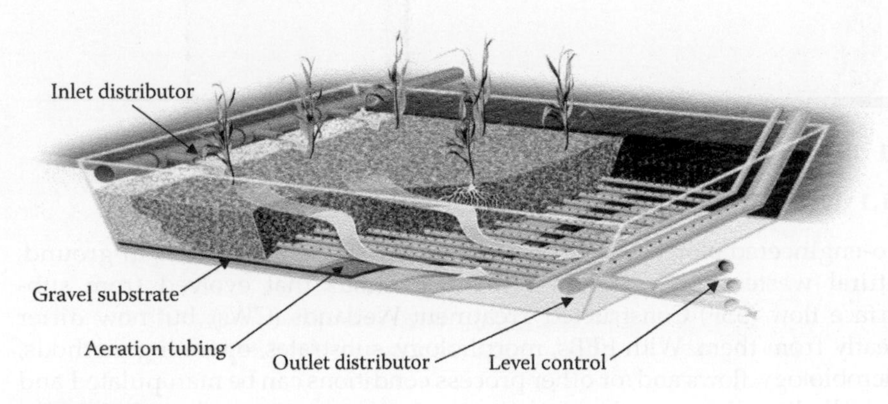

FIGURE 1.1
An early engineered wetland (BBR) cell.

While a variety of aggregate materials (e.g., crushed rock and slate) can be used as substrate materials, alone or in combination with sand, the most common substrate material in BBRs is gravel.‡ Generally, the materials used as these substrates are available from suppliers in ready-to-use (e.g., screened and washed) forms, and can be put in place immediately§ (see Section 7.3 in Chapter 7 for further information on aerobic EEB substrates).

BBR morphology has evolved considerably since the early 2000s, and currently these bioreactors have much thicker substrates (1–2 m+ vs the 0.7–1.0 m of the early varieties) and no longer need to be vegetated with wetland plants at all (although some still are), instead often being unplanted or vegetated with terrestrial plants.

Figure 1.2 shows an operating BBR cell at BNIA.

* Physical–chemical sorption and other reactions can (and usually do) also occur in aerobic EEBs.

† The treatability tests described in Sections 6.4 and 6.5 in Chapter 6 involved EW cells of this sort.

‡ There are a variety of kinds of gravel used in EEBs including quartz gravel; relatively pure calcite gravel (calcium carbonate, often called agricultural limestone); dolomite gravel (calcium magnesium carbonate); and various kinds of crushed rock and concrete referred to by the name.

§ See Chapter 7 for more details on EEB substrates, their logistics and their placement.

FIGURE 1.2
A BBR cell in operation at BNIA.

This particular cell (one of four similar ones) is vegetated with terrestrial road mix grasses, and these grasses have no involvement in the treatment process. Indeed, the EEBs at BNIA do not resemble conventional ideas of what "wetlands" should look like. They have no open water and are vastly smaller in surface area (by an order of magnitude or more) than what CWs treating the same amounts of WW would be. Accordingly, although such EEBs at some airports are still sometimes referred to as EWs, it is more appropriate to refer to them as BBRs to avoid mistaken impressions.

Another variety of aerated EEB is the submerged attached growth reactor (SAGR bioreactor, see Section 3.9 in Chapter 3), a kind that is used for nitrification and operated in series with lagoons for the upgrade of municipal lagoon WWT systems for northern communities.

There are also now several kinds of anaerobic EEBs, and in this book they are referred to as anaerobic bioreactors (ABRs, see Chapter 4). Besides two kinds of biochemical reactors ([BCRs] passive and semi-passive treatment ones, see Chapter 5), other kinds of EEB ABRs include denitrification bioreactors (DNBRs, which, as the name suggests, remove nitrates), and successive alkalinity producing systems (SAPS bioreactors, which are used to safely neutralize iron-rich, acidic WWs such as acid rock drainage [ARD]). There are also a variety of anaerobic specialty BCRs including several kinds used to target specific CoCs (e.g., arsenic and selenium), and electro-biochemical reactors (EBRs) that make use of electrical potentials across bioreactor beds to enhance performance.

EEB ABRs are usually un-vegetated and buried in the ground, or located at the bottom of ponds to maintain anaerobic conditions in them.*

* Again, describing them as "wetlands" is inappropriate, although where the latter (location below ponds) is the situation, the peripheries and surfaces of these impoundments may be colonized by opportunistic wetland plants and algae.

As was mentioned, in contrast to the active operating mode of conventional WWT facilities and the semi-passive mode of aerated EEBs such as BBRs, there are kinds of BCRs that can operate in either passive or semi-passive modes (see Section 5.3 of Chapter 5).

The latter, semi-passive treatment BCRs, have relatively simple aggregate substrates similar to those in BBRs and do not have aeration tubing in them as do BBRs. In these cases, the carbonaceous materials (active media) that the anaerobic microorganisms in the biofilms require for their metabolisms do not form part of the substrates. With them, the active media are injected into the BCR as liquids (e.g., methanol and molasses) along with the wastewater influent.

In contrast, passive treatment BCRs contain substrates of which a large part is carbonaceous materials (solid active media such as biosolids), whose biodegradation by various kinds of microbes provides the metabolites needed for CoC transformations. The microorganisms in passive treatment BCRs include anaerobic heterotrophs such as cellulose-degrading bacteria (CDB), acid-producing bacteria (APB) and fermenters, and these provide breakdown products (e.g., ethanol, acetates, lactates, and H_2) that other kinds of bacteria and archaea use in their metabolisms (MEND 1999).

Despite EEBs being identified with one or other kinds of oxic conditions (e.g., aerated gravel bed ones will be bulk aerobic and BCRs bulk anaerobic), there will always be microenvironments in their biofilms where the opposite conditions are found. For example, despite the turbulent and highly mixed aerated conditions in BBRs, there will still be places in their biofilms (see below) where anaerobic conditions (and anaerobic bacteria) are found. The reverse will be true for anaerobic EEBs such as BCRs where aerobic microenvironments will be found.

Initially, all EEBs (then referred to as EWs) were wetland plant-vegetated, HSSF* aerated gravel-bed BBRs but as the ecotechnology developed, the types available were expanded to include vertical sub-surface flow (VSSF†) varieties of BBR, as well as several kinds of anaerobic EEBs such as BCRs, including ones with substrates dominated by solid organic materials. As a result, the generic name has been changed from EWs to EEBs.

EEBs are designed (see Chapter 8) using standard chemical engineering Reaction Kinetic methods based on the results of treatability tests that allow the definition of the parameters needed for scale-up.

Microorganisms‡ are largely present in EEBs in bacteria-generated biofilms attached to substrates and other surfaces under the water level, and

* With HSSF EEBs, the WW being treated flows parallel to the cell surfaces within the substrates.

† With VSSF EEBs, the WW being treated flows vertically through the substrates.

‡ The microbes in EEB biofilms are mostly bacteria but they may include archaea as well as opportunistic, symbiotic and entrapped fungi, algae and other micro-eukaryotes. Biofilm microorganisms are referred to herein as bacteria unless it is necessary to distinguish other kinds.

the bacterial metabolic functions in them dominate CoC transformations.* Gradients of oxygen, nutrients and the products of bacterial activity determine the structure of the biofilms (see Section 1.11).

In addition to the physical and chemical parameters determined during treatability tests to define the scale-up factors needed to design full-scale EEBs, it has been found that the collection of information on genetic material (genomes) from the mixed communities of microbes in the biofilms can be used to improve their design. Early on, the genomic characterizations of the bacteria and archaea in biofilms focused on amplifying specific regions within segments of DNA critical for the production of proteins by their cells, and using literature data to identify the microbes present (taxonomic analyses). Now these kinds of microbial analyses have been supplemented with "shotgun" and whole genome sequencing of all of the DNA in a sample (metagenomic analyses).

These advances in genomics (see Section 1.11) allow rapid and detailed analyses of the biofilm microorganisms and their metabolisms during WWT in EEBs. Once the main consortia of microbes have been identified for a specific situation, testing can be carried out to determine which of them have the greatest impact on treatment in an EEB. When this information is known, it can be exploited to improve the entire treatment process (biocontrol).

Early BBRs (then called EWs) were vegetated with wetland plants and many of the initial treatability tests (see Chapter 6) and EEB Systems involved cells vegetated with Reeds (*Phragmites* spp), Cattails (*Typha* spp) and other hydrophilic wetland plants. However, it soon became clear that other than aesthetic and possibly insulation values, the plants contributed little to EEB operations, and currently, unless a client wishes it or some project aspect necessitates it, the surfaces of BBR cells are either vegetated with terrestrial plants or left un-vegetated. ABR cells are usually buried or located at the bottom of impoundments and wetland plants are not a factor with them.

Most kinds of EEBs are now fully proven, and active development continues aimed at making them even more efficient and versatile.

Table 1.1 summarizes the main kinds of EEBs.

* Such biotic and abiotic transformations in EEBs include redox reactions (i.e., oxidation where CoCs change by losing electrons, and reduction where they gain electrons); degradation by mineralization, biodegradation by hydrolysis; biosynthesis (microbial biomass growth); isomerization, conjugation, precipitation (physical, chemical and biological removals by flocculation, coagulation, and settling); and sorption (physical and biological adsorption, absorption and mechanical entrapment in biofilms). In this book, unless there is a need to be specific, microbiotic transformations will be referred to as biodegradation.

TABLE 1.1
Types of EEBs

	Characterizing Bacteria	Main Purpose	Morphology
Aerobic EEBs			
BBRs (BREW bioreactors)	Aerobic heterotrophs including nitrifiers	The removal of CoCs susceptible to aerobic biodegradation	Open gravel substrate beds, un-vegetated or vegetated surfaces (terrestrial or wetland plants)
SAGR bioreactors (submerged aerated growth reactors)	Nitrifying bacteria	Nitrification of the effluents from upstream Aerated Lagoons forming parts of lagoon/SAGR bioreactor EEB Systems. Provide functional upgrades of facultative lagoon WWT systems used to treat municipal wastewater in small- to medium-sized northern communities	Non-vegetated gravel substrate beds
Anaerobic EEBs			
DNBRs (denitrifying bioreactors)	Denitrifying bacteria	The removal of nitrates from WWs and the ultimate conversion of them into nitrogen gas	Buried or water-covered VSSF BCR-like bioreactors containing solid (often wood chip-based) substrates
SAPS bioreactors (successive alkalinity producing systems)	Iron-reducing bacteria	The conversion of ferric iron in ARDs to ferrous iron in a manner which limits ferric iron hydrolysis and allows the ARD to be neutralized by an underlying limestone layer	Water-covered downflow VSSF bioreactors containing upper, organic-based substrates underlain by limestone
Passive treatment BCRs	Sulfate-reducing bacteria	The removal of many dissolved metal(loid)s, largely as insoluble sulfides	Buried or water-covered upflow or downflow VSSF bioreactors containing organic-based, amended substrates
Semi-passive treatment BCRs	Sulfate-reducing bacteria	The removal of many dissolved metal(loid)s, largely as insoluble sulfides	Buried or water-covered downflow VSSF bioreactors containing gravel substrates

(*Continued*)

TABLE 1.1 (*Continued*)
Types of EEBs

	Characterizing Bacteria	Main Purpose	Morphology
EBRs (electro-biochemical reactors)	DNB, SRB and SeRB	The removal of many dissolved metal(loid)s, largely as insoluble sulfides	Similar to BCRs but with layers of conducting material upstream and downstream of the substrate with a small battery connected between them
Specialty BCRs	SRB, SeRB, AsRB, CrRB, etc.	BCRs designed to remove specific CoCs such as As, Cr and Se	

1.1.2 EEB Systems

EEB cells usually make up only part of EEB Systems that consist of one or more parallel sets of basins (cells) connected in series (trains). In such trains, the secondary (2°) treatment EEB cell(s)* may be preceded by† primary (1°) treatment cells/components (e.g., rock filters, cascades, vaults, ponds, grit removal chambers, oil and grease removal equipment, oil–water separators [OWSs], lagoons, CW cells, and limestone drains) that may be physical, chemical and/or biological in nature. EEB cells also may be followed by tertiary (3°) treatment cells/components that may also be physical, chemical and/or biological in nature (e.g., carbon filters, sand filters, CW cells, phosphorus removal equipment, and ponds), and sometimes final disinfection methods and equipment (e.g., chlorination and UV).‡

Besides their 1°, 2° and 3° treatment process cells per se, EEB Systems usually include a variety of ancillaries such as diversion chambers/cells, control structures, ditching, piping, pumps, blowers, instrumentation and controls, heaters, cleanout facilities, electrical equipment (e.g., motor controllers,

* The secondary treatment phase of a WWT facility is usually (but not always) biological in nature.

† Such 1° treatment cells/components are also usually up-gradient of the 2° treatment cells, and will be referred to herein as being "upstream" of them. Similarly, 3° treatment cells/components (usually down-gradient of the 2° treatment cells) will be referred to as being "downstream" of the 2° treatment cells.

‡ For many wastewaters that are microbially contaminated (e.g., sewage), treatment in an aerated EEB cell (e.g., a BBR) will itself disinfect the effluent, although regulatory authorities are often reluctant to approve sewage treatment projects without downstream disinfection. In any case, treatment in a BBR will result in an effluent that is clear enough so that UV disinfection is relatively easy, and this disinfection method is favored with them over other disinfection processes such as chlorination.

transformers and power lines), equipment covers and blower- and pump houses.

In addition to 2° EEB cells, EEB Systems sometimes can include other secondary treatment components such as Aerated Lagoons or membrane systems (e.g., reverse osmosis [RO]) in advance of the EEB cells. However it should be clear that, unlike the case with well-designed aerated EEBs (which do not produce excess sludge), Aerated Lagoons do, and membrane systems produce concentrate streams, so including them in an EEB System will involve operators in by-product management. This may or may not be a problem depending on the WWs involved; if they are high in biochemical oxygen demand (BOD) (e.g., as are many leachates and airport WWs*), considerable amounts of sludge may be generated requiring management.

EEB Systems often do not look or perform at all like one would expect a natural WWT system should and, as was mentioned, should not be categorized with CWs or other passive natural treatment systems. An example of a modern EEB System is the one at BNIA (see Section 3.6 in Chapter 3, also Higgins et al. 2010a) at which the four BBR cells appear to be grassy fields near the airport's main runway (see Figure 1.2 for a picture of one of the BBR cells at BNIA). Indeed, increasingly, BBR cells are either not vegetated at all (Higgins et al. 2007a,c,d), or are vegetated with ordinary terrestrial vegetation (as in Figure 1.2), further emphasizing how far the EEB ecotechnology has evolved from that of SSF CWs.

Figure 1.3 illustrates the relationship between CWs, EEB Systems and mechanical WWTPs.

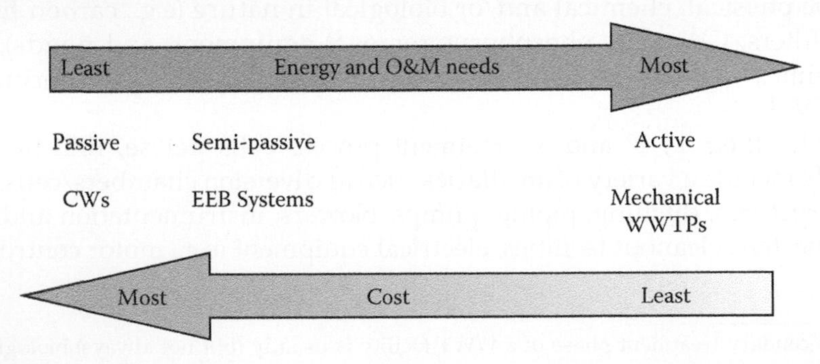

FIGURE 1.3
With EEB Systems there is a trade-off between land and mechanical complexity.

* Aerated and un-Aerated Lagoons are also usually inappropriate at airports as they involve open water that may attract waterfowl and thus represent bird strike hazards, although there are sometimes ways to deal with this as is seen in Figure C.1 in Appendix C where the aerated pond in front of BBR cells at LHA is covered with floating rafts of vegetation.

It is noted that, while EEBs are very much more sophisticated than CWs, they still share the latter's low energy and O&M requirements while allowing almost un-attended operation much of the time. As is addressed in more detail later, EEB Systems can be designed to operate in the tropics or in northern areas (e.g., the Arctic). EEB Systems are licensed in the same manner as are active WWTPs in a number of jurisdictions in Canada, the United States and elsewhere. EEB Systems allow the treatment of acidic, neutral or alkaline wastewater streams in warm weather and under the most extreme winter conditions.*

An important aspect of EEB Systems is their modularity. They often involve two or more trains of cells, and while these trains may all be constructed at once, there is an opportunity with this kind of WWT to phase their construction over a period or periods.

By "mixing and matching" the cells and other components of EEB Systems, these advanced treatment methods can be designed to allow the clean-up of WWs containing even the most recalcitrant (otherwise difficult-to-treat) CoCs in one or more trains that can operate almost unattended at high efficiencies whatever the ambient air temperature, and in water at temperatures as low as 0.1°C. EEB Systems allow the removal from WWs of CoCs efficiently and effectively, reaching levels of 90%–99%+ in most cases.

The major markets for EEB Systems are at airports (where they are used to treat residual, glycol-contaminated streams resulting from the de-icing of aircraft in cold weather); for municipalities (where they treat sewage and other domestic wastewater streams); and for the mining industry (where they treat a variety of mining impacted waters [MIWs], such as ARD). Other important markets include those for the treatment of landfill leachates, process waters from petroleum and petrochemical plants, and waters from agricultural operations.†

EEB Systems allow the treatment throughout summer and winter in economical systems that can consistently meet stringent wastewater discharge criteria over extended periods.

EEB Systems do not require continuous operator presence or attention, and only need be periodically inspected and monitored. In many cases, this ecotechnology is the best available technology economically available (BATEA) for treating WWs, stormwaters and groundwaters contaminated with otherwise recalcitrant contaminants.

The cells and other components of EEB Systems can be "mixed and matched" to allow the treatment of WWs containing any combination of CoCs. EEB Systems are highly modular, allowing easy expansion (just add a

* The properties of EEBs and EEB Systems are summarized in Section 9.1 of Chapter 9.

† Section 9.1 in Chapter 9 summarizes the EEB ecotechnology while Section 9.2 describes the current "State-of-the-art for BBRs and BCRs." Section 9.3 suggests current, planned, proposed and envisaged areas for EEB-related R&D and treatability testing that will enhance the ecotechnology.

train or cells), and should more stringent criteria ever be required in future, these systems can be relatively easily upgraded to meet them (Higgins 2014).

1.1.3 Seminal EEB Systems

In addition to detailing the R&D and many of the treatability tests* that were carried out to develop the EEB ecotechnology, this book presents full descriptions of two seminal EEB Systems that defined its scope.

One is the currently operating full-scale BBR-based EEB System at BNIA in upstate NY, USA where four aerated, gravel-bed BBRs, each as big as a football field, successfully treat both the highly concentrated Spent Glycols that collect below aircraft being de-iced and glycol-contaminated stormwater run-off (GCSW) from around airport property. The BNIA EEB System demonstrates that very large, aerobic VSSF EEAs (BBRs) can be used on-site to treat large volume, cold, intermittent flows of airport wastewater throughout northern winters economically and efficiently. Like all new technologies, the BNIA BBRs had their "teething" problems but these have all been successfully resolved and planning is underway to expand this EEB System with a few more BBR cells, ones which will incorporate some of the advances outlined herein. The BBR ecotechnology can be described as fully proven and demonstrated.

In 2010, for the design of the EEB System project at BNIA, the Lead Author† (who was the project director for its treatability testing, design, engineering, construction supervision and start-up), on behalf of the NFTA, Stantec, Stantec NAWE, Urban and other members of the project's design team, was awarded the Diamond Award in the Environmental Category by the NY branch of the American Council of Engineering Companies (ACEC) and a very prestigious Honor Award by the US National ACEC. The design also won the 2010 Project of the Year Award (Environmental Category) from the American Public Water Works Association of Western NY, the APWA NY Environmental Project of the Year Award.

The second seminal EEB System described herein was a BCR-based demonstration-scale project (the Teck Demo) that operated for many years, summer and winter, near the site of the lead–zinc metal refinery and smelter of Teck Cominco in Trail, BC, Canada, successfully removing lead, cadmium, zinc and arsenic from landfill leachates from hundreds and even thousands of mg/L to trace levels. The Teck Demo showed that passive treatment BCRs need to be coupled with downstream tertiary treatment components to produce EEB Systems that can treat even the most recalcitrant MIWs for maximum efficiency and effectiveness. Never in the past had BCRs been tested

* A treatability test is a set of experimental activities undertaken at a reduced scale (at an indoor or outdoor pilot facility, or at a larger outdoor demonstration facility) whose purpose is to define kinetic and other scale-up parameters needed to allow the design of a full-scale EEB System.

† One of the other Co-Authors (Stiebel) was Jacques Whitford's (and later Stantec's) Project Executive for the project.

to the high level that the ones at Teck Cominco were subjected to, and the advances in "know-how" and "know-what" found during the Teck Demo project will set the standard for designing, building and operating BCRs and their associated EEB System components in future. In addition to the larger Teck Demo at Trail, a smaller but still demonstration-scale test unit was also constructed and operated there to assess the specifics of arsenic removal in BCRs. As with the BBR ecotechnology, the BCR ecotechnology can now be said to be fully developed and proven.

1.2 Bioreactors

Bioreactors are biological WWT* systems in which contaminant transformations in waters passing through them are dominated by microbial (biotic) processes.† Bioreactors may have in them globules/aggregates of suspended growth microbial biomass (e.g., floc, microbial accumulations circulating freely in wastewater under turbulent conditions due to aeration and/or mechanical mixing), or fixed films where the wastewater being treated flows over attached biomass. Varieties of the activated sludge process are the most common suspended growth, active WWT (conventional fossil fuel-based) methods and the large mechanical WWT facilities (WWTPs) treating municipal sewage in many cities rely on these processes.

Fixed-film biological WWT involves the use of attached microbial communities in biofilms to biodegrade contaminants (WEF 2010). Consortia in most bioreactors treating WWs may involve several different kinds of microbes but are dominated by bacteria. These carry out a variety of processes and functions. Gradients of oxygen, nutrients and the products of biological activity determine the structures of biofilms (Ivanow 2016).

Besides EEBs such as BBRs and BCRs, there are various other kinds of fixed-film "biochemical reactors" used for WWT, including the following (Qureshi et al. 2005; WEF 2010):

- Trickling filters
- Rotating biological contactors (RBCs)
- Moving-bed biofilm reactors
- Integrated fixed-film activated sludge units
- Biological filters (anammox and sludge blanket)

---

* Unless there is reason to define a particular stream, herein the term WW will generically mean any domestic wastewater, septage, leachate, groundwater, seepage, percolate, process water, pore water, mine water, cooling water, blowdown, backwash, run-off, aqueous sludge or any other wastewater that it is desired to treat in an EEB System.

† Abiotic processes also occur in EEBs but are ancillary.

These technologies represent conventional mechanical (active) WWT methods. The alternative to such active methods are passive WWT methods whose energy source is ultimately solar (although gravity can also be involved). For example, constructed wetlands and many lagoon-based municipal WWT systems are largely passive in nature.

In truly passive WWT systems, once built, no further inputs should be required. Indeed, some (Gusek & Wildeman 2002) define as a "passive" system only "...any wastewater treatment system that requires no power after construction and does not involve a mechanical WWTP." However, there are many kinds of natural WWT systems which are not active but cannot be accurately defined as fully passive either, as inputs are required continuously or periodically of: chemicals (e.g., adding liquid carbonaceous materials to some types of BCRs); energy (e.g., electricity for air blowers) and/or materials (e.g., additional limestone that needs to be added periodically to limestone drains). These hybrid kinds of systems can be defined as semi-passive WWT systems. These kinds of systems share many of the aspects of fully passive WWT treatment systems but also have some "active" aspects (e.g., air being added continuously to a BBR).

EEB CoC transformations often are classified and "characterized" (made distinctive) by one or a few types of bacteria that are said to carry out specific, targeted reactions at specific ranges of pH and oxidation-reduction potential (ORP). The characterizing bacteria may only form a small part of the microbial consortia present in an EEB but nevertheless the distinctive process or processes that they carry out reflect the purpose of the designated bioreactor. This characterizing group (and a number of different kinds of microbes may be involved in a group) will have optimal ranges of pH and redox (ORP or Eh) at which their members thrive (Bass Becking et al. 1960; Metcalfe & Eddy 2004).

Table 1.2 illustrates some of the approximate ranges of conditions germane to BCRs and other anaerobic EEBs.

TABLE 1.2

Optimal ORP Ranges for Anaerobic EEBs

Flow through an EEB System	Characterizing Microbe(s)	Final Electron Acceptor	Approximate ORP Range (mV)	Optimal pH Range
↓	Aerobic respirators	Oxygen	+50 to >+200	
	DNB	Nitrate	−50 to >+50	2–8
	DNB	Nitrite	−50 to −200	
	SeRB	Selenate	−50 to −200	
	SRB	Sulfate	−100 to −250	4–10
	SeRB	Selenite	<−150	
	APB	Carbon	−75 to −300	
	Methanogens	Carbon	<−300	4–9

In Table 1.2 (which is a work in progress), DNB are denitrifying bacteria and SeRB are selenium-reducing bacteria.

From Table 1.2, it may be seen that denitrification is a more energetically favored form of microbial respiration than sulfate reduction, and where a wastewater being treated contains comparable amounts of both sulfates and nitrates, sulfate reduction may only occur after all of the nitrates have been reduced.*

1.3 Bioreactor Engineering

In the mid-to-late 1990s, an initiative (the BREW Project—see Section 1.4) led to the evolution of SSF CW technology into a new, modern kind of natural system WWT ecotechnology based on innovative, advanced forms of constructed wetlands which, as was mentioned, were initially called EWs (Higgins 1997; Higgins et al. 1999a; Higgins 2003) and later engineered bioreactors (EBs) to emphasize the difference between them and CWs. They are now referred to as EEBs to show that they represent sustainable, ecological engineering-based ecotechnology.

Natural WWT processes may be "ecologically engineered" if they involve one or more of the following aspects:

Design Modifications

- Aeration
- Engineered substrates
- System ancillaries

Process Additions/Modifications

- Energy
- Chemicals
- Dilution

* Too much should not be made of this. In cases where the sulfate concentration is much larger than the nitrate concentration, as was the case for the Teck Demo where the influent sulfate concentration into the first BCR was in thousands of mg/L (see Section 5.6 in Chapter 5) and the nitrate concentration was only 40–50 mg/L, sulfate reduction overwhelmed nitrate reduction. Although passive treatment BCRs can be designed to allow comparable amounts of both nitrate reduction and sulfate reduction to occur at different locations in the same BCR cell, such is not recommended for full-scale facilities. Rather, a cell or cells dedicated to denitrification should be followed in an EEB System treatment train by separate ones dedicated to sulfate reduction, etc.

Vegetation Modifications

- Stress-resistant species
- Harvesting for nutrients removal
- Phytoremediating plants

Advanced Systems Operations

- Stream control
- Recycle
- Computerized operation
- Biocontrol

As was mentioned, adding air to a gravel bed such as a SSF CW is one way to turn it into an EEB (Higgins 1997). Many of the transformations of pollutants in a natural WWT system treatment are dependent on microbially mediated aerobic transformations. In SSF CWs, most of the needed oxygen for such reactions comes from that dissolved in the influent (feedstock) or that supplied by wetland plants such as Cattails that "pump" air to aerobic microbes in their root zones. This limits the degree of oxidation reactions such as nitrification that can occur. For example, ammonia removals of 30%–70% are typical with CWs. Oxygen availability also limits the thickness of SSF CW substrates to the maximum that plant roots can penetrate, (0.4–0.8 m) and hence also reduces the potential for passive heat retention since thicker substrates allow this. One way to overcome this limitation is to oxygenate the wastewater being treated. This may be accomplished in various ways (see Section 1.5). Once this is done, substrate thickness can be substantially increased and for aerated EEBs such as BBRs, values as high as two meters and more have been used (Liner & Kroeker 2010).

SSF CWs pass the wastewater they are treating through beds of permeable substrate of high hydraulic conductivity such as gravel. Another way by which SSF CWs can be "engineered" is by replacing all or part of their substrates with suitable engineered substrates capable of chemically adsorbing, precipitating, causing the volatilization of or otherwise altering the form of a wastewater's CoCs (Higgins et al. 2004; Hussain et al. 2014, see Section 7.6 in Chapter 7). By this means, the removal levels of certain contaminants (e.g., phosphates and arsenates), which can be relatively low otherwise, can be increased to 99% or more. For example, steel slag will sorb phosphorus from WWs.

Energy can be added to enhance performance. For example, many industries and other kinds of businesses that are sources of WWs requiring treatment have to manage, dispose of or otherwise handle large volumes of cooling water and other streams containing significant amounts of low-grade heat. Some of the main, size-dictating reactions occurring in treatment wetlands and bioreactors are highly temperature-sensitive, and they ordinarily require design for operation under the coldest conditions. This might result in systems which are overly large (and expensive) and which are over-designed for operating conditions during most of the year. Energy addition, either by mixing in or heat exchanging with relatively warm streams, or by using cooling coils in primary EEB System cells, can greatly reduce the size of a system, allowing enhanced treatment (Higgins et al. 1999a).

In addition to adding energy, another way to engineer a treatment wetland cell is to modify it so that heat is retained. Many microbially mediated wetland metabolisms slow as wastewater temperatures drop, and may become so reduced at the coldest water temperatures encountered (i.e., in winter) that the treatment system would have to be very large to achieve desired effluent CoC concentrations. However, some WWs (e.g., landfill leachates and municipal sewage) are naturally warm when generated, and if heat loss from them can be reduced, CoC transformation processes can be continued to be carried out at higher temperatures, requiring corresponding smaller systems. SSF EEB cells retain heat much better than do other kinds, and their sides and surfaces can be insulated to even further retain heat. In addition, in some cases, Sedimentation Pond and lagoon cells in front of EEB cells in EEB Systems can even be fitted with insulated floating covers to retain heat. By using such process modifications, EEBs smaller in surface area can be used (Higgins 2000b).

Another major way that EEBs may differ from ordinary constructed wetland systems is the practice with them of adding things to the process. Additions need not be limited to energy and air. Chemicals can be added as well (Higgins et al. 2006a). For example, a carbon-rich wastewater stream from another source can be added to a very low BOD, high nitrate stormwater to allow denitrification to occur, or alum can be added to a primary or tertiary treatment Sedimentation Pond cell of an EEB System to precipitate phosphorus, to name just two. In addition, the pH of a wastewater stream in an EEB System may be adjusted by adding acid or caustic to enhance certain removals and/or reduce toxicity. Similarly, in some cases, dilute streams may be added as well (e.g., stormwater streams added to process streams being treated).

Still another way by which vegetated EEBs may sometimes be engineered is by manipulating their vegetation (if there is any). Often, the WWs being treated in a natural WWT system such as a CW contain contaminants or have conditions such that they will adversely affect the plants used, negatively stressing or killing them. WWs with excessively high or low pHs, or ones containing toxic pollutants or salt can be very hard on the common types of aquatic vegetation. However, if vegetated EEBs are being considered, plants can be selected that are more stress resistant, better allowing their use under such conditions.

All emergent wetland macrophytes provide some level of phytoremediation (see Section C.6 in Appendix C), be it only adsorption on surfaces of their root systems). With vegetated EEB cells, the possibility exists of enhancing this, and/or of even selecting wetland plants that have phytoremediating properties which allow them to take up and metabolize, sequester and/or otherwise remove heavy organics, inorganics and heavy metals from WWs passing through their root systems.

Although it is well known that at certain times of the year wetland plants can take up from 10% to 15% of a wastewater's influent nitrogen, and 40%–60% of its phosphorus (Kadlec & Knight 1996), the harvesting of vegetation from CWs to remove these contaminants is usually impractical and/or uneconomic, and is not normally carried out. However, in certain kinds of EEB Systems, there may be special situations that may prove exceptions to this rule. With vegetated EEB cells, plant harvesting and disposal can be practised. Also, the "dry" surfaces of some kinds of vegetated EEB cells present opportunities for harvesting equipment access to occur much more easily than with other kinds of natural treatment systems.

Still another way to "engineer" a constructed wetland system is to operate it in an advanced manner. In EEB Systems, wastewater feed rates may be monitored and controlled using SCADA systems to maximize performance (e.g., lower feed rates and hence longer retention times may be used in colder weather to compensate for temperature effects). Another way is to recycle streams among EEB System cells in ways that improve performance.

Finally, the microbiome in a biological system may be modified and/or controlled using genomic methods to enhance performance (Higgins & Mattes 2014, also see Section 1.11).

CoC transformations in passive treatment BCRs are more complex than they are in semi-passive BCRs (see Chapter 4), and usually involve the microbial metabolization of the complex carbonaceous materials of the active media (e.g., carbohydrates, cellulose, hemi-cellulose, lignins, lipids, and proteins) from the transformation of solid media (e.g., wood chips, composts, and biosolids) via processes such as biodegradation, fermentation, enzymatic hydrolysis, biotransformation, bio-oxidation and bioreduction and/or biomineralization.* This produces progressively simpler compounds that are metabolized by the characterizing and other bacteria.

In addition to the active media, the substrates of most passive treatment BCRs and other anaerobic EEBs usually also contain inert support material(s) that provide permeability (e.g., sand and crushed rock) and/or reactive material(s) that provide support, permeability and/or alkalinity (e.g., limestone gravel). Some have materials specifically added to their substrates initially to provide all or part of the microbial consortium (e.g., a manure). In addition to these substrate materials, a passive treatment

* In addition to microbially mediated processes, a variety of other biotic as well as abiotic reactions/processes also occur in a BCR's substrate including filtration, sorption, co-precipitation, stabilization, ion exchange and anaerobic degradation as well as microbially enhanced chemisorption and chelation.

BCR or other anaerobic EEB may contain separate layers of non-substrate alkalinity addition/buffering material(s) such as limestone, and relatively thin layers (lenses) of separator and bedding materials such as sand. In some cases, crushed sandstone may be added to a substrate to improve precipitate sorption.

If it has not already been removed in advance, energy-generating, microbially mediated aerobic respiration* in a BCR cell or other anaerobic EEB will remove dissolved oxygen from an influent wastewater as follows:

$$C_6H_{12}O_{6(aq)} + 6\,O_{2(aq)} \rightarrow 6\,H_2O + 6\,CO_{2(aq)} \tag{1.1}$$

where $C_6H_{12}O_6$ represents a bioavailable carbonaceous material. This kind of oxygen scavenging is important for all anaerobic EEBs, and if it does not occur adequately because of inadequate upstream treatment, channelization or some other design limitation, their functions may be jeopardized. Other major microbial processes that occur in the cells of EEB Systems besides aerobic respiration include fermentation (anaerobic), methanogenesis (anaerobic), sulfate oxidation/reduction and iron oxidation/reduction.

While many types of active media can be considered for passive treatment, anaerobic EEBs such as BCRs (e.g., wood waste, composts of various kinds, straw, and organic soils), wood chips and the biosolid by-products from paper recycling plants are among the most desirable kinds for them (see Section 7.4 in Chapter 7).

Preparing substrates for passive treatment BCRs is much more complex and expensive than the situation for aerated EEBs and semi-passive BCRs. Not only are the carbonaceous active media in passive treatment BCRs and other anaerobic EEBs more difficult to manage (e.g., composts and biosolids are harder to handle than granular aggregates), but they usually also involve more than one kind of material (e.g., support materials as well as two or more kinds of active media) and each of these materials must be separately accessed, brought to a large preparation site, stored, mixed homogeneously and only then moved to the bioreactors for placement. The preparation area may be on-site near the bioreactors or at some off-site location.

Wherever the substrate preparation site is, provision has to be made at it to protect the local environment from dust, excess traffic, odours and contaminated run-off from the piles of component materials stored and mixed there† (see Chapter 7).

* Readers are reminded that Equation 1.1, like all such equations, is an artefact and at best only represents an overall condition. More accurately, electron flow using ATP via the microbes produces the oxygen while the parallel metabolism of carbon results in the production of carbon dioxide during the biosynthesis (growth) of more microbes. Nevertheless, if one is aware of their limitations, equations are valuable characterizations of processes.

† This fact gives mining and waste management firms, companies that are used to handling large volumes of solid materials, advantages in building EEB Systems.

1.4 The BREW Project

Although the use of aerated gravel-bed "wetlands" for sewage treatment has been reported in the past (e.g., Hays 1935; Davies & Hart 1990), the systematic development of the current aerated EEB ecotechnology dates from the multi-party Bioreactor Engineered Wetland (BREW) Project (the BREW Project) carried out in Ontario, Canada led by Soil Enrichment Systems (SESI) in the mid-to-late 1990s (Higgins 1997; IRAP 1999, see Glossary for details on SESI).

The 3-year BREW Project had a number of participants besides SESI, including four universities, several government agencies (e.g., the Canadian National Research Council, Environment Canada and two branches of the Ontario Ministry of the Environment), NGOs (e.g., the Toronto and Region Conservation Authority), consulting engineering companies (e.g., CRA, Jacques Whitford, NAWE and others) and other organizations (e.g., Dofasco Steel, 3M Canada Ltd. and the International Minerals Corporation [IMC]). The two pre-eminent wetlands engineering experts in the world were advisors to the BREW Project (Dr. Robert Kadlec and the late Sherwood Reed, the Lead Author's mentor).

The results of the BREW Project proved that aerated SSF gravel-bed bioreactors (then called EWs, now called BBRs) gave superior contaminant removals compared to those in similar, non-aerated CWs for the treatment of WWs such as sewage, septic tank overflow and a municipal/commercial landfill leachate (Higgins 2014).

After its completion (IRAP 1999), a version of the ecotechnology was first patented as Forced Bed Aeration™ by NAWE (Wallace 2001a) and the name "engineered wetlands" was adopted to distinguish them from CWs.

When Stantec NAWE was closed in 2009, some of the patents relating to Forced Bed Aeration were transferred to Naturally Wallace, a new firm involving some of the principals of the old NAWE, with full rights to these patents being retained by Stantec and some of the developers.

Naturally, Wallace later sold the rights to some of these patents for North America to Nexom,* although most of the aerated EEB (BBR, formerly EW) ecotechnology as now practised is not covered by patents and the original patents have mostly expired. (In any case, "know-what" is as or more important than "know-how" in designing, building, and operating BBRs as well as other kinds of EEBs.)

* Formerly Nelson Environmental Limited (Nelson). In addition to designing SAGR bioreactors (see Section 3.9 in Chapter 3), Nelson was an equipment supplier (blowers and aerations systems) for EEB System projects involving BBRs (e.g., that at BNIA). Nexom has advanced the SAGR Bioreactor ecotechnology and several of its innovations are covered by patents.

At about the same time as the BREW Project, Jacques Whitford carried out a major CW project* at Edmonton International Airport (EIA, or YEG using its IATA code) aimed at treating GCSW resulting from cold weather aircraft de-icing activities (Higgins & MacLean 1999; Higgins et al. 2001; Higgins 2002a,b). As part of the original EIA CW project, the first indoor and outdoor pilot-scale treatability testing of the use of aerated EEB ecotechnology to treat glycol-contaminated WWs was carried out and this supplemented the testing of the BREW Project.

1.5 Comparison of Constructed Wetlands and EEBs

In general, despite having much smaller surface areas than CWs, EEBs are much more efficient at removing CoCs from WWs passing through them. Table 1.3 outlines typical removal efficiency ranges.

TABLE 1.3

Relative Removal Efficiencies of CWs and EEBs

Contaminant	CWs	EEBs
BOD	50%–90%	70%–99%+
TSS	60%–80%	70%–99%
TKN	40%–60%	90%–99%+
TP	30%–60%	95%–99%+
Soluble organics	80%–90%	95%–99%+
Dissolved metal(loid)s	40%–90%	90%–99%+
Pathogens	3–6 log	4–9 log

1.6 Horizontal versus Vertical Flow in EEBs

Whether aerobic or anaerobic in operation, for some EEB cells there is the option of operating them in HSSF modes or in VSSF modes.

Table 1.4 compares well-designed HSSF and downflow VSSF EEB bioreactors.

It is emphasized that the comparisons in Table 1.4 are relative and will vary depending on the specific situation. For example, both VSSF and HSSF EEBs often can achieve very good removals of residual suspended solids, but HSSF ones do a somewhat better job of doing so. (In any case, bulk removals

* The Lead Author was the designer and project manager for this project, and another of the Co-Authors (Stiebel) was project executive. The EIA CW system has now (2012) been partially upgraded to an EEB System (Liner 2013, see Section 3.9 in Chapter 3).

of most of the grit and suspended solids should occur upstream [before] the EEB cells of an EEB System.) Flow in most BBRs constructed recently is downflow VSSF, while that in SAGR Bioreactors is HSSF. Generally, the floors of HSSF EEBs are flat and parallel to their substrate surfaces, but in a few cases terrain and other considerations may allow sloped bottoms (1/2% to a few per cent).

TABLE 1.4

Comparison of VSSF and HSSF EEB Modes

Parameter	VSSF	HSSF
Cell shape	Can be any shape, but are usually rectilinear	Rectilinear cells usually required
Hydraulic capacity	Higher	Lower
Wastewater stratification	Cannot occur	Can occur
Compaction	Can occur	Regional compaction highly undesirable
WW channelling	Cannot occur	Can occur
SS removal	Better	Best

1.7 EEB Treatability Testing

To design EEB Systems treating a different kind of WW than ones that have been treated before, or proposing ones to be located at a new site where different conditions than those that have been encountered before (e.g., climate, soils and environmental constraints) are extant, treatability testing has to be carried out to provide scale-up information needed to design the full-scale EEBs to be constructed.

Figure 1.4 (see later) illustrates the morphology of a pilot unit for treatability testing involving two downflow EEBs: an ABR cell (in this case a BCR that might produce H_2S, hence the vent) and a vegetated BBR cell in series (with grow lights over it and referred to as an "Engineered Wetland" cell as this was made for an early treatability test). This test unit also includes a feed tank (in this case another tote, although larger vessels are often used, as may be seen in Figure 6.4 in Chapter 6) and a mixing tank. Flow is from left to right.

A picture of the actual pilot unit that is sketched in Figure 1.4 is found as Figure 6.4 in Chapter 6.

In the past, WWT facilities such EEB Systems were usually custom-designed to treat waters having deemed worse-case CoC concentrations and flows and the only data collected during their pilot- or demonstration-scale treatability tests were those required to define kinetic and associated scale-up parameters (see Chapter 8). For these systems, it was assumed that textbook descriptions

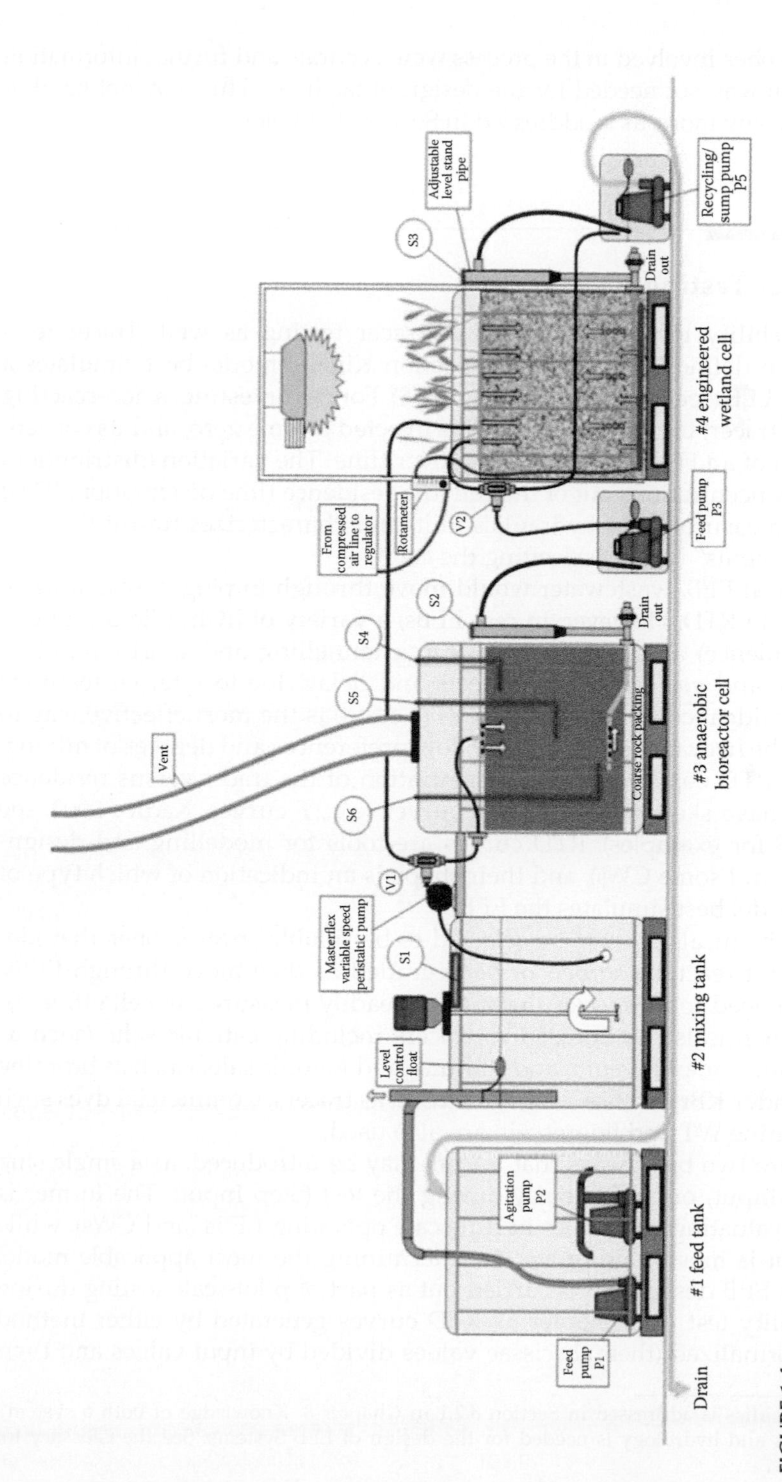

FIGURE 1.4
A two-cell EEB System pilot unit.

of the microbes involved in the process were accurate and further information about them was not needed for the design of facilities. This may not be altogether true any more, as is addressed in Section 1.11 below.

1.8 Tracer Testing

EEB treatability tests usually involve tracer testing as well. Tracer tests are used to define which kind of Reaction Kinetic model best simulates a particular EEB (see Section 8.2 in Chapter 8). For tracer testing, a non-reacting chemical (tracer) of concentration [C] is injected at time zero, and its concentration out of an EEB cell is measured over time. The variation (distribution) of tracer concentrations out of the cell (the residence time distribution, RTD) will be a measure of the hydraulics in it, and characterizes transit times of "water elements" (aliquots) leaving the cell.*

In an ideal EEB, wastewater would move through in plug flow and there would be no RTD. However, in real EEBs, a variety of hydraulic conditions (e.g., turbulence) and inefficiencies (e.g., channelling and short circuiting, transverse and vertical mixing effects and delay due to internal features) lead to nonideal conditions, and an RTD curve is the most effective way to quantify the hydraulics and identify flow preferences and degrees of mixing. Typically, RTD curves plot the concentration of the tracer versus residence time and have skewed, bell-shaped curves (e.g., F curves, Kadlec 2003, see Chapter 8 for examples). RTD curves are tools for modelling and designing EEBs (and some CWs), and their shape is an indication of which type of reactor model best simulates the EEB.

Tracer chemicals need to be selected to be soluble species: ones that ideally will not react, be sorbed or be degraded as they move through EEBs. They also need to be species that can be readily measured in cell effluents. Various chemicals can be used as tracers including cationic salts (such as those of lithium, potassium, and sodium) and anionic salts (such as bromine and chloride). KBr and NaCl are often used as tracers. Commercial dyes such as rhodamine WT and fluorescein are also used.

There are two basic ways that tracers may be introduced: as a single slug (Impulse Input) or continuously during the test (Step Input). The former is best for evaluating hydraulics in full-scale operating EEBs (and CWs), while Step Input is most appropriate for determining the most applicable model for use in EEB design and is carried out as part of pilot-scale testing during a treatability test (see Chapter 6). RTD curves generated by either method can be normalized (their abscissae values divided by input values and their

* EEB hydraulics is addressed in Section 8.2.1 in Chapter 8. Knowledge of both a system's hydraulics and hydrology is needed for the design of EEB Systems. See the Glossary for definitions.

ordinate values divided by the nominal residence time) (see Equation 8.4 in Chapter 8). A normalized RTD curve of this sort is called an E (exit age) curve.

In designing EEBs, E curves from step input tracer addition found from the tracer tests forming part of pilot-scale treatability tests are used to determine the Reaction Kinetic model most applicable for a particular substrate/wastewater/other conditions combination. This determination, coupled with the kinetics and other design parameters determined in other parts of a treatability test, can then be used to "size" (help design) new full-scale EEBs.

Figure 1.5 shows two tracer curves obtained from a sodium bromide tracer test carried out as part of pilot-scale treatability testing to determine the applicability of using an EEB System to treat polluted groundwater at a contaminated former chemical facility site. In this case, the pilot-scale EEB System used consisted of BCR cell followed by a BBR cell.

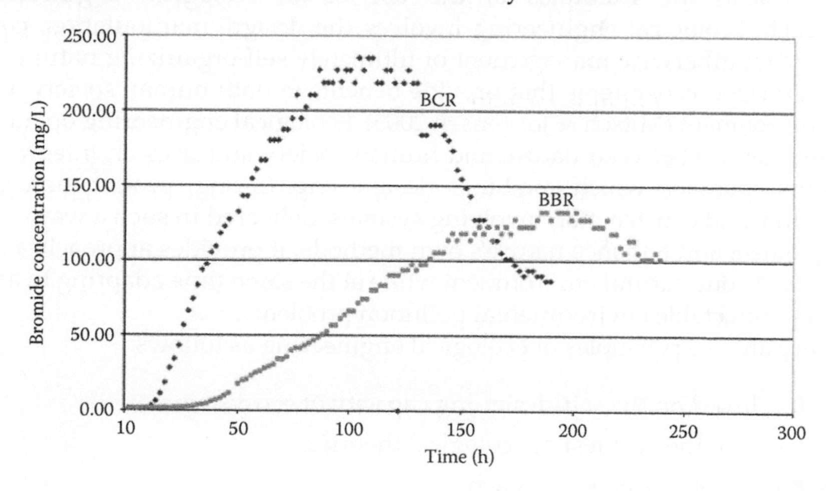

FIGURE 1.5
RTD E-curves for a BBR and a BCR in series during the treatability testing of contaminated groundwater.

In the normalized graph (Figure 1.5) in which the tracer test results for the two EEBs have been superimposed, the lower right-hand curve is for tracer testing in a BBR that had a mixed gravel substrate, and indicated that the design of a full-scale unit could be facilitated (using kinetic results determined from the other part of the treatability test) using a multiple tanks-in-series (TIS) with delay model (Kadlec & Knight 1996, see Equation 8.7 in Chapter 8). It also showed (upper left-hand curve) that the BCR cell in this case could be modelled as a multiple TIS one (see Equation 8.16).

It is emphasized that, depending on the wastewater being treated, the kind of substrate being used and certain other conditions, different models may be determined to be applicable for the same kinds of EEBs. For example, the tracer test carried out for the BNIA Project (see Section 3.6 in Chapter 3) indicated that the best design model for that pilot unit's BBR cell treating glycol-contaminated waters was a continuously stirred reactor (CSR) model, not the kind of model suggested above for treating a groundwater contaminated with many CoCs.

More information on the design of EEB Systems is provided in Chapter 8, and the state-of-the-art for two kinds: a BBR-based EEB System at an airport and a semi-passive treatment BCR-based EEB System at a remote northern base metal mine are found in Section 9.2 in Chapter 9.

1.9 Ecological Engineering

EEB Systems are examples of the use of an ecological engineering approach. Ecological engineering involves the design, manipulation, control and/or otherwise management of ultimately self-organizing natural or bioengineered ecosystems that provide benefits to both human society and the environment (Mitsch & Jørgensen 2003). Ecological engineering operates at the interface between nature and human society and uses an integrated systems approach which combines design, engineering, project management and field construction involving systems, delivered in such a way as to complement and enhance nature's own methods. It provides approaches for conserving our natural environment while at the same time adapting to and solving intractable environmental pollution problems.

There are five principles of ecological engineering as follows:

1. It is based on the self-designing capacity of ecosystems.
2. It can be the acid test of ecological theories.
3. It relies on systems approaches.
4. It conserves non-renewable energy resources.
5. It supports biological conservation.

Ecological engineering is not environmental engineering (although all ecological engineers can also be called environmental engineers, the reverse is not true); it is not aimed at energy conservation ecotechniques such as those involving gravity and solar energy (although it usually involves them); it is not "green" engineering based on eco-ethics or other green philosophies; and it is not the same as Industrial Ecology,* although it can be related to each of them (Higgins 2002a,b).

Ecotechnology is not just another name for biotechnology. It is more than that. Ecotechnology intimately involves not just an environmental focus, but

* Industrial ecology is the practice of trying to treat/set up/operate industrial and technological systems in the same manner as natural ecological systems whereby the uses of energy and materials are minimized and material cycles are closed looped with an overall goal of sustainable development.

also environmental values and the principles of sustainable development. Table 1.5 illustrates the differences between ecotechnology and biotechnology.

Ecotechnologies, by definition, are interactive and holistic. Their use implies not only a shift to biologically based systems approaches but also a shift to decision-making that is also ecosystem based.

TABLE 1.5

Comparison of Ecotechnology and Biotechnology

Characteristic	Ecotechnology	Biotechnology
Basic unit	Ecosystem	Cell
Basic principles	Ecology	Genetics; cell biology
Control	Forcing functions, organisms	Genetic structure
Design	Self-design with human help	Human design
Biotic diversity	Protected, enhanced	Changed
Costs	Reasonable	Enormous
Energy basis	Solar based	Fossil fuel-based

Ecotechnologies are used in all parts of the environmental field but find particular use in ecological engineering, WWT, composting, soil bioremediation, land management and some types of hazardous waste treatment. The use of ecotechnologies in WWT is ancient but it is only recently that their use in other parts of the environmental industry has become more widespread, often through the use of integrated systems approaches.

Ecological engineering seeks to provide lower cost solutions to environmental problems by using natural systems assisted by engineering methods. Combining ecosystem function with human needs is the essence of ecological engineering. Ecological engineering bridges ecology and engineering as is illustrated in Figure 1.6.

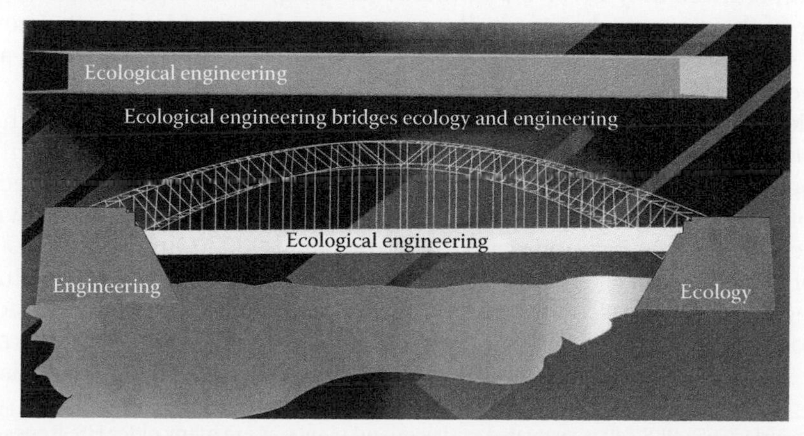

FIGURE 1.6
Ecological engineering bridges ecology and environmental engineering.

Dealing with environmental stress is an important part of ecological engineering. Ecosystems may be stressed by a variety of factors ranging from overuse by humans, compaction of soils, diseases, insect pests, drought and other changes in environmental conditions. Ecological engineers recognize the symptoms of such stresses and deal with their underlying causes. With ecological engineering, optimizing desired biological outcomes is achieved by identifying and dealing with rate-limiting conditions (Higgins 2002a).

The energy and economic values of ecological engineering projects such as those involving EEB Systems can be quantized using Emergy Analyses which evaluate system components on common bases to determine their net value to human society (Brown & Herendeen 1996; Odum 1996), and the new science of systems ecology (Jørgensen 2012) describes how ecosystems work and how they respond to disturbances such as WWT in them.

1.10 Biofilms

1.10.1 Biofilm Formation

Early in the development of life on earth, prokaryotic microorganisms (archaea and bacteria) evolved the ability to associate in biofilms in highly structured communities (microbial consortia) (Stoodley et al. 2002, Wikipedia). The purpose of biofilms is to contain bacteria (and, to some extent, archaea) and provide them with protected environments where they may thrive and grow.

The CoCs that it is desired to transform in EEBs can be present in soluble (dissolved), colloidal and suspended solids forms, as well as part of entrained solids (grit). When wastewater containing them passes through an EEB, many CoC transformations are by microbially mediated processes. A small portion of these may involve planktonic (floating) microorganisms in open water (if there is any), but the bulk occurs on and in the biofilms that cover most solid surfaces below the water level. In EEBs, biofilms penetrate and permeate bedding materials and porous substrates.*

Wikipedia defines biofilm-forming microorganisms as any group in which cells stick to each other on a surface. However, more rigorously, biofilms can be defined as: "…structured communities of microbial cells enclosed in self-produced (secreted) polymeric matrices, and adherent to inert or living surfaces in aqueous environments" (Stoodley et al. 2002; Qureshi et al. 2005).

Biofilm formation is a general attribute of bacteria and these dense cellular aggregates may be based on single or multiple kinds of them. Usually, bacterial consortia are involved rather than a single type of bacterium. In EEBs and other kinds of fixed-film bioreactors used for WWT, bacterial consortia in

* If the surface of a BBR cell is vegetated with wetland plants, as are many older HSSF ones, the part of the substrate bed that is penetrated by plant roots may forms an even more complex substrate particles/biofilm/root(let)s matrix.

biofilms dominate CoC transformations. In biofilms, the metabolic activities of the different kinds of bacteria involved are integrated wherever they occur. Additionally, the relatively high biomass concentrations represented by biofilms have the effect of significantly reducing hydraulic retention times in bioreactors of all sorts.

EEB biofilms usually consist of thick yet permeable layers containing complex communities of mostly aerobic, anaerobic and facultative bacteria (both Gram positive and Gram negative), although some opportunistic, predatory, symbiotic and/or entrapped archaea, fungi, algae, micro-eukaryotes and other microorganisms may be present as well. These microbes are held together and protected inside a bacterial-secreted mucous, high-molecular-weight gel, often referred to as extracellular polymeric substance (EPS), and for EEBs is a slimy, glue-like substance that adheres to submerged interfaces in aqueous environments.* This EPS constitutes the biofilm (Nguyen et al. 2012; Ivanow 2016). The microbes in a biofilm are physiologically and behaviorally different from the same organisms when in their planktonic states, and in their biofilm modes, different suites of genes are regulated.† Figure 1.7 shows biofilm (green) growing on the gravel particles of an operating, pilot-scale BBR cell.‡

FIGURE 1.7
Biofilm growing on gravel substrate in a pilot-scale BBR cell.

* Biofilms are common and do not always involve gels like those in EEBs. For example, the plaque on teeth is a kind of biofilm.

† More details of biofilm formation, development and properties can be found in Wikipedia.

‡ The test cell shown in Figure 1.7 had Plexiglas windows glued to openings in the side of the one-cubic meter pilot-scale test vessel (tote, see Chapter 7). Although they provided an excellent view of the development of the biofilm, the windows leaked badly and the test cell had to be scrapped.

In natural WWT systems such as EEBs, the biofilm is largely made up of exopolysaccharide alginate but also includes nucleic acids, proteins, lipids, metal cations (in some cases) and entrapped materials such as colloids and suspended solids. This biofilm binds the microbes together in a three-dimensional matrix that affects local physiochemical characteristics such as mass transfer, adsorption capability and stability. There are various kinds of bound EPS (sheaths, capsular polymers, condensed gels and loosely bound polymers) and soluble EPS (soluble macromolecules and colloids) and these affect a biofilm's ability to bind to organics and nutrients in a wastewater in contact with it. As a diffusive barrier, the biofilm protects the largely bacterial cells from phages, other microbial predators (everything from rotifers to bloodworms) and from toxic compounds in the water. Bacterial cells in a biofilm release enzymes that allow them to move freely through it.

Wikipedia defines biofilm formation as involving five stages including initial adhesion to a surface by planktonic microbes, subsequent irreversible attachment via cell pili and other methods, two stages of maturation involving cellular growth and recruitment (proliferation and maturation), and dispersion as the biofilm spreads to colonize adjacent surfaces.

Biofilm formation depends on a variety of cellular, surface and environmental factors. There is considerable cooperation and interactions among the bacteria in a biofilm and these microbial accumulations have many charged groups (e.g., carboxyl, phosphoric, hydroxyl, phenolic, and sulfhydryl) that impact adhesion and reactive characteristics (Qureshi et al. 2005; Nguyen et al. 2012). Porous and/or rough surfaces (e.g., gravel particles in an EEB) enhance biofilm formation, as does the availability of adequate amounts of nutrients, especially nitrogen and phosphorus.

EPS is distributed in various layers of varying depth through a biofilm, establishing its structural and functional integrity and mechanical stability (Nguyen et al. 2012). EPS at a submerged surface promotes microbial adhesion by altering the surface's physiochemical properties such as charge, hydrophobicity and roughness, and creating scaffolds with physical characteristics and pore structures that promote microbial cell attachment. The oxygen content (and hence the kinds of microbes present) in biofilms in aerobic bioreactors typically progresses from aerobic conditions to anoxic conditions to anaerobic conditions (De Beer et al. 1994).

1.10.2 Clogging

While biofilms are the essence of all EEBs, care must be taken in design so that excessive amounts of EPS do not occur. If organic loading to an EEB (usually expressed as BOD loading) becomes too high, the EPS may expand to bridge the gaps between the substrate particles (clogging), and this will cause plugging, channelization and even, in some cases, the formation of "biomats" and "hardpan" layers' in the substrate.

1.10.3 Sludging

Under certain situations, an almost "explosive" growth of EPS, called "'sludging," can happen. It is manifested by the generation of sludges that may be described as "rubbery" or "mushy"; may be present as continuous accumulations of sludge or "gelatinous" globules; and may be of various colors (yellow, orange, black and whitish-gray) depending on the situation and the kinds of contaminants present (e.g., the presence of iron may color sludges black or orange depending on the form involved*). EEB Systems can be designed to prevent the impact of any sludging on them. Figure 1.8 shows an accumulation of sludge flushed from a well on the central de-icing pad at a Canadian airport.

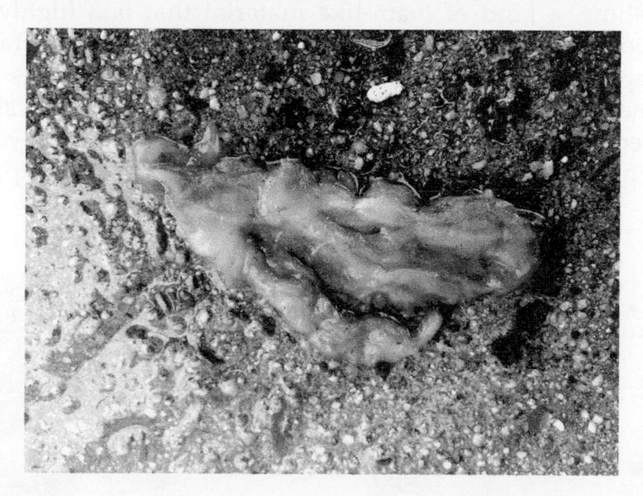

FIGURE 1.8
Sludge at an airport.

Sludging is not limited to WWT systems, but can occur in any collection, storage, transport or other management systems where a wastewater, run-off or stormwater contains an adequate amount of organic material and sufficient nutrients, and when certain other conditions occur. Sludging has become a particular problem at airports where the reformulation of the glycol-rich fluids using for de-icing in cold weather has led to relatively nutrient-rich WWs.

Since most sludges are insoluble in water (and in most solvents), their formation may be a problem in the design of WW management systems. With proper EEB design, sludging will not be a problem in EEB Systems (which can easily cope with the kinds of WWs prone to sludging) but where there is a potential for them, the design of their EEBs may have to be adapted to prevent them from being so.†

* However, beware. Certain species of bacteria are also able to color WWs and sludges red to orangish-red without excess iron being involved.

† Sludging is discussed in more detail in Section 7.5 of Chapter 7.

1.10.4 Sliming

When the microbial cells in biofilms uptake nutrients, they channel much of their metabolic energy towards producing EPS and less towards cell growth/division. However, if nutrients (e.g., phosphorus and especially nitrogen) become scarce, the bacteria in a biofilm face the prospect of either being trapped in an unfavorable environment or escaping the biofilm. They do the latter by producing enzymes capable of breaking it down and overproduce the (lipo)polysaccharides normally found outside of bacterial cell walls. This situation is referred to as sliming (Glymph 2015).

If this happens, it results in the detachment of pieces (sheets) of the biofilm as slime, a kind of foam-like material that is a highly undesirable condition. Slime formation most often occurs when certain industrial WWs are treated, as these can be deficient in the nutrients and other conditions necessary for biofilm maintenance. Figure 1.9 shows slime floating on the receiving waters at an airport EEB System treating glycol-GCSW (which was nutrient-deficient) after manually added nutrient supplies were cut off.

FIGURE 1.9
Undesirable slime formed after nutrient deficiency.

This sliming incident occurred early in the operation of the BBR-based EEB System when the operators responsible for adding needed nutrients (bags of urea and phosphate fertilizer) to the nutrient-deficient influent stream of the airport's EEB System were diverted to snowploughing duties during a large winter storm and stopped adding the needed nutrients. This led to extreme nutrient deficiency for the biofilm bacteria in the system's BBRs. The result was large-scale slime formation that led to literal small "volcanos" of foam forming on the surfaces of the BBR cells where aeration air was surfacing. Effluent BOD concentrations rose to very high levels because of the slime that was then reported as BOD. Regulatory discharge levels for BOD were far exceeded, the visible contamination of the receiving waters (a small stream) occurred, and a costly clean-up was required.*

It is emphasized that the terms "clogging," "sludging," and "sliming" as used in this book are definitive, and should not be confused with similar terms used in other publications and industries. For example, in conventional activated sludge-based WWTPs, by-product and recycled waste activated "sludge" (WAS) streams are usually black liquids containing 0.5%–2% solids, while the materials referred to as sludge herein are mushy semi-solids. Similarly, many use the terms "biofilm" and "slime" interchangeably but in EEBs, biofilms in place are desirable (and essential) while slimes resulting from biofilm detachment represent undesirable conditions.

Both sludges and slimes can clog WWT treatment facilities and infrastructure (e.g., piping). While slimes, like sludges, are insoluble in water and most solvents, and can lead to unsightly foams in receiving waters (as is shown in Figure 1.9) because they consist of sheets of detached biofilm that continue to separate, slimes will eventually disperse and disappear (but may cause major disruptions until they do). Such is not the case with sludges. These can, as was mentioned, often be flushed from piping and other facilities but will not disperse over time and may have to be mechanically collected and disposed of.

1.10.5 Biofilm Environments

As was mentioned above, in the biofilms of aerobic EEBs (e.g., BBRs), there will always be anaerobic microenvironments, and hence anaerobic processes caused by anaerobic microbes can and will occur, albeit to limited extents. Similarly, aerobic microenvironments (and hence aerobic microbes and processes) will also be found in the biofilms of anaerobic EEBs (e.g., BCRs). The importance of these seemingly anomalous microenvironments should not be underestimated, and genomic techniques are now available to identify their aspects (see the next section); this can suggest methods to exploit them to enhance the performance of EEBs.

For example, during the EW Test Project (see Section 4.5 in Chapter 4), it was found that the expected anaerobic sulfate reduction process in a pilot-scale BCR (which was expected to be characterized by sulfate-reducing bacteria, SRB) was supplemented and complemented in the bioreactor by a simultaneous sulfur oxidation process (which involved sulfur-oxidizing bacteria, SOB), and that the relative influences of these processes varied along with changing process conditions (temperature in that case). This finding has had ramifications on the design of ABRs.

* The airport now has dedicated staff EEB Systems operators, while other airports with such systems sub-contract these operations to firms whose staff do not get diverted to other duties

1.11 Genomics

1.11.1 Bacteria

In order to grow in a laboratory or a bioreactor such as an EEB, a microbe such as a bacterium must have an energy source; a source of carbon; sources of required nutrients (and certain micronutrients); and a permissive range of environmental conditions such as temperature, pH and other factors (e.g., adequate oxygen content for aerobic bacteria).

Sometimes bacteria are categorized based on their patterns of growth under various chemical (nutritional) or physical conditions (e.g., anaerobes are ones that grow under anoxic conditions and thermophiles are ones that can grow at relatively high water temperatures). Other examples are phototrophs (microbes that use radiant energy [light] as an energy source), heterotrophs (ones that use organic chemicals with a C–H bond as a carbon source), autotrophs (ones that use CO_2 as a carbon source) and lithotrophs (ones that use [oxidize] inorganic carbon). Textbooks on WWT provide descriptions of the bacteria supposedly involved in the process(es) occurring, and designers often assume that these descriptions are accurate. Indeed, generally during treatability testing carried out to provide scale up information needed to design full-scale bioreactors such as EEBs, the only data collected are those required to define the kinetic and associated physical and chemical parameters, and that further information on the bacteria involved was thought not to be needed. This may not be altogether true any more.

This is because there now is a developing capability, Genomics, available to supplement data on a wastewater's physiochemical properties with data on the microbial communities present allowing the determination of how they influence treatment.

This new capability, some aspects of which are referred to as synthetic biology,* can be used to determine which types of bacteria and other microbes are associated with various kinds of WWs, soils, sediments, slimes and sludges. Having information of this sort allows ecological engineers to design commercial systems better able to treat WWs more efficiently and effectively. The new capability involves the study of a collection of genetic material (genomes) from a mixed community of organisms.†

Genomics can be illustrated using the biological oxidation (nitrification) of ammonia to nitrate in WWs by microbes as an example. Nitrification is defined in textbooks as a two-step aerobic process involving two groups

* Synthetic biology builds on genomics and is the science of using microorganisms to make industrial processes more green, sustainable and robust. Some aspects are alternatively referred to as genomic engineering.

† Herein, Genomics refers to the study of microbial communities. Where environmental samples are involved, it is often referred to by its sub-discipline, environmental genomics (envirogenomics).

of nitrifier bacteria (Metcalfe & Eddy 2004; Hill et al. 2015). During the initial step, ammonia-oxidizing bacteria (e.g., *Nitrosomonas, Nitrosococcus, Nitrosovibrio, Nitrosospira,* and *Nitrosolobus* spp.) are involved and nitrite is produced:

$$NH_4^+ + 1.5O_2 \rightarrow NO_2^- + 2H^+ + H_2O \tag{1.2}$$

This is quickly further oxidized to nitrate by nitrite-oxidizing bacteria (e.g., *Nitrobacter, Nitrococcus,* and *Nitrospira* spp.):

$$2NO_2^- + O_2 \rightarrow 2NO_3^- \tag{1.3}$$

In this way, highly toxic ammonia is converted to less toxic nitrate (Kinsley et al. 2002). Using genomic testing, much more can be determined regarding the bacteria that catalyze/facilitate the above reactions.

Genomic testing seeks to answer questions such as: Do the above equations (and bacteria mentioned) really describe what is happening during nitrification in a bioreactor? Other questions that may be able to be answered by Genomics include whether the aforementioned bacteria are the only kinds present; how do they interact with any other microbes also present; what affects them; how do bacterial populations change over time and/or when process conditions change (e.g., loading and temperature); and, most importantly, can the microbes be manipulated to enhance performance (biocontrol).

Early on, genomic characterization of microbial consortia involving prokaryotes (bacteria and archaea) focused on the 16S rRNA gene,* a segment of DNA critical for the production of proteins by their cells (Fraser et al. 2015).

For cultured microbes, genomic information on a bacterium can come from a single clone, making sequence assembly and annotation tractable. However, most natural WWT systems involve communities of many kinds of bacteria present in biofilms attached to substrate and other materials. One way to describe these microbial consortia is to use specific components of their DNA that are amplified via the polymerase chain reaction (PCR) (Valentini et al. 2009).

These amplified components of DNA (reference genes) from microbes found in WWs or sludges being treated can be compared to descriptions of the genes of various kinds of microbes found in the scientific literature and/or listed in curated databases of known annotated sequences (e.g., BOLD, GenBank and MG-Rast) and these can be used to identify the species present. The curated databases are particularly instructive where medically important microorganisms are involved, as for these species not only have the pathogens among them been identified and studied, but so have been

* Corresponding 18S rRNA can similarly be used for micro-eukaryotic and fungal cells.

any related species which are not pathogenic. For example, many Legionella species can be identified in such databases even though most of them are not pathogenic and some of the nonpathogenic ones even occur in the biofilms of fixed film WWT facilities. Other kinds of less intensively studied microbes (e.g., magnetotactic bacteria [MTB], see below) have yet to be included but may be soon as the databases are being continually updated as new information becomes available.

1.11.2 Biostimulation and Bioaugmentation

In a few cases, supplies of specific microbes or microbial consortia needed for biocontrol may already be available commercially. In other cases, microbes already present in soils and WWs can be identified and prepared for use (cultured) in quantity in fermentation tanks or other vessels, and such bacteria can be used to inoculate* EEB cells. The process is known as biostimulation.

In still other cases, microbes already identified at another site as having desirable qualities can be imported for use in inoculants. The process is known as bioaugmentation and can be controversial.

The uses of biostimulation and bioaugmentation to enhance treatment in EEBs are examples of biocontrol.

1.11.3 Taxonomic Studies

Taxonomic studies (16S Analyses†) are a kind of microbial community genetic testing involving sampling a wastewater, sediment, soil, slime or sludge; extracting DNA from it; sequencing the microbes‡ in it from which the DNA came by PCR amplification of their V3–V4 regions using conserved primers; and determining the microbes from which the amplified DNA sequences came. These data are then analyzed, assigning the classifications present using information from the literature and/or curated databases. The result of taxonomic studies is the classification of the microbial communities found in a wastewater, soil or sludge at several levels: by kingdom, by phylum, by class, by order, by family, by genus and by species, the latter two of which are most germane. Figure 1.10 illustrates a classification at the kingdom level of results from the taxonomic study of a sample of sludge found coming out of piping for an airport stormwater collection system.§

* For example, samples of arsenic-reducing bacteria [AsRB] identified at a mine site in one jurisdiction could be brought to another to facilitate the treatment of arsenic-contaminated wastewater from a different mine site there.

† The corresponding 18S RNA gene can similarly be used for fungal and micro-eukaryotic cells.

‡ For the genomic testing mentioned below, an Illumina MiSeq instrument located at the Biology Department of the UoW was used for sequencing.

§ As was mentioned, airports are places where EEB Systems are often used for WWT. See Appendix D for more information on the formation of sludges at airports and how they can affect the design and operation of EEBs.

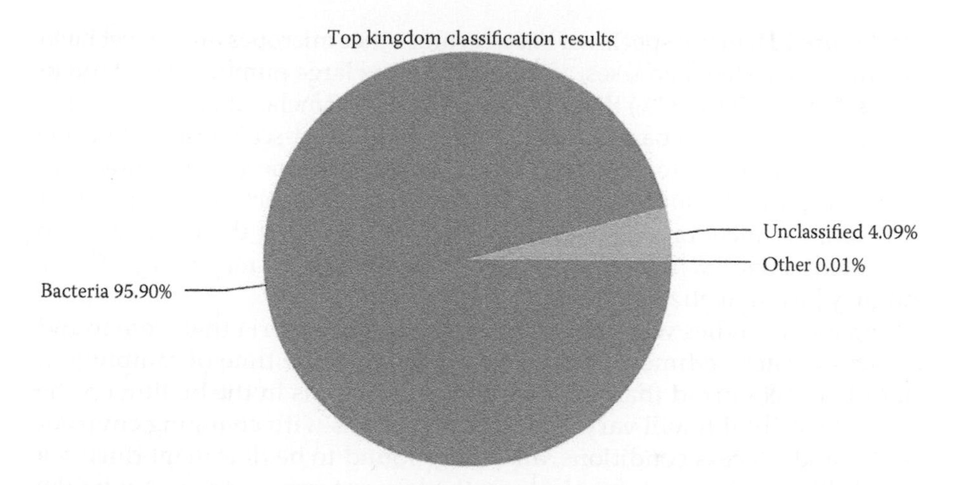

FIGURE 1.10
Taxonomic results of a sludge at the kingdom level.

As may be seen from Figure 1.10, not surprisingly almost all of the microbes associated with the sludge in this case were bacteria. Similar pie charts can be derived from the results of taxonomic studies in progressively more detail at the phylum, class, order, family, genus and species levels.

Figure 1.11 graphs species results for a taxonomic study of a sample of slime found coming out of piping of the same airport stormwater collection system from which the sludge was found.

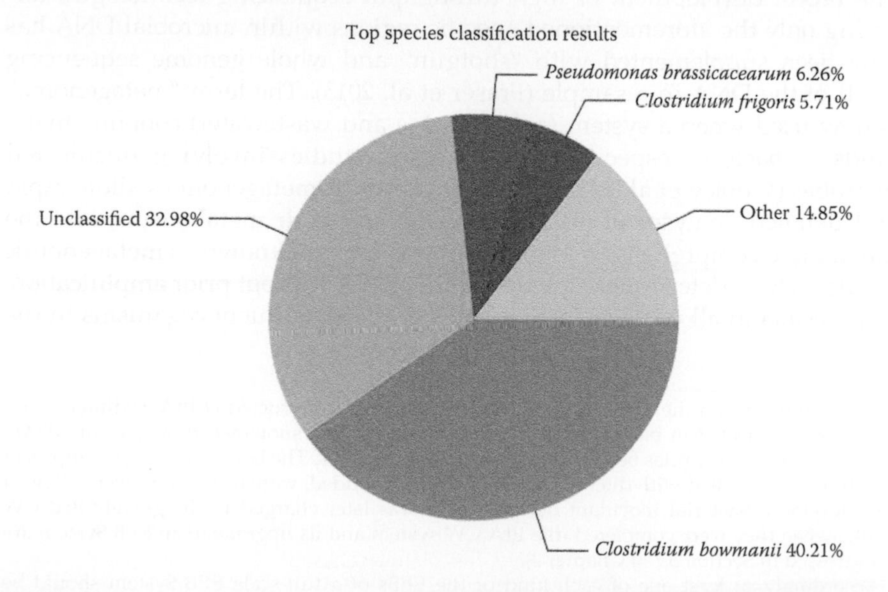

FIGURE 1.11
Taxonomic results for a slime at the species level.

In Figure 1.11, many species of bacteria and other microbes are not yet been included in curated databases, accounting for the large number of bacteria in Figure 1.11 pie chart (33%) that are categorized as "Unclassified."

Generally, when preparing an EEB, for either a pilot-scale test unit or one for a full-scale bioreactor, it is preferable to "prime" the bioreactor's substrate* with an inoculant solution made up by biostimulating bacteria in samples of local soil, sediment or sludge from around the source of the wastewater to be treated as these will have already been exposed to the major target CoCs, and may be acclimatized to them.

Taxonomic studies will only identify the major bacteria that were found in a wastewater, sediment, soil, slime or sludge at the time of sampling. It should be recognized that the microbial populations in the biofilm on the substrate of an EEB will vary with time as well as with changing environmental and process conditions, and those found to be dominant during a treatability test may not remain so in the longer term, and may not be the same as those in a subsequent full-scale EEB, even if they are primed with the same inoculant solution. Follow-up taxonomic studies should be carried out on samples from an operating full-scale EEB after it has been running for a period to determine whether the same microbes that dominated target CoC biodegradation during the earlier treatability test continue to do so, so that the right kind of biocontrol can be considered.†

1.11.4 Metagenomic Studies

The recent development of high throughput sequencing technologies targeting only the aforementioned specific regions within microbial DNA has now been supplemented with "shotgun" and whole genome sequencing of all of the DNA in a sample (Fraser et al. 2015). The term "metagenome" is now used when a system (e.g., a sludge and wastewater) contains many kinds of bacteria, especially for genomics studies involving uncultured microbes (Wooley et al. 2010). These advances in metagenomics allow rapid and detailed analyses of microorganisms and their metabolisms, and the much more comprehensive methods of this sort are known as metagenomic studies. These determine sequences of the DNA without prior amplification. This results in all parts of the genomes of the constituent organisms in the

* For example, when the very large SSF CW system was constructed at EIA, sediment samples were collected in barrels from the ditches into which snowmelt flowed from ADAF-contaminated snow piles beside the airport's de-icing pads. The barrels for these samples of sediment were filled with ditch water; had fertilizers added; were mixed; and were allowed to develop a bacterial inoculant decant which was later charged to the gravel-filled CW cells when they were completed (the EIA CW system and its upgrade to an EEB System are addressed in Section 3.7 of Chapter 3).

† Accordingly, at least one of each kind of the EEBs of a full-scale EEB System should be designed with sample points so that samples can be obtained from within the bioreactor's substrate bed during operations

community being sequenced and this in turn results in very large amounts of genomic data being generated.

A rapidly developing sub-discipline of metagenomics provides the tools that allow insight into these resulting data. Bioinformatics is used to obtain information about metabolism and functions of the various species that make up the microbial communities involved. Bioinformatics uses computational procedures to extract meaningful data from the very large data sets produced by such analyses (Fraser et al. 2015). Such evaluations allow a better definition of complex bacterial communities and a more precise determination of what is taking place.

Metagenomic studies can reveal not only which microbes are responsible for some desired result (e.g., ethylene glycol biodegradation in a BBR when treating an airport wastewater), but also the metabolic pathway (or pathways) responsible can now be identified. In some cases, the normal functional roles they play in interactive consortia of microbial species responsible for the contaminant transformation functions associated with a properly functioning treatment system can also be defined. However, the complexity of the interrelationships between various species is not yet well understood, and scientists and engineers are now only beginning to gain insight into some of these relationships for applications such as treating WWs.

If it turns out possible to identify not only specific bacteria responsible for a desired result (e.g., the aforementioned EG biodegradation), and the metabolic pathway or pathways they use, the potential for biocontrol then exists, as conditions in a bioreactor may then be manipulated by genomic engineering methods to enhance the desired outcome. For example, the metabolic pathways for the biodegradation of ethylene glycol by the bacterium, *Pseudomonas putida*, have been reported (Mückshel et al. 2012). If this bacterium was found in the substrate of a BBR treating airport wastewater using a taxonomic analysis (and it has not been so far), and its metabolism was defined and deemed important during the bioinformatic analysis of data from a subsequent metagenomic analysis, biocontrol steps could be taken to biostimulate it. If not, it could be bioaugmented.

As was mentioned, the amount of data produced by a typical metagenomic study of a sample is enormous, and it is useful to have means to zero in during the bioinformatic analysis to data on specific metabolic pathways that are likely to be involved. This may be achieved by (1) using information from treatability tests; (2) assessing EEB operating results; (3) having determined what species of bacteria are present from the results of an earlier taxonomic study; and/or (4) using knowledge of what the major CoCs are involved and what their biodegradation products are or have been separately measured to be. For example, for the proposed treatment of an ethylene glycol-contaminated airport wastewater, the determination by chromatographic methods of what volatile fatty acid (VFA) biodegradation products are present (see

Section 2.3.2 in Chapter 2) will be very useful, as knowledge of the presence, absence and concentrations of the various VFAs may materially assist design.

1.11.5 Magnetotactic Bacteria

Often, sludges and slimes are encountered that are colored or stained black or orange. This may be due to the presence of iron and/or to specific bacteria that cause such colorations. Figure 1.12 shows a black-colored material due to the former flowing out of a stormwater collection system's overflow pipe into receiving waters at an airport. Genomic testing of the material showed that it was a kind of slime formed under anaerobic conditions and testing of the water flowing out of the pipe with the slime showed that it most likely originated in upstream concrete piping in the collection system which was deteriorating, resulting in rubble on which biofilm temporarily formed at oxic/anoxic interfaces until nutrient deficiencies and hydraulic surges displaced it as slime.

FIGURE 1.12
Black material coloring a slime.

A Fourier transform infrared (FTIR) spectrum was carried out at UoG on the dried solids from the centrifugation of a sample of this black slime and the results showed a pattern similar to a published spectrum for magnetite. It is well known that magnetite can be produced by biomineralization by kinds of bacteria called MTB.* MTB will take up to 100× more iron than is usual for bacteria, and are particularly active when ferrous iron is present

* See Arakaki et al. 2008 and "Magnetotactic Bacteria" in Wikipedia.

(as was found to be the case here). That the black solids once separated from the slime were probably magnetite was further indicated by the fact that a very strong ceramic magnet was able to rapidly coalesce and move the dried solids, indicating their magnetic properties.

Indeed, magnetite formation (black fines) has not only been observed in the effluents of airport WWT systems (both those using EG-based ADAFs and those using PG-based ones), but it has also been observed in waters from other process systems where glycols are involved (e.g., where EG is used to dehydrate natural gas at compressor stations). The black fines that have been observed from time to time in the effluents from EEB Systems do not seem to adversely affect their operations (the fines only rarely appear in any case) and R&D is being considered to better determine when and how they form.*

* See Section 9.3 in Chapter 9.

2

Wastewaters Treatable in EEBs

2.1 Types of WWs Treated in EEB Systems

There are many kinds of WW streams that can be/have been successfully treated in EEB Systems and they involve wide ranges of both *labile* (easy-to-biodegrade) and recalcitrant contaminants (Higgins 2014). These WW streams include: *municipal wastewaters* (Wallace 2004; Higgins 2008; Liner & Kroeker 2010); *landfill leachates* (e.g., drainage, percolate, pore water and seepage from municipal, commercial, industrial and hazardous waste landfills, Higgins 1998, 2000b; Kinsley et al. 2000; Nivala 2005); *Mining influenced wastewaters* (Duncan et al. 2004; Higgins et al. 2006a,b; Higgins & Liner 2007; Mattes et al. 2010; Higgins & Mattes 2014); *industrial wastewaters* (e.g., process waters, blowdowns, washwaters, backwashes, aqueous sludges, WWTP effluents, petroleum and petrochemical plant site runoffs, Higgins 2003; Wallace & Kadlec 2005); *meteoric water* (a general term for stormwater runoff and/or snowmelt that has been or may be contaminated, Higgins et al. 2012); *water from ASGM* areas (Anderson et al. 2005; Higgins & Anderson 2014); *impacted groundwater* (that which has picked up pollutants either via contact with contaminated materials, soils or waste accumulations and/or via the influx of other kinds of WWs, Higgins 2003); *agricultural wastewaters* (e.g., WW from animal exercise, rearing and barn areas, milkhouse waters, percolates and leachates from manure piles and compost production areas, agrichemical operations' WWs; water from greenhouses and field runoff at farms, Hurd et al. 1999; Higgins 2009) and *other contaminated waters* such as the glycol-contaminated stormwaters at airports resulting from cold weather aircraft de-icing operations (Higgins et al. 2007c,d; Wootton et al. 2010). Any of these WWs if released to the environment untreated or inadequately treated can result in adverse impacts to surface waters and groundwaters.

The parameters measured in WWs that are most germane to the design of EEBs include those that measure *carbon or its oxygen demand* (e.g., biochemical oxygen demand [BOD: five-day, ultimate and carbonaceous]; chemical oxygen demand [COD] and total carbon [TC] (total organic carbon and total inorganic carbon [TIC]); *phosphorus* compounds (organic phosphorus [Org-P], ortho-phosphate [o-PO_4], and total phosphorus [TP]); *nitrogen* compounds (organic nitrogen [Org-N], ammonia-, nitrite-, and nitrate-nitrogen [NH_3-N, NO_2-N, and NO_3-N] as well as total Khjedahl nitrogen [TKN, the sum of Org-N and NH_3-N], total nitrogen [TN], and nitrogenous BOD); *suspended solids* (usually measured as total suspended solids [TSS]); *ions* (e.g., sulfides, cyanides, chlorides, hydroxides, and oxyanions such as sulfates and arsenates) and *metal(loid)s* (both dissolved [filtered] and total [unfiltered] and particularly involving ones such as Al, B, Ca, Cd, Cu, Co, K, Hg, Mg, Mn, Mo, Na, Ni, Sb, Sn, Pb, and Zn, with special attention to *iron* [Fe_T, ferrous, and ferric]).

Also important are the measurements of *pH; biodegradation products* such as *volatile fatty acids* (VFAs); *acidity and alkalinity* (both usually measured as mg $CaCO_3$/L); the *dissolved oxygen* content (DO usually measured in mg/L); the *redox* condition (the oxygen reduction potential [ORP] or Eh, usually measured in millivolts, mV); *turbidity* (NTU), and *conductivity* (a measure of salinity and TDS, μS/cm) (Bass Becking et al. 1939).

For specific WWs, parameters germane to EEB treatment may also include concentrations of various kinds of organic compounds such as aliphatics, aromatics (e.g., BTEXs, PAHs, and PCBs), dioxins and furans, other chlorinated aromatics, oil and grease, and VOCs of various kinds.

2.2 Discrete and Lumped Parameters

While some contaminants in WWs exist in *discrete* forms (dissolved single species such as ammonia or many dissolved metals and metalloids), many others occur in a range of forms (e.g., dissolved, colloidal, and particulate) and, for the latter (e.g., the particulates of suspended solids), may exist in ranges of particle sizes and morphologies. In a natural WWT system, these components may be each removed at different rates. Such types of CoCs are deemed *lumped parameters*. Examples of lumped parameters include TSS, Org-N, Org-P, BOD, TN, and TP.

In addition, some other CoCs may be preferentially associated with certain parts of a lumped parameter. For example, many metals, metalloids, pathogens, and organic materials are often found associated with mineral and organic particulates in WWs (see below), and the removal of these will also remove that part of them. Their removal leaves the non-particulate parts of the lumped parameters (e.g., the colloidal and dissolved components) to be removed more slowly later in the system. To design EEBs, it is important to appreciate the form of the CoCs in the WWs to be treated.

2.3 The Treatment of WWs in EEBs

2.3.1 The Removal of Organics in EEBs

Organics may be present in WWs in or as part of solid particulate forms (e.g., entrained grit), as suspended solids, as colloidal material (only some of which may be filterable) and as dissolved species. These organics may be aliphatics, aromatics, organic acids, chlorinated compounds and solvents (e.g., PCP and CCl_4), waxes, polymers of many kinds, energetics (TNT, RDX, and nitrobenzene), sludges (e.g., sewage sludge and industrial WWTP biosolids), slimes, biocides (herbicides and pesticides), oxygenated compounds (e.g., alcohols, glycols, ethers, and aldehydes), plant materials and their degraded products (e.g., cellulose and lignin), as well as a host of other chemicals and materials and the degradation products thereof. EEB Systems have to be capable of managing any and all of them.

Bulk removal rate constants for lumped parameters such as BOD, Org-N and Org-P available from the literature may reflect the rapid and easy removals of the particulate parts, and may not be indicative of the (different) rate constants for the removals of the colloidal and dissolved components. This may be important for bioreactor design for advanced systems if such a component is the object of the treatment. In general, in well-designed EEB Systems, it is preferable that the bulk removals of any entrained material and the particulate portions of contaminants occur in upstream primary treatment cells, leaving the EEB cells *per se* to deal with the more recalcitrant components.* Accordingly, treatability tests for a WW (see Chapter 6) may have to simulate a primary treatment step (e.g., bulk removal of suspended solids) so that data from the secondary, EEB treatment step, more accurately reflect what would be expected in a full-scale EEB System.

BOD is a measure of the quantity of oxygen that microbes use to oxidize organics, and BOD tests for WWs usually are for a 5-day period (BOD_5), even though after this period there will still be biologically oxidizable organics left in the sample being oxidized. The BOD_5 test† is used since allowing biological degradation to proceed to its culmination, ultimate BOD (BOD_u), usually would take too long to be useful as a practical analytical test on a regular basis.

TOC, COD and BOD are the organic parameters usually of most interest in the design and operation of EEB Systems.

COD is a measure of the capacity in water to consume oxygen during the transformation of carbonaceous compounds and materials (including the

* EEB cells alone can remove entrained materials and suspended solids by a variety of often interrelated processes such as settling, filtration and physical–chemical precipitation and sorption near their inlets but this, over time, may lead to the formations of mats and "hard pan" that will impede influent flows.

† A BOD_7 test is often used in Continental Europe. It is ~1.15 × BOD_5 for domestic wastewaters.

decomposition and biodegradation of plant and other organic materials) plus the amounts of oxygen needed to oxidize any ammonia and nitrite present.

Accordingly, COD is higher than BOD since a WW may contain nitrogen compounds and organics not amenable to biological degradation. Typically, for domestic WWs, BOD_5/COD varies from 0.4 to 0.8 (averaging about 0.6 for municipal WWs) while BOD_5/TOC varies from 1.0 to 1.8 (averaging about 1.6, Metcalfe & Eddy 2004). There are two levels of BOD: that caused by the microbial degradation of carbonaceous material such as cellulose, sugars and glycols to CO_2 and water (cBOD), and that caused by the further degradation (oxygen demand) of nitrogenous material (nBOD) such as proteins and other Org-N compounds.

BOD is approximately the sum of 1.5 times c-BOD plus 4.6 times TKN. BOD may then be defined as

$$BOD = cBOD + nBOD \tag{2.1}$$

nBOD is important as its impact on total BOD can be many times higher than its concentration would indicate. For example, the nBOD of ammonia is about 4.5 times the ammonia concentration present.

With many municipal WWs treated in EEB Systems, typically 75% of suspended solids and 40% of the filterable solids are organic compounds, and a large proportion of these are removable during primary treatment before EEB cells (Metcalfe & Eddy 2004). Some WWs (e.g., landfill leachates), in addition to containing suspended and filterable organics, will contain varying amounts of colloidal and soluble organics that may be made up of proteins, carbohydrates, fats and oils. These are susceptible to biological removal by microbes in the fixed films of EEBs.

The very rapid *aerobic biodegradation* of dissolved and soluble organics by the heterotrophic bacteria of biofilms in aerobic EEBs such as BBRs may be illustrated by:

$$(CH_2O) + O_2 \rightarrow CO_2 + H_2O + \text{new bacteria} \tag{2.2}$$

where (CH_2O) represents an organic of any sort. Organics may also degrade or otherwise transform abiotically in a bioreactor. The transformation (biotic or abiotic) may not be direct and may proceed via several intermediate steps, forming by-products that in turn are oxidized into the final product. Two main groups of bacteria are involved in aerobic biodegradation: aerobic *chemautotrophs* that effect the oxidation of the organics (including the ammonification of organic nitrogen compounds) and the other, nitrifying chemautotrophs, that can also effect the oxidization of ammonia.

The anaerobic biodegradation of organics in EEBs (and there will always be anaerobic micro-environments in the most aerobic EEBs and vice versa) will be carried out by facultative bacteria and by obligate anaerobic heterotrophic bacteria in the biofilms.

For example, the following equations illustrate the *anaerobic biodegradation* by the heterotrophic bacteria of biofilms in anaerobic EEBs such as BCRs of an indicative organic compound in a WW (here represented by $C_6H_{12}O_6$) to acetic acid, lactic acid and ethanol, respectively.

$$C_6H_{12}O_6 \rightarrow 3CH_3COOH + H_2 \tag{2.3}$$

$$C_6H_{12}O_6 \rightarrow 2CH_3CHOHCOOH + H_2 \tag{2.4}$$

$$C_6H_{12}O_6 \rightarrow 2CO_2 + 2CH_3CH_2OH \tag{2.5}$$

The most common partial anaerobic biodegradation product in an EEB ABR is acetic acid. The acids, etc. produced by the first stages on anaerobic biodegradation are then further biodegraded by methane-generating bacteria and archaea to form the ultimate products (e.g., CH_4, CO_2, NH_3, and H_2S). In addition to biodegradation, organics (especially higher molecular weight ones and xenogenic ones) in a WW being treated in an EEB ABR may be removed by precipitation, sorption, biofiltration and biostabilization.

For the operation of EEB Systems, there are two kinds of artefacts associated with the testing of BOD and COD as carried out in commercial analytical testing laboratories that must be considered in assembling data to be used to design EEBs.

For BOD, standard testing includes inoculating a WW sample with a defined kind of microbial consortium and measuring oxygen demand for the biodegradation of organics over the defined period (i.e., 5 days). This is fine where the microbial inoculants involve bacteria "acclimatized" to (having been previously exposed to and having "learned how" to metabolize the kind[s] of the organic material[s] present in the WW [e.g., those in sewage]), so many of them can begin metabolizing the organics as soon as the BOD test begins. However, where the organics in the WW being tested are dominated by a compound with which most of the standard formulation bacteria are not already acclimatized (e.g., propylene glycol in an airport WW), there may be a period before those bacteria in the commercial inoculant that can readily metabolize the unfamiliar compound become dominant and take over its biodegradation. This results in a "lag" or delay period before biodegradation really gets underway. In some cases, this may use up part of 5-day BOD test period, and as a result, the standard BOD test may seriously underestimate concentrations (by up to 50% or more) (the ultimate cBOD will not be affected).

The lag period may be eliminated by including in the BOD_5 test's microbial inoculant bacteria already acclimatized to the expected organic. However, since regulators monitoring the performance of WWT systems such as those involving EEBs usually want BOD performance testing to be carried out at "arm's length" by accredited commercial analytical labs, such is usually not an option. For those having EEB Systems constructed to treat their WWs, it is recommended that they negotiate in advance with their regulators to have performance monitored using COD instead, after presenting to them the results of correlation studies on the relationships between ThOC, BOD_u, COD and TOC for the specific WW involved.

There are sometimes problems with the measurement of COD by commercial analytical labs using standard methods in cases where very concentrated samples are involved (e.g., de-icing fluids and residual streams from their use at airports containing more than about 30% glycol, see the next section). Measurements with highly concentrated samples often turn out to be higher than what should be expected based on the concentration of the dominant pure organic in them, and the small amounts of impurities and additives in the commercial products usually cannot explain the differences.* In this case, the discrepancy may be due to the fact that to measure the COD of samples containing a very concentrated organic, the sample must be first diluted, then the diluted sample must have its COD measured, and finally the COD of the undiluted sample is estimated by back-calculation.

Where the amounts of dilution are measured by weight (which can be measured much more accurately) rather than by volume (which is harder to measure as accurately), such discrepancies often disappear.

2.3.2 The Treatment of Airport WWs in EEBs

As was mentioned, airports are one of the most important areas of use for BBR-based EEB Systems. Many contaminants are found in stormwater runoff and other WW streams at airports (e.g., they may include grit, dust, sand, lubricant residuals, oil and grease, sewage from leaks, spilled fuel, firefighting chemicals,† chemicals used to remove rubber from runways,‡ washing product chemicals [e.g., detergents] and fertilizers used on grassed areas around taxiways and runways). In addition, at certain times of the year, these streams will also contain residuals from surface (pavement) de-icing chemicals and glycol-based freezing point depressant solutions (fluids) used in de-icing operations to control ice and snow build-ups on aircraft fuselages during freezing weather.

* For example, pure EG has a COD of about 1.1 MM mg/L (see Equation D.3 in Appendix D) but measurements of the COD of de-icing fluids containing 92% of it by commercial analytical labs using standard methods often indicate values above 1.3 MM mg/L.

† Many airports, especially the larger ones, have test areas where firefighting and other emergency response methods are practised regularly. The foams and retardants used in such testing may enter airport runoff collection systems, and in some cases can represent significant pollution sources.

‡ Tire rubber is left on airport runways during landings and may be periodically removed by washing with high-pressure water containing chemicals that soften and de-bond the rubber, allowing its hydraulic removal. The rubber and the chemical solution get washed onto adjacent grassy areas from which they and their degradation products can enter stormwater runoff at the airport (Findall & Basran 2001).

The biological treatment of glycol-contaminated streams from airports is especially challenging since doing so involves dealing with cold, nutrient-poor WWs of variable flows and strengths (Murphy et al. 2014). Since they usually involve large areas, runoff from airport watersheds during and after large precipitation events can cause upsets to any on- or off-site WWT facilities, and this necessitates the provision for them of bypass measures and large volume balancing ponds and/or vessels in front of them. Even so, severe disturbances in their operations are common in such situations. One of the few technologies capable of coping with these large volumes of runoff at high efficiencies are BBR-based EEB Systems.*

Glycol-contaminated WW at airports can involve up to three streams: (1) *Spent Glycols*, relatively concentrated solutions (up to many per cent glycol) which are collected from beneath aircraft at de-icing areas and pads and/or at central de-icing facilities (CDFs); (2) *glycol-contaminated stormwater* (GCSW) where de-icing fluids are blown by jet or prop wash onto areas adjacent to de-icing areas, drip off onto aprons and runways during aircraft movement, or are collected with removed snow and ice, ending up in stormwater sewers and ditches and (3) glycol-contaminated *snowmelt* from snow and ice ploughed up from runways, aprons, taxiways and de-icing pads.

Spent Glycols are often vacuumed up around de-icing areas, pads and CDFs using Glycol Recovery Vehicles (GRVs, vacuum sweeper trucks or trailers) and/or enter dedicated sewers (this is the case at BNIA). Spent Glycols can be treated on-site, processed for recycling on- or off-site, or sent off-site to local WWTPs (usually municipal ones). Indeed, sending at least some of their Spent Glycols to a local municipal WWTP is the most common method of glycol management at northern airports.

Generally, municipal WWTPs charge Airport Operating Authorities (AOAs) and/or their airport's Sprayers (firms contracted by AOAs, airlines and others to carry out de-icing operations) for both the volumes of Spent Glycols (and, in some cases, some GCSW as well) sent to them as well as (in many cases) "BOD surcharges" for organic concentrations above some defined limit. Such WWTPs usually want Spent Glycols diluted to <10% glycol before acceptance (Strong-Gunderson et al. 1995). As a result, often Spent Glycols received at a WWTP cannot be sent directly into its facilities undiluted, and there has to be storage at their facilities (tanks and impoundment basins) to receive this concentrated material and introduce it in controlled amounts into municipal sewage being treated. This is often due to the fact that even small amounts of airport glycols fed into an activated sludge unit without control can cause a deterioration in floc structure and enhance the formation of *Nocardia* scum in an aeration basin. In some cases, WWTPs are less than keen to receive glycol-contaminated streams from local airports.†

* Aircraft and surface de-icing chemicals and methods are addressed in more detail in Appendix D.

† Indeed, such reluctance was one of the reasons for the EEB project at BNIA (see Section 3.6 in Chapter 3).

The other major destination for Spent Glycols is recycling by firms that specialize in doing so (Recyclers). These companies collect Spent Glycols at some airports and send them to off-site recycling facilities directly or via on-site *concentrators* (the latter are distillation units that concentrate the glycols to typically about 50%–70% by volume before they are sent for upgrading for use in materials such as automotive anti-freeze*). Generally, recovery of Spent Glycols becomes economic when the glycol content is above 12% (Zitomer 2001), so usually only Spent Glycols are concentrated (although if there are convenient, impounded and richer glycol-contaminated stormwater volumes available, some of them may be sent to concentrators as well). Both the overheads (distillate) and bottoms streams from any on-site concentrators have to be managed. The distillate from a concentrator may still contain ~0.5% glycol (5000 mg/L) and exert a BOD_5 of 3000–4000 mg/L. Any bottoms stream from a concentrator will likewise still contain not insignificant amounts of glycol and has to be managed at airports having them as well. Generally, in addition to the revenues they receive from selling the recycled glycols from their concentrators, Recyclers often charge AOAs and their clients for the volumes of glycols managed.†

As was mentioned above, despite the best efforts of AOAs to deal with all of their ADAFs by managing *Spent Glycols*, a significant portion of the sprayed fluids will end up in glycol-contaminated stormwater. Since airports cover large areas, they generate large volumes of GCSW, volumes that are usually too large to send off-site to local WWTPs, except in limited amounts and in specific cases.

There are two kinds of glycol-based fluids used at airports. *Aircraft de-icing fluids* (ADFs) are used to melt and remove accumulated snow, ice and frost on aircraft surfaces just before departures, and *aircraft anti-icing fluids*(AAFs) used to inhibit snow and ice build up on aircraft at areas and other locations where they are parked for periods‡ (collectively, ADFs and AAFs are referred to as *aircraft de-icing and anti-icing fluids*, ADAFs).

In North America, ADAFs are based on either *ethylene glycol* (EG, $C_2H_6O_2$) or *propylene glycol* (PG, $C_3H_8O_2$). ADAFs also contain water and small amounts

* At Montreal-Trudeau Airport (YUL), the AOA, Aéroports de Montreal and a recycling firm have installed sophisticated distillation and processing equipment to collect and upgrade Spent Glycols from the airport's CDF and imported from other airports back into airport-grade fluids that are re-used for de-icing. Despite this, a considerable portion of the ADAFs sprayed at YUL still reports to GCSW.

† One of the motivators for an EEB System being considered for a northern airport is the fact that once that airport installed a CDF, the amounts of glycol being recovered increased substantially compared to those from the previous at-gate de-icing operations, and the amounts that were charged to the AOA for collecting the Spent Glycols, concentrating glycols from it in an on-site concentrator facility and trucking the concentrated material to off-site upgrading facilities increased by a very large amount.

‡ See Appendix D for more details on them.

of proprietary "additive packages" (*ADPACs**). Industrial grade glycol-based fluids also normally contain some impurities from their manufacturing process, as well as very small amounts of their polymeric analogues (e.g., polyethylene glycol in the case of EG).

During de-icing, ADFs are diluted by the hot water they are sprayed with and the water and the glycols end up in Spent Glycols along with water from the snow/ice/frost they flush from the aircraft surfaces being de-iced, plus snowmelt and any precipitation occurring. As a result of de-icing operations at airports, glycol-contaminated WW streams will be generated and these may enter local ditches and ponds where they begin to biodegrade microbially both aerobically and anaerobically (Dwyer & Tiedje 1983; Clifton et al. 1985; Rossiter et al. 1985; Strong-Gunderson et al. 1995; Higgins & MacLean 1999). This can result in the contamination of receiving waters and noxious odours† if their discharge is impeded in any way (e.g., hold up in ponds or ditches).

As was mentioned, in addition to the glycol itself, ADAFs contain varying proportions of water, contaminants, impurities and the ADPACs. Usually, hot water makes up about half of the volume of the ADAFs sprayed in aircraft, and glycol concentrations in the sprayed fluids can range from 30% to 70%. For the approximately 50% glycol ADAFs commonly used for aircraft de-icing and anti-icing, BODs are in the hundreds of thousands of mg/L‡ (Higgins & MacLean 1999). The Spent Glycols from sprayed ADAFs during de-icing can have BOD concentrations as high as 50,000 mg/L or more at some airports, and these are hundreds of times the BOD concentration of raw municipal sewage (EPA 2000a,b). Such high concentrations, combined with the inherently variable nature of winter storms and the low temperatures of the glycol-contaminated streams, challenge the use of conventional stormwater treatment technology. A large slug of such glycol-contaminated water can quickly deplete the dissolved oxygen in receiving waters, killing aquatic organisms that need aerobic conditions. In addition, some glycols such as EG can be quite toxic if they get into the environment without treatment.

GCSW at airports represents the destination for a significant part of the total volume of ADAFs used (up to 90% at some airports but typically 40%–60%) and these streams used to be rarely treated. Concentrations of glycols and their daughter products (VFAs, see Section D.1 in Appendix D) in GCSW at airports can range from several tens of mg/L to many thousands of mg BOD/L.

Snowmelt results from the melting of glycol-contaminated snow and ice that is ploughed up from runways, taxiways and other airside paved surfaces

* The ADPACs are chemicals added in small quantities (often ~1%) to meet insurance and other requirements and to provide desired properties (e.g., corrosion protection, see Section D.2 in Appendix D).

† Indeed, it was the odours from such degradation once temperatures rose (especially anaerobic degradation) that led to the construction of facilities for treating glycol-contaminated streams, both for a CW system at EIA, see Section 3.7 in Chapter 3) and for an anaerobic digester-based WWT plant at Albany International Airport (ALB) in New York State, USA (aerated EEBs such as BBRs do not generate unpleasant odours).

‡ See Table D.3 in Appendix D for some typical values

in winter and piled in various areas around an airport. Often, airports have dedicated snowmelt pads onto which the most contaminated ploughed up snow and ice (e.g., that collected from on and around de-icing areas) is accumulated. During warming periods (e.g., the Spring freshet), the collected snow and ice melt and the glycol-contaminated material in it enters the drains at/under pads or a CDF, often into the Spent Glycol stream.*

An evaluation of ploughed snow at a medium-sized airport showed that even after most of the glycols in melting snow had been removed, ADPAC chemicals (see Section D.2 in Appendix D) remained in the snow, and the snowmelt from them can result in toxicity in receiving streams well after the end of a de-icing season if the meltwater is not directed into streams that are treated (Corsi et al. 2006). The results indicate that at airports, just measuring the glycol concentrations in a stream or BOD alone will not be an indication of the fate and transport of toxic ADPAC components.

EEB Systems for airports will always involve secondary treatment in BBRs and normally do not need to include other secondary treatment components such as ABRs or Aerated Lagoons. EEB Systems at airports may range from ones treating only Spent Glycols and a limited amount of snowmelt (as is planned for YYT) to ones treating a range of airport WWs including Spent Glycols, GCSW, snowmelt, surface de-icing chemicals, runoff from paved and grassed areas and a host of other CoCs.†

Table 2.1 illustrates the composition of two EG-contaminated streams from two different airports: one a diluted Spent Glycol stream which contains little else (Stream 1), and the second, the aforementioned stormwater runoff stream (Stream 2).

TABLE 2.1

Some Analyses of Two Different Glycol-Contaminated Residual Streams

Parameter	Stream 1	Stream 2
Ethylene glycol	13%	41 mg/L
COD (mg/L)	26,000	–
BOD (mg/L)	21,000	35
TOC (mg/L)	7100	19
NH_3-N (mg/L)	2	0.2
NO_3-N (mg/L)	0.1	0.1
TKN (mg/L)	20	1
o-PO_4(mg/L)	10	0.03
TP (mg/L)	12	0.1
ORP (mV)	180	130

* For example, such is the case at YYT and BUF.

† The treatment of airport wastewater is seminal to the EEB ecotechnology and, as a result, an appendix (Appendix D) is devoted to reviewing the background and related information about the glycol-containing fluids that are used at them.

As may be seen from the Stream 1 analyses in Table 2.1, this WW would definitely require treatment before discharge to receiving waters by the airport (or need to be diverted to a local WWTP). Also, most of its nitrogen is in the form of Org-N (TKN less ammonia and nitrate, see Section 2.4) that would most likely be ammonified in an aerated EEB, and the bulk of its phosphorus is in a reactive form at relatively high levels.

If the analyses of Stream 2 in Table 2.1 were indicative, it might be discharged to the environment without treatment. However, seasonal variations cause its organic content to vary from that indicated to thousands of mg/L, and it too would require treatment. The point here is that ostensibly similar streams (glycol-contaminated residual streams resulting from airport de-icing operations) can be vastly different in composition, and their treatment might require completely different morphologies for EEB Systems. In each case, a treatability test would be essential if EEB treatment were to be considered (see Chapter 6).

The complete oxidation of EG* (either biotically and/or abiotically) may be given by:

$$\underset{\substack{\text{ETHYLENE GLYCOL}\\ MW = 62\,g/mol}}{C_2H_6O_2} + \underset{\substack{\text{OXYGEN}\\ MW = 32\,g/mol}}{2.5O_2} \rightarrow \underset{\substack{\text{CARBON DIOXIDE}\\ MW = 44\,g/mol\\ MW\ Carbon = 12\,g/mol}}{2CO_2} + \underset{\text{WATER}}{3H_2O} \tag{2.6}$$

This equation may be used as a basis for understanding the relationships of the parameters for glycols at airports (Higgins & MacLean 1999; Switzenbaum et al. 2001; ACRP 2012). As may be seen from the above equation, it requires 2.5 moles of oxygen to completely oxidize one mole of ethylene glycol, producing two moles of carbon dioxide and three moles of water in the process.

The amount of oxygen required is known as the *theoretical oxygen demand* (ThOC). For pure EG, the ThOC for aerobic degradation is 2.5 mol $O_2 \times 32$ g O_2/mol O_2 divided by 62 mol EG/g EG = 1.29 g O_2/g EG. For EG, the ThOC concentration is essentially the same as the COD concentration because the analytical procedure for determining COD in the laboratory essentially destroys virtually all of the EG in a sample.† Also, the EG concentration times 2×12 (the moles of carbon as part of the CO_2 product) divided by the molecular weight of EG (62) gives the total organic carbon‡ (TOC) concentration.

* Equation D.2 in Appendix D illustrates the equivalent aerobic transformation of PG.

† The same is true for PG, see Equation D.4 in Appendix D

‡ TOC is an important parameter used in the design of BBRs for treating glycol-contaminated streams at airports, one that can be correlated to cBOD and COD and measured by on-line instruments. ThOC, BOD, COD and other physical parameters for EG, PG and diethylene glycol (DEG, another glycol sometimes used for aircraft de-icing in Europe) are summarized in Table D.1 in Appendix D.

For pure EG, it is 0.39 times its concentration in mg/L. While Equation 2.6 represents the complete oxidation of EG, in many cases, transformation by biodegradation many involve intermediate steps that form VFAs*. For example, when the dihydric alcohol, ethylene glycol, is oxidized, the first VFA formed is a hydroxy acid, *glycolic acid* (also called hydroxyacetic acid), and the next before the carbon–carbon bond breaks is *oxalic acid* (also called ethanediotic acid).

$$\begin{array}{ccccc} CH_2OH & & CH_2OH & & COOH \\ | & \rightarrow & | & \rightarrow & | \\ CH_2OH & & COOH & & COOH \\ \text{EG} & & \text{Glycolic acid} & & \text{Oxalic acid} \end{array} \quad (2.7)$$

Formic acid (HCOOH) and glyoxylic acid (COCOOH) may also be formed, and the presence or absence of certain metal ions may influence VFA formation. Together with acetic acid, glycolic acid (the conjugate base is glyoxylate†) and oxalic acid are known as C_2 carboxylic acids.‡

VFAs can accumulate in vessels and ponds where glycol-contaminated residual streams from airports are stored or held up, and their presence will be indicated by lower than expected pHs of the WW in them. Such can negatively impact some kinds of downstream treatment facilities (e.g., nitrification is inhibited at lower pHs). All of the VFAs are readily removed in BBRs and in them concentrations rarely exceed one mg VFA_T/L. Nevertheless, there are ramifications to VFA formation for the storage and handling of glycol-contaminated residual streams at airports.

As is discussed above, common analytical tests (e.g., those for BOD or COD) used in the monitoring of CoCs at airports do not distinguish between the glycol and the VFA degradation products if they persist. If such is not taken into account, it may effect de-icing operations in general.

The State-of-the-art for BBR-based EEB Systems at airports is addressed in Section 9.2.2 in Chapter 9.

* While the term VFA refers specifically to "fatty" acids and such are the bulk of the degradation products, other glycol breakdown products such as alcohols, ketones, ethers and aldehydes may also be present when glycols degrade. Analytical techniques to identify total VFAs (VFA_T) measure the sum of the concentrations of all of these, including the nonacid breakdown products, and calibrated gas chromatographs (GCs) have to be used instead to identify specific VFAs when more accurate results are desired.

† The glyoxylate cycle is important as it enables bacteria, fungi and plants to convert fatty acids to carbohydrates.

‡ The equivalent C3 cycle involves oxidative biodegradation of the trihydric alcohol, 1,3 propylene glycol, first to the hydroxyl acid, ethylene lactic acid, then to the diprotic acid, malonic acid. There are many more kinds of VFAs possible when de-icing using PG-based fluids and their biodegradation may result in the presence of the C2 carboxylic acids as well.

2.3.3 The Treatment of Municipal WWs in EEBs

Municipal WWs include sanitary sewage, septic tank overflow, sewage sludge (biosolids), septage, combined sewer overflow (CSO), municipal landfill leachates, and municipal stormwater. All of these types of domestic WWs can be readily treated in EEB Systems, mostly ones based on 2° BBR cells.

One of the most common uses for small HSSF BBRs has been as the main treatment step in on-site WWT Systems (where they replace the leach bed of septic systems for single homes) and in larger de-centralized WWT systems for communities of a number of homes (and often other facilities) which, although connected by sewers, are not accessed by major trunk sewers to large central WWTPs (the latter are often also called "cluster" systems).

Treated water from BBR-based decentralized WWT systems may be infiltrated into local soils (if such are suitable and regulations permit) or disposed into receiving waters. The following picture shows how a 64 home, cluster-type EEB System in the community in MI, USA involving an HSSF BBR with effluent disposal via an infiltration bed can be designed into the landscape (Figure 2.1).

FIGURE 2.1
A cluster-type EEB System.

Table 2.2 compares the typical ranges of concentrations of some CoCs in septic tank overflow, primary sewage and municipal facultative WWT lagoon water.

TABLE 2.2

Typical Domestic WW Composition Ranges (mg/L)

Parameter	Septic Tank Overflow	Primary Sewage	Lagoon Effluent
BOD	120–250	30–2000	10–40
TSS	40–60	50–400	20–80
NH_3-N	20–40	15–40	0.5–20
NO_3-N	0–1	0	0.1–1
TP	10–15	10–50	3–4
Fecal coliforms[a]	5–6	5–7	1–6

[a] log/100 mL

In EEB Systems, as with active mechanical WWTPs, treatment may be categorized as being to primary, secondary or tertiary levels with, as was mentioned, EEB cells providing the secondary treatment.

Table 2.3 illustrates typical treatment levels for a variety of CoCs for municipal WW.

TABLE 2.3

Typical WWT Levels (mg/L)

CoC	Primary	Secondary	Tertiary
TSS	30–100	20–40	10–20
BOD	100–200	20–40	10–20
NH_3-N	n/a	5–10	<5
NO_3-N	n/a	20–40	<10
TP	>10	0.3–0.5	<0.1

In municipal WWs such as sewage, significant proportions of the CoCs are associated with particulates, and although these can be removed in the 2° treatment EEB cells, it is more appropriate that they be mostly removed upstream in the primary treatment cells/components of an EEB System that precede them. In municipal sewage, most of the suspended solids, 65%–85% of the organics, 30%–60% of the phosphorus, 20%–40% of the nitrogen, 20%–80% of the heavy metals and all *Cryptosporidium* oocysts can be removed in this manner. Other pathogens (bacteria and viruses) may sorb to particles in the sewage and can be removed with them. This will leave the BBRs (and, in some cases, anaerobic EEB cells such as BCRs as well) to deal with the remaining (and more recalcitrant) soluble and colloidal CoCs.

The original development of "Engineered Wetlands" following the *BREW Project* led to the design, construction and successful operation of more than 100 of these BBRs, mostly for municipal WWT. Most of these early EWs (BBRs) are vegetated with wetland plants and are relatively small, ranging in treated WW flow rates from a few m^3/d to a few hundred m^3/d.* Municipal WW streams generally contain enough bioavailable nutrients (phosphorus and nitrogen) such that additional amounts need not be added to EEB System influents (in contrast to the case with airport GCSW).

In the case of highly polluted municipal WW streams such as septage (which can have typical CODs of 25,000 mg/L, TSS values of 40,000 mg/L, ammonia concentrations of 200 mg/L and TPs of 250 mg/L), considerable upstream (primary) treatment is required (e.g., filtration, settling, and chemical dosing) and the 1°-treated stream still may have to be metered into other, more dilute, higher flow influent streams before 2° treatment in EEB cells.

Conventional WWT systems in most municipalities involve extensive sewer systems to a series of pumping stations that pump the collected sewage streams from local areas to one or more large central WWTPs where sewage is treated, disinfected and discharged to receiving waters. The operation, maintenance and other upkeep (chemicals and electrical power) of these sewage systems is complicated and expensive, require large staffs to operate 24/7, are subject to periodic disruptions (e.g., broken sewer pipes) and necessitate the regular replacements of various components (e.g., pumps, motors, pipes, and instruments). EEB Systems for municipal WWT are well-proven technology and they offer (to those municipalities where sufficient land and other infrastructure is available) the option of replacing (and/or supplementing as a community expands) central WWTPs with a series of unconnected local EEB Systems that would operate independently of each other, significantly reducing CAPEX and OPEX.

2.3.4 The Treatment of MIWs in EEBs

Another major use for EEB Systems is in the treatment of *mining influenced waters*. Much of the treatment of MIWs in EEB Systems involves removing suspended, colloidal and dissolved metals and metalloids using EEB Systems that involve BCRs.

Mining at underground mines, open pit mines, heap leach facilities, quarries, etc. produces a variety of WWs containing suspended, colloidal and dissolved metals, metalloids and other CoCs, many of which are especially challenging to treat. In addition to the kinds of WWs found at mine sites (e.g., waste rock and tailings pile leachates and percolates, mill process waters, site runoff and ore pile runoffs), reclaim water and other streams may also contain significant levels of ammonia and nitrates resulting from the use of ammonium nitrate-based blasting explosives.

* But not all. The Casper EEB System has a design flow rate of 6000 m^3/d (See Section 3.4 in Chapter 3) but it was designed to treat contaminated groundwater at a closed oil refinery site, not municipal wastewater.

At base metal mines and their associated mill facilities, the primary metals found in the WWs are iron, aluminum, and manganese, while minor metals and metalloids (collectively metal[loid]s) also present may include As, Cd, Cr, Cu, Ni, Pb, Sb, Se, and Zn. Anions and oxyanions such as sulfates, carbonates, chlorides and a host of others are usually present as well.

At precious metal mines (e.g., those mining for gold, silver, and platinum group metals) and their associated mill facilities, WWs may also contain ammonia, cyanides, Hg, Mo, Sn, Sb, and other CoCs.

Graphite mines and quarries for phosphate rock and other industrial minerals will each produce WWs with unique qualities and CoCs. Some WW streams from some kinds of mines may be highly acidic, while others may be quite basic. Surface and underground blasting to access ores may lead to the contamination of WWs at mines with fuel oils, ammonia and nitrates. All of these WW and runoff streams can be categorized as mining influenced waters.

MIWs then are any waters whose chemical compositions have been affected by mining or mineral processing (Wildeman & Schmiermund 2004). MIWs include highly acidic WWs such as ARD; alkaline *mineral processing waters* (MPWs) that often contain various other oxyanions as well as sulfates; *marginal waters* (any other MIWs impacted by mining/mineral processing that have alkalinity and pHs of about 7); and *residual waters* (basic waters that contain high levels of TDS in the form of sodium, potassium, chloride and sulfate ions).

The most recalcitrant MIWs are (1) *ARD* (pHs of <3.5, acidity of >500 mg $CaCO_3$/L, containing dissolved metals such as iron, aluminum and/or manganese at concentrations of more than 50 mg/L, and with elevated levels of sulfates) and (2) *MPWs* (high pH, and often containing soluble oxyanions such as arsenates and selenates). Mining influenced waters also include *neutral mine drainage* (NMD) (Higgins et al. 2006a–c). EEB Systems can be designed to treat any of these MIWs.

MIWs may also contain metal and other ion complexes (e.g., those with cyanides). Indeed, continuing long-term releases of metal and sulfate-contaminated waters such as ARD and other metal-contaminated MIWs occur at an estimated 70% of the world's mine sites (Lopez et al. 2009), and regulations for the concentrations of them in WWs discharged to the environment are becoming progressively more stringent in most jurisdictions.

Conventional (active) WWT facilities at mine sites are often used to manage MIWs during operations and can continue to be used to treat such streams after mines are closed down. However, as some MIWs may continue to flow for extended periods (decades to centuries) after mining has ended and as conventional WWT facilities are costly to operate and maintain, their economic viability over such extended periods is questionable. The mining industry is seeking more "passive" methods to manage such MIW streams, especially ones that can, over the longest periods, allow full walk-away closures. EEB Systems can meet this need.

Owing to the nature of MIWs, EEB Systems for treating them at mine sites tend to be much more complex than those at airports, or even those for treating for municipal landfill leachates. Typically, they will consist of relatively large primary treatment components (e.g., Sedimentation Ponds and limestone drains), followed by secondary treatment parts involving both aerobic and anaerobic EEBs. These in turn will be followed by tertiary treatment components such as polishing CW cells. In addition, effluents from EEB Systems may be directed through Stormwater Wetlands also treating site runoff and/or into nearby local Natural Wetlands (if such are conveniently available) before discharge to receiving waters.

Figure 2.2 illustrates the Teck Demo, a demonstration-scale BBR-based EEB System that operated for many years at the Teck Cominco lead–zinc refinery and smelter in Trail, BC, Canada (see Section 5.6 in Chapter 5 for a detailed description). The Teck Demo treated an MIW contaminated with arsenic, cadmium and zinc. For the Teck Demo, the MIW was collected in French Drains around an old arsenic scrubber pond and old landfill and pumped to the EEB System which consisted of two passive treatment, upflow VSSF BBR cells in series followed by four Constructed Treatment Wetland cells in series, two HSSF CW cells, a FWS CW cell and a Pond Wetland cell (see Section B.3 in Appendix B for background on such CWs). Treated effluent from the final one, the Pond Wetland cell, was disposed of by spraying it on a grove of hybrid poplar trees (see Section C.6 in Appendix C for a review of phytoremediation methods such as the phytoirrigation method illustrated).

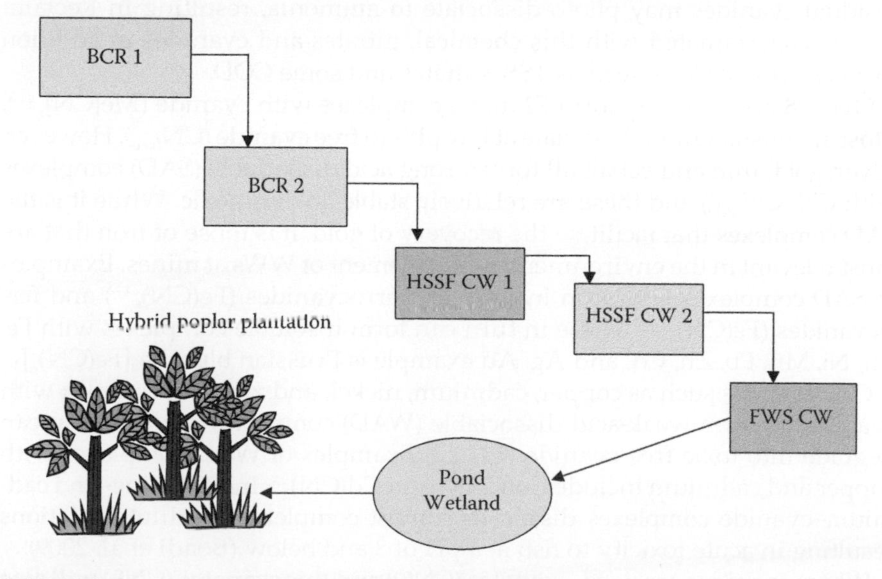

FIGURE 2.2
The Teck Demo EEB System.

2.3.5 The Treatment of MIWs Containing Cyanides

Cyanides deserve special mention (Palmer et al. 1988; Higgins et al. 2006c). Sodium cyanide is used as a lixivant in the recovery of silver and gold. About 90% of gold recovered worldwide relies on the use of cyanide (Mudder & Botz 2004). For gold mining, two larger mechanical plant-based cyanidation processes (the carbon-in-pulp process and the carbon-in-leach process) are generally used for higher (but still low) Au-containing ores, and heap leach (or valley fill) cyanidation which is used to *beneficiate* ores containing less than 0.04 Troy ounces of gold per ton of ore (EPA 1994, see Section 6.10 in Chapter 6 for an example of the treatment of MIWs from a heap leach facility using an advanced kind of EEB).

The cyanidation of process for extracting gold from ore using cyanides can be given by Eisner's equation:

$$4Au + 8CN^- + O_2 + 2H_2O \rightarrow 4Au(CN)_2^- + 4OH^- \tag{2.8}$$

The process is adversely affected by some other metals as well as free sulfur and sulfides. In the carbon-in-pulp or the carbon-in-leach processes used at gold mines and heap leach facilities, complexed gold is often removed from slurried crushed ore by carbon adsorption. At surface and underground gold mines, the remaining materials (tailings, often >99% of the total) are pumped as a slurry to a tailings pond where the solids settle out.

Decanted water from such tailings ponds (reclaim water) may (and usually is) be pumped back to the gold mine's mill to slurry more ore. In such WWs, residual cyanides may photo-dissociate to ammonia, resulting in Reclaim Waters contaminated with this chemical, nitrates and cyanides in addition to conventional CoCs such as TSS, sulfates and some COD.

Over 28 elements can form 72 metal complexes with cyanide ($Me[CN]_x^{y-}$). Most are unstable and dissociate at low pHs to free cyanide (CN_{free}). However, silver, gold, iron and cobalt all form strong acid dissociable (SAD) complexes with CN (CN_{SAD}) and these are relatively stable and nontoxic. While it is the SAD complexes that facilitate the recovery of gold, it is those of iron that are most relevant in the environmental management of WWs at mines. Examples of SAD complexes with iron include the ferrocyanides ($Fe(CN)_6^{4-}$) and ferricyanides ($Fe(CN)_6^{3-}$). These in turn can form insoluble complexes with Fe, Cu, Ni, Mn, Pb, Zn, Cd, and Ag. An example is Prussian blue, $Fe_4[Fe(CN)_6]_3$.

Other metals such as copper, cadmium, nickel, and zinc can complex with cyanides to form weak-acid dissociable (WAD) complexes (CN_{WAD}) that can degrade into toxic free cyanide (CN_{free}). Examples of WAD complexes with copper and cadmium include $Cu(CN)_2^-$ and $Cd(CN)_2^-$. Indeed, zinc- and cadmium–cyanide complexes dissociate almost completely in dilute solutions resulting in acute toxicity to fish at a pH of 8 and below (Boadi et al. 2009).

Where cyanides are used, cyanates (CNO) and thiocyanates (CNS) will also often be present and these can hydrolyse to ammonia. Analytical techniques are available to measure them as well as total cyanides (CN_T), WAD cyanides and free cyanides, but SAD cyanides are usually determined by difference as follows:

$$CN_{SAD} = CN_T - CN_{WAD} - CN_{free} \tag{2.9}$$

The chemistry of cyanides is important because an important area for the use of EEBs is in the management of cyanide-contaminated Reclaim Waters recycled from gold mine tailings ponds (Higgins 2006; Higgins et al. 2006c; also see Section 6.6 in Chapter 6).

EEB Systems can be designed to treat acidic, neutral and alkaline MIWs and have already been demonstrated to be effective in treating many of them. EEB Systems for mines (and associated facilities such as smelters and metal refineries) tend to be more complex than those for treating municipal and airport WWs, and usually involve both aerobic (e.g., BBR) and anaerobic (e.g., BCR) cells. They can be designed for almost unattended operations, whatever the ambient temperatures, and in many cases can be designed to evolve over time into more passive treatment systems.

A state-of-the-art, semi-passive treatment BCR-based EEB System designed to treat acidic MIW at a northern base metal mine is discussed in Section 9.2.3 in Chapter 9.

2.3.6 The Treatment of Leachates in EEBs

Leachates may involve internally generated CoCs and/or the emanations of pore water, runoff, seepage or drainage from landfills and other waste accumulations (e.g., fly ash and coal piles at power stations, bark and sawdust piles at sawmills and pulp and paper facilities, and manure piles at livestock production facilities). Depending on the source, leachates (including those from municipal, commercial, industrial, hazardous waste landfills) usually by their natures contain wider slates of CoCs and other species than do many other kinds of WWs. In addition, CoC concentrations in leachates can: vary in pH from acidic to alkaline; continue to be generated for extended periods (although their concentrations often decline over time); be difficult to assemble for treatment; and vary in temperature from warm to near freezing.

Before the treatment of any leachate in an EEB System can be considered, comprehensive sampling and CoC data analyses plus one or more treatability tests will always be required (see Chapter 6 for some examples of the results of treatability tests for leachates) (Nivala et al. 2012). The scope and morphology of any EEB System to treat a leachate will be different for each situation. It has already been demonstrated that EEB Systems involving BBRs and BCRs can effectively treat municipal solid waste (MSW) landfill leachates, and indeed the treatment of the leachate from a municipal/commercial landfill was one of those treated during the seminal *BREW Project* (see Section 1.4 in Chapter 1).

EEB Systems for treating WWs will tend to range in complexity from those needing to be as sophisticated as those for treating MIWs to relatively simple ones involving only one or more smaller BBRs. In some cases, complex leachates such as those from highly contaminated hazardous waste landfills may be involved and these can include toxic and potentially bioaccumulator CoCs such as mercury and chlorinated aromatic chemicals. In some cases, the sub-surface disposal of effluents from leachate treatment EEB Systems will be possible (e.g., for the treatment of smaller domestic WW streams and, in a few cases, some airport WWs), but in most cases treated effluents from them will need to be surface-discharged to wetlands, lakes, and rivers (receiving waters).

Table 2.4 presents the concentration ranges of just a few CoCs found in four kinds of typical leachates, those from: municipal solid waste (MSW) landfills, kinds of industrial landfills (in this case gypsum stacks, see Section 7.6 in Chapter 7), bark piles at sawmills and coal piles at power stations. As may be seen from it, the kinds of CoCs and their concentrations vary widely, and accordingly so will the designs of any EEB-based facilities to treat them.

TABLE 2.4

Typical Ranges of CoC Concentrations in Four Leachates

Landfill Type	MSW	Gypsum Stack	Bark Pile	Coal Pile
pH	5–8	3–5	3–6	2–7
BOD (mg/L)	100–1000	n/a	1000–15,000	n/a
COD (mg/L)	100–50,000	n/a	1000–15,000	0–10
NH_3-N (mg/L)	1–3000	1–2	1–15	0–2
o-PO_4 (mg/L)	0.5–2	200–1000	1.5–5	0–2
SO_4^{2-} (mg/L)	20–2000	1000–2000	n/a	500–20,000
Turbidity (NTU)	1000–2000	50–150	n/a	2–500
Conductivity (mS/cm)	30–200	3000–4000	1000–3000	2000–3000
VFAs (mg/L)	5–15	n/a	200–3000	n/a
Cl^- (mg/L)	1–3000	15–25	n/a	2000–3000
F^- (mg/L)	n/a	30–100	n/a	n/a
Other ions (mg/L)				
Cu	0.1–1.0	n/a	n/a	0.1–10
Fe	1–2000	10–20	n/a	10–6000
Mn	n/a	0.5–1	n/a	2–500
Ni	0.1–1.0	n/a	n/a	1–2
Zn	0.1–30	100–300	0–0.2	2–30

Specific leachates may require a specific EEB Systems design to treat them (e.g., a gypsum stack leachate with CoCs in the range of those illustrated in Table 2.4 would require extensive primary and tertiary treatment).

2.3.7 The Treatment of Other WWs in EEBs

In addition to the four categories of WWs discussed in previous sections (those from airports, municipalities, mines, and leachates), EEBs have great potential for treating a great variety of other WWs. WWs requiring treatment may also be generated by landfarms, by compost production facilities and by a host of feedstock and by-product storage piles at commercial and industrial facilities. Petroleum and petrochemical facilities, as well as petroleum pumping stations and natural gas compressor stations, all produce oily and other WWs that are amenable to treatment in EEB Systems. The burgeoning field of hydraulic fracking is also an area where treatment EEB Systems may prove advantageous and R&D on their use for this potential market is recommended (see Section 9.3 in Chapter 9).

2.4 Nitrogen in WWs

2.4.1 Nitrogen Transformations in EEB Systems

Many WWs are contaminated with significant amounts of nitrogen compounds, often in the form of organic nitrogen compounds from degrading plant matter and other detritus, and from other organic materials in WWs (e.g., organics in sewage or added chemicals in airport glycols). In aqueous environments, such Org-N compounds may be mineralized to ammonia (*ammonified*, see the next section) aerobically and anaerobically by facultative bacteria. In the aerobic zones of a treatment wetland (see Section B.2 in Appendix B), providing that sufficient alkalinity is present, ammonia will be microbially oxidized (nitrified) to nitrites and nitrates (see Equations 1.2 and 1.3 in Chapter 1).

In anaerobic zones, providing sufficient organic matter is present, nitrates will be microbially reduced (denitrified) to nitrogen gas, which bubbles out of the water. There is a hierarchy of pollutant removals in wetlands, and the heterotrophic microbes (hetrotrophs) that remove nitrogen will first consume organics before turning to nitrogen compounds as energy sources. This means that to clean up a WW containing both BOD and nitrogen, there has to be sufficient residence time to allow these sequential reactions.

Airport runoff streams are usually deficient in the nutrients needed to ensure the viability of the bacteria making up a BBR's biofilms. Experience has shown that nutrients need to be present in bioreactor influent streams at a ratio of 100/5/1 for C/N/P. For BBRs, maintaining this ratio of C/N/P will avoid either oversupplying the microbes in its biofilms with nutrients (which might lead to excess sludge formation and the clogging of gravel beds) or not providing them with enough (which would lead to highly undesirable slime formation).

Ammonia is classified as a nonpersistent, noncumulative toxic substance. In water, ammonia may exist in either an ionized form, the ammonium ion (NH_4^+), or a dissolved molecular form (NH_3, the un-ionized form), depending on the water temperature and its acid–base state (its pH):

$$NH_3 + H_2O <=> NH_4^+ + OH^- \quad (2.10)$$

The ionized, NH_4^+, form predominates under normal conditions. For example, at 25°C and a pH of 7, only 0.6% of the ammonia in water exists in the un-ionized form. However, at higher temperatures and pHs, the proportion increases. By a pH of 9.5 and 30°C, the proportion in molecular form increases to 72%. This is important because concentrations of as little as 0.1 mg/L of un-ionized (free) ammonia in water can be acutely toxic to many forms of aquatic life. In addition, after mixing into receiving waters, as little as ~0.02 mg NH_3/L can cause chronic toxicity to many marine organisms. Further, the presence of certain co-contaminants such as copper can exacerbate toxicity.

It should be noted that where generators of ammonia-contaminated WW streams depend on the on-site volatilization of ammonia gas in impoundments, vessels or ditches prior to discharge to receiving waters, this process is highly temperature dependent and will be greatly reduced in cold weather.*

Ammonia compounds in WWs are usually (but not always) measured and reported as their equivalents in terms of nitrogen (e.g., ammonia nitrogen [NH_3-N], nitrate nitrogen [NO_3-N]), and care must be taken not to confuse the reporting of the molecular concentration and that as its nitrogen equivalent.†

2.4.2 Ammonification

Ammonification (mineralization) is the biological transformation of the organic nitrogen (Org-N) into ammonia by hydrolysis and microbial biodegradation. Under aerobic (oxygen-rich) conditions, the effective reaction is

$$\text{Org-N} \Rightarrow NH_4^+ \quad (2.11)$$

Some anaerobic bacteria can also mediate the transformation, but at very much lower rates. In fields, vegetated ditches, ponds and wetlands, nitrogen compounds in runoff are readily taken up by the plants, fauna and microbial

* As is noted elsewhere, aerated EEBs such as BBRs (see Section 3.2 in Chapter 3) are designed to allow significant nitrification even with the coldest water, and in addition their aerated natures facilitate ammonia volatilization in warmer waters if pHs are right.

† Confusion over such matters is common and being clear about this is critical when designing WWT facilities. For example, a wastewater's reported nitrate concentration of 20 mg/L might mean 20 mg NO_3/L or 20 mg NO_3-N/L (an over fourfold difference in the amount of nitrate present). Other species such as phosphorus compounds may also be reported in such a manner.

communities. However, once these organisms die, their tissue nitrogen is released to the water again as organic nitrogen compounds, so such take-up is of little value in nitrogen removal unless vegetation harvesting is practised (which is rarely the case).

Ammonification is dependent on temperature, pH, available nutrients and the C/N ratio. Generally, most Org-N compounds in WWs are readily ammonified if treated in BBRs.

2.4.3 Nitrification

Nitrification is the major transformation by strict anaerobic bacteria that converts (oxidizes) ammonia nitrogen in a WW to nitrate nitrogen. Equations 1.2 and 1.3 in the previous chapter illustrate the nitrification reactions. Equation 1.2 is the rate-limiting step. The oxidation of the ammonia (nitrification) consumes oxygen (4.57 g O_2/g NH_4^+) and alkalinity while releasing water and nitrate, creating biomass (the mediating bacteria) and adding acidity (more detail on nitrification is found in Section C.4 in Appendix C, particularly Equation C.8).

The *Nitrosomonas* and *Nitrobacter* bacteria are relatively sensitive and are greatly influenced by the alkalinity, pH and dissolved oxygen concentration in a WW. Temperature and nominal residence time are also fundamental factors that control performance. The growth rate of *nitrifiers* has been simplified to the following equation (Parker 1975):

$$\mu = \mu_m \frac{N}{K_N + N} \frac{O_2}{K_{O_2} + O_2} [1 - 0.83(7.2 - pH)] \tag{2.12}$$

where μ is the specific growth rate (L/d), μ_m maximum specific growth rate (L/d), N concentration of ammonia nitrogen (mg/L), O_2 concentration of dissolved oxygen (mg/L), K_N half-saturation constant for ammonia (mg/L), and K_{O_2} half-saturation constant for dissolved oxygen (mg/L).

Values for μ_m and K_N can be estimated with the following (Knowles 1965):

$$\mu_m = 100.0413.T - 0.944 \tag{2.13}$$

$$K_N = 100.051.T - 1.158 \tag{2.14}$$

where T is the design temperature in °C.

Typical aeration equipment recommended is specified to supply a minimum dissolved oxygen concentration of about 2.1 mg DO/L to water in a pond or lagoon to nitrify 1 kg of ammonia (4.6 lb. O_2/lb. NH_3), and this probably applies to aerated EEBs as well (the minimum DO concentration for nitrification needs to be about 2 mg/L, but it proceeds best if DO is >5 mg/L*).

* Such is usually not a problem in aerated EEBs such as BBRs.

A value of 1.2 mg/L is suggested for K_{O2} (Rich 1999). Given the inputs for the variables above, a limiting growth rate can be determined for *Nitrifiers* that can be subsequently used in sizing aerated EEBs (and lagoons) for nitrification. The inverse of the calculated growth rate provides a close estimate of the minimum detention time to establish resident *Nitrifier* populations.

For Aerated Lagoons, the inverse of the growth rate provides the basis for selecting the hydraulic detention time. The caveat to application of the equation for design is that the biomass (the nitrifying bacteria) must be kept in suspension; that there is sufficient alkalinity and oxygen; that there are no substances toxic to *Nitrifiers* in the WW being treated; and that flow conditions do not vary significantly.

In order for nitrification to proceed, alkalinity is required (and used up). About 7.1 g of alkalinity (as $CaCO_3$) are consumed per gram of ammonia nitrogen converted to nitrate.* If the amount of alkalinity in a WW is limited, and if there is sufficient ammonia being nitrified, all of the buffering capacity of the WW may be removed as a result of nitrification, leaving an un-buffered WW capable of varying quickly over a range of pHs. This may be exploited in chemical phosphorus removal (e.g., in a following tertiary treatment cell after an EEB cell) whereby the addition of very small doses of neutralizing chemicals can cause large changes in pH with accompanying large amounts of phosphorus removal by precipitation.

Since *Nitrosomas* and *Nitrobacter* do not compete well with the heterotrophic bacteria that consume BOD, if it is there in excess, BOD must be removed before nitrification can get underway. In an EEB System, this can be carried out by placing a BOD removal cell upstream of an ammonia removal cell.†

Nitrification can be inhibited by some heavy metals (e.g., copper) and certain organic toxins, and such may influence the design of an EEB System (e.g., a BBR for removing ammonia might be located after a BCR in a treatment train).

Nitrifying bacteria are very sensitive to toxic substances, and that during WWT that they must compete with heterotrophic bacteria metabolizing organics in the WWs‡; ones that also require nitrogen compounds for their growth (Wallace et al. 2006).§

Ideally, microbial nitrification requires a pH in the 6.5–7.8 range to proceed well. Nitrification is pH sensitive and its rate falls quickly in waters with

* Practically for conservative design, about 10 g of alkalinity should be present per gram of ammonia nitrogen in a wastewater being treated.

† This is the case with Nexom's's facultative lagoons upgrade process (see Section 3.9 in Chapter 3) where an Aerated Lagoon is located in advance of nitrifying SAGR Bioreactor cells.

‡ Practically, the competition between nitrifiers and the organics-metabolizing heterotrophs in a wastewater being treated and containing nitrogen compounds and organics is such that nitrification usually will not commence until the concentration of the organics has been reduced to a few tens of mg/L. This does not mean that nitrifiers will not be present as many of them can also metabolize organics.

§ Generally, for bioreactors such as EEBs to provide for the requirements of the bacteria in the biofilms, at least one-tenth the amount of nitrogen (from ammonia, nitrate or organic nitrogen in compounds) must be present as is the amount of carbon (from organics).

pHs below 6.5. (At a pH of 5.8, the rate is only 20% of that at pH 7, EPA 2002.) In some cases where ammonia-containing airport WWs are of low pH, the ability of nitrifying bacteria to convert ammonia to nitrate is significantly reduced (Transport Canada 1999) and low pH streams discharged to the receiving waters may be much more toxic than anticipated.* If it is desired to remove ammonia from a low pH WW, an EEB System train must include some means to raise pH first (e.g., a limestone drain in front of a BBR for ammonia removal or caustic solution injection as a 1° treatment step.)

Nitrification is a well-understood biological treatment process at water temperatures above 10°C, but little to no data have been available for systems operating below 1.0°C. Such conditions are typical of lagoon-based treatment systems in winter for small-to-medium size communities in a variety of cold-climate locations. However, the real situation is that nitrifying bacteria do not proliferate at low water temperatures, but those already present can and do continue to metabolize ammonia if their populations are large enough, even in very cold water.†

Ammonia-contaminated streams are readily treated in BBR-based EEB Systems and removal levels can exceed 99% (see Table 6.1 in Chapter 6, and Higgins et al. 2006b).

2.4.4 Denitrification

Denitrification is a biological reduction process that converts (reduces) nitrate to nitrogen gas under anaerobic conditions. A variety of common facultative bacteria such as *Bacillus, Micrococcus,Pseudomonas,* and *Spirillum* are said to carry it out. Denitrification requires a labile carbon source and energy, and the former may come from vegetation, suspended solids, or an added material such as methanol. The effective denitrification reaction is

$$NO_3 + \text{Organics} \rightarrow N_2 + OH^- + 2CO_2 + H_2O + \text{Biomass} \tag{2.15}$$

As with nitrification, the effective denitrification reaction also involves two steps as follows:

$$NO_3^- + 2e^- + 2H^+ \rightarrow NO_2^- + H_2O \tag{2.16}$$

$$NO_2^- + 3e^- + 4H^+ \rightarrow 0.5N_2 + H_2O \tag{2.17}$$

---

* Where the low pHs are the result of glycol degradation to VFAs (see Section D.1 in Appendix D), caustic can be added to the influents of BBRs to add alkalinity and increase pH. At some airports such as Halifax Stanfield International Airport (YHZ) where ARD results from the disturbance of pyritic slate underlying the airport, any EEB System would have to involve pH adjustment in a SAPS Bioreactor (see Section 4.3 in Chapter 4) before nitrification could occur.

† BBRs and SAGR Bioreactors now routinely and effectively treat wastewaters at temperatures down to just above freezing.

The NO_x^- (nitrites plus nitrates) ions are the electron acceptors and the reactions consume acidity. Denitrification is energetically favored over other anaerobic reactions such as sulfate reduction, and where both ions are present in a WW, it may inhibit the latter until nitrate concentration is low. The majority of the facultative heterotrophic bacteria involved in denitrification are able to use both oxygen and nitrate, and will use the former if it is present, so DO concentrations for this process proceed best if DO is <0.5 mg/L (i.e., oxygen inhibits denitrification).

Denitrification will readily occur in both FWS and SSF CWs, but removals are limited and greatly negatively impacted by low water temperatures. Enhanced biological nitrogen removal (EBNR) plants can be used when large volumes of WW need to be denitrified. These are extensions of the activated sludge process and supplement that processes' aerobic treatment (nitrification) steps with anoxic ones of various sorts (Constantine 2008). Where more moderate volumes of water need to be denitrified, this can be accomplished using a denitrification bioreactor (see Section 4.2 in Chapter 4).

2.5 Phosphorus in WWs

2.5.1 Phosphorus Species under Aqueous Conditions

Phosphorus compounds in a WW, as with nitrogen compounds, come in inorganic and organic forms, the latter often sourced from organic matter in the water or, in the case of an industrial WW, from specific phosphorus-containing chemicals in it. Phosphorus is very mobile in aquatic systems, and there are many complex pathways, sinks and sources for it. Phosphorus is released as/becomes: (1) *organic phosphorus* (Org-P) compounds that may be in dissolved, suspended solid, colloidal or particulate forms, some of which may be mineralized in somewhat the same manner as Org-N (see Section 2.4.2); (2) *soluble inorganic phosphorus* species called ortho-phosphates or reactive phosphorus (o-PO_4 involving PO_4^{3-}, HPO_4^{2-}, $H_2PO_4^-$, and H_3PO_4) and (3) *condensed phosphates* (e.g., pyrophosphates, tripolyphosphates, trimetaphosphates, and hexametaphosphates, to name a few of the phosphorus-containing compounds often found in WWs: those sourced from detergents, industrial cleaning agents, soft drinks, toothpaste, and baking powder).

Unlike nitrogen compounds with nitrogen gas (see Equations 2.15 and 2.17), there is no gaseous sink for phosphorus in natural systems. Phosphorus removals from WWs are by the following often inter-related processes: (1) *sorption* (adsorption onto, absorption into and/or mechanically entrapment in); (2) *uptake by biota* (e.g., aquatic plants, algae, and bacteria); (3) *abiotic chemical precipitation*; and (4) certain *biological removal* processes that occur in active

WWTPs and in semi-passive and passive natural treatment systems such as EEBs. Phosphorus is stored as polyphosphates in bacterial cells called *phosphorus accumulating organisms* (PAOs) and this "excess" or *luxury* uptake can be converted back to ortho-phosphates as needed. This mechanism can be exploited for the removal of phosphorus in some sorts of EEBs such as BBRs.

While phosphorus can be sorbed by many surfaces, media and even by biota (see Section 2.5.5), its sorption in certain tertiary treatment EEB cells is addressed in more detail in Section 7.6 of Chapter 7.

2.5.2 Condensed Phosphate Hydrolysis

When released into WWs, condensed phosphates will convert (*hydrolyse*, the splitting of the chain or ring involved) into ortho-phosphates, but their rates for doing so are very slow and are influenced by: (1) water temperature; (2) concentration; (3) the exact species of condensed phosphate involved; (4) the presence of certain multivalent metal ions (e.g., species such as Ca^{2+}, Al^{3+}, Fe^{2+} which can somewhat accelerate hydrolysis; and (5) pH (hydrolysis is faster – but still slow – at pH >7 and <11). Many WWs will contain some condensed phosphates (e.g., as much as 10 mg/L of sewage's typical TP of about 40 mg/L can be due to them). The most common method of removing ortho-phosphates from WWs involves chemical precipitation (see below) and some chemical precipitants will also remove all or part of some kinds of condensed phosphates as well, but many of them will pass through a chemical precipitation process without reaction, resulting in effluents that might not meet total phosphorus discharge concentration criteria.

It is also known (Lawson et al. 2015) that in a kind of advanced active treatment WWT method called enhanced biological phosphorus removal (EBPR) processes, hydrolyzing PAOs such as the bacterium, *Candidatus "Microthrixparvicella,"* will sequester large quantities of phosphorus intracellularly (luxury uptake) and mediate the breakdown of condensed phosphates and other macromolecules containing phosphorus. It is to be expected that PAO will also be present in BBRs, and given their highly aerobic environments, more accessible biofilms (fixed film on substrate particles in EEBs vs suspended growth floc in WWTPs based on the activated sludge process) and longer nominal residence times (days in a BBR vs hours in a conventional WWTP), it is not surprising that condensed phosphate hydrolysis will be accelerated in the EEBs. This was indeed what appeared to happen during a pilot-scale R&D test (the *Phosphorus Augmentation Project*) when various chemical precipitation methods were tested using tripolyphosphate-spiked septic plant overflow as feedstock when comparing precipitation alone and precipitation following pretreatment in a BBR cell. Further R&D is being proposed (see Section 9.3 in Chapter 9) to assess this phenomenon.

2.5.3 The Uptake of Phosphorus by Biota

Plants will take up nutrients such as phosphorus into their tissues but in natural systems such as CWs and vegetated EEBs, these nutrients will be returned to the water column as Org-P when the plants die and decay. While plant harvesting to remove such nutrients is possible in SSF aerated EEBs such as vegetated BBRs, the amounts that can be removed in this way are not large and doing so is usually impractical as a phosphorus control method. Microbes such as algae and bacteria will also take up ortho-phosphate to support cell growth, but again this is released when the cells senesce. However, many microbes including bacteria, yeasts and algae are also able to accumulate large amounts of polyphosphates internally and at least for some algae, this storage (up to 20% of dry weight) is said to occur in acidic vacuoles (Pick & Weiss 1991). Such uptake into vacuoles may also occur in the kinds of bacteria forming biofilms in EEBs.

2.5.4 The Chemical Precipitation of Phosphorus

Chemical precipitation has been the most common abiotic method of phosphorus removal from WWs and a variety of precipitants are used including iron compounds (e.g., ferric chloride, ferrous sulfate, ferric sulfate and some steel industry by-product liquors), calcium compounds (e.g., lime, limestone), aluminum compounds (e.g., alum) and mixtures (e.g., lime and ferric sulfate) (Brooks et al. 2000).

As an example of phosphorus removal by an iron compound, that by ferric chloride can be given by:

$$FeCl_3 + PO_4^{3-} \rightarrow FePO_4\downarrow + 3\,Cl^- \qquad (2.18)$$

Many iron precipitants work best in certain pH ranges: that above the 4.5–5 range, although it will still work in the 6–9 range (the precipitation of phosphorus by hypervalent iron is addressed in Section 2.5.2).

Phosphorus removal by a calcium compound can be illustrated for lime to give a hydroxyapatite precipitate by:

$$5Ca(OH)_2 + 3HPO_4^{2-} \rightarrow Ca_5(OH)(PO_4)_3\downarrow + 3H_2 + 6OH^- \qquad (2.19)$$

This works best over a pH range of 9–10.

Phosphorus removal by alum can be illustrated by:

$$5Al_2(SO_4)_3.14H_2O + 2PO_4^{3-} \rightarrow 2AlPO_4\downarrow + 3SO_4^{2-} + 14H_2O \qquad (2.20)$$

This works best over a pH range of 6–9 and is widely used to remove phosphorus from domestic WWs.

Other kinds of basic materials such as sodium hydroxide can also be used to treat o-PO_4^--contaminated WWs, but many of these form soluble reaction products rather than precipitates and are not usually useful for chemical phosphorus removal.

Chemical precipitation can be used in EEB Systems to polish effluents from upstream EEB treatment, reducing effluent phosphorus concentrations in the ppm range to very low levels in some cases (<0.5 mg TP/L).

Active WWT plants based on the activated sludge process treat WWs primed with recycled waste activated sludge from a downstream clarifier mixing with feedstock in an aerated vessel in which bacteria react with contaminants as suspended growth (see Section 1.2 in Chapter 1). Such processes are particularly keyed to removing BOD. There are now a variety of active treatment EBPR (and biological nitrogen removal [BNR]) processes available (Metcalfe & Eddy 2004) in which the aerobic vessels of an activated sludge process plant are supplemented by anoxic and/or anaerobic vessels, and recycle streams and clarification steps are more complicated.

2.5.5 Enhanced Phosphorus Removal in EEBs

As was mentioned above, PAOs in the biofilms of BBRs will use ortho-phosphates for cellular growth and will remove and store in their vacuoles significant amounts of polyphosphates as well. Thus, BBRs readily function as semi-passive kinds of biological phosphorus removal systems.

With EEB Systems, there is even the potential to further optimize (enhance) phosphorus removal using genomics (see Section 1.11 in Chapter 1). An example of the potential inherent in the use of genomics in understanding and providing directions for change in a system was revealed in a recent study that examined the change in bacterial population in an EBPR system (Lawson et al. 2015). The study monitored temporal changes in microbial community structure and potential activity across the zones in the active treatment plant. By sampling and using genomics to study the samples on a daily basis, the authors were able to determine changes in the microbial community structure by examining the ratio of small subunit ribosomal RNA. During the course of the study, the removal of phosphate slowed down, with results for removal that were significantly different from those observed at the beginning. The study showed that many of the bacteria that were for the most part responsible for accumulating the phosphate, and hence for its removal, diminished in numbers but that other rare microbes present in much lower numbers increased in number to compensate for and take over phosphate removal.

Further, it was noted that these time-based patterns could not be explained by measuring physiochemical parameters that are associated with the normal processes taking place in the system. This suggested that there were other heretofore unrecognized factors that were in fact ecological drivers. These include grazing and viral lysis that effectively modulated the bacterial community structure. The authors noted an increase in amoeboid protozoan activity during the period when phosphate removal was at its lowest that decreased when phosphate removal returned to higher levels. Increased predation has been shown to stimulate bacterial activity (Berdjeb et al. 2011). The study illustrates the kind of complex interactions that can take place in a biological/bacterial-based treatment system.

It is believed that by understanding more of the dynamics of these changes in bacterial community structure, they can be modified and hence used to improve and control the overall remediation systems functioning. Modifying substrate supply, or adding additional microorganisms when needed, might ameliorate the effects of grazing and viral lysis. Since the compact beds of substrate particles in EEBs inhibit the ingress of and provide refugia from predators such as rotifers and bloodworms, bacteria in their biofilms are not usually subject to the destruction of the fixed films that sometimes occurs in the vessels of active WWTPs with more accessible biofilms.

2.6 Metals and Metalloids in WWs

2.6.1 General

EEB Systems have to be able to remove or otherwise manage many kinds of metals and metalloids in WWs being treated in them. These WWs can be highly diverse and can involve in them varying concentrations of toxic heavy metals (e.g., Ag, Cd, Hg, Pb, and Tl), macro-nutrient metals (e.g., Ca, Fe, K, and Mg), micro-nutrient metals and metalloids (e.g., B, Ba, Be, Co, Cr, Cu, Mo, Mn, Ni, Se, and Zn), alkali earth metals (e.g., Mg, Mn, Mg, and Na), and, in some cases, radioactive and transuranic metals (e.g., certain isotopes of Ce, Sr, and U). Metals and metalloids (collectively, metal[loid] s) can be divided into those that form cations in water (e.g., Al, Ca, Co, Cu, Hg, Mg, Mn, Pb, and Zn); those that form anions in water (e.g., As, B, Cr, Mo, Sb, and Se); and those that form soluble complexes (e.g., chelates, Me–organic complexes, some carbonates, and chlorides). For example, many metals in MIWs exist as suspended colloids, or are sorbed by organic and mineral substances.

The exact speciation of metal(loid)s in WWs will be impacted by their original forms (e.g., they can be labile or recalcitrant, and can exist in dissolved or suspended states). Speciation is also impacted by pH, WW organic content and types, salinity, ionic strength, DO, ORP and acidity/alkalinity. In EEBs and the other associated cells/component of EEB Systems, interactions may occur with WW metals and metalloids by ion exchange, sorption (physical–chemical sorption and biosorption), surface precipitation, chelation/complexation, redox reactions and uptake by plants (translocation) and with microbes.

The designs, kinds and treatment cells/components of EEB Systems will also be impacted by the toxicities of any heavy metals in the WWs being treated. Al, B, Ba, Be, Ca, Co, Mg, Mn, Ni, and Si are relatively low in toxicity, while Ag, As, Cd, Cr, Pb, Hg, and Tl are high, with Cu, Fe, Se, and Zn having more moderate and variable toxicities.

Dissolved metals and metalloids can be removed in anaerobic zones of EEBs by a variety of microbially mediated precipitation reactions as sulfides, oxides, hydroxides, and carbonates, and/or may be sorbed to parts of substrates. In aerobic zones, for some dissolved metals (e.g., iron), microbially mediated oxidation reactions may be supplemented/replaced by chemical oxidation before metals are removed by precipitation following hydrolysis. Water pH will dictate whether microbial or chemical oxidation reactions dominate. Iron, manganese, and aluminum hydroxide precipitates in a bioreactor's effluent will readily sorb and co-precipitate significant fractions of other dissolved metals.

2.6.2 Iron and EEBs

Many of the most challenging applications of EEBs involve the treatment of WW streams that are highly contaminated with iron. This is particularly true for mining influenced waters. Table 2.5 compares iron and a few other CoC concentrations in some typical MIWs: two ARDs and NMD.

TABLE 2.5
Iron and Other CoCs in MIWs (mg/L)

CoC	Base Metal Mine Drainage	Coal Mine Drainage	Neutral Mine Drainage
Iron	<10–30,000	50–500	0.1–50
Magnesium	5–500	10–100	10–100
Aluminum	1–4,000	10–100	0–3
Copper	0–800	0–50	50–500
Sulfate	500–20,000	10–5,000	200–25,000
pH (unitless)	0–3.5	3–6	6–9

In the ARDs (which have acidities of >500 mg/L as $CaCO_3$), iron is usually (at least initially) in its ferrous form (Fe^{2+}) which, if oxygen is present, will oxidize to ferric iron (Fe^{3+}) biotically and abiotically.

$$4Fe^{2+} + O_2 + 4H^+ \rightarrow 4Fe^{3+} + 2H_2O \quad (2.21)$$

Ferric iron is only soluble below a pH of about 3 and if the pH is above this value under aqueous conditions, the ferric iron will then hydrolyse abiotically to a ferric oxyhydroxide sludge (ochre), which is a mixture of $Fe(OH)_3$,

$FeOHSO_4$ and FeOOH. Other dissolved metals and some salts may co-precipitate with the sludge. For simplicity, the ferric iron hydrolysis reaction is usually represented as follows*:

$$Fe^{3+} + 3H_2O \rightarrow Fe(OH)_3\downarrow + 3H^+ \quad (2.22)$$

Iron-contaminated streams such as ARD and NMD can be readily treated in EEB Systems, and those for ARD treatment may include primary treatment limestone drains (see Section C.3 in Appendix C) or secondary treatment SAPS bioreactors (see Section 4.3 in Chapter 4) followed by BCRs (see Chapter 5) followed in turn by polishing BBRs or other aerobic treatment cells (Higgins et al. 2004).

Although iron is typically found in its ferrous (Fe II or Fe^{2+}) and ferric (Fe III or Fe^{3+}) oxidation states, hypervalent Fe (IV), Fe (V), and Fe (VI) valence states are known, and the latter (ferrate) is of special interest (Lee et al. 2004). In the past, the manufacture of this aggressive oxidant has been very expensive as stable salts were only available with potassium, but a US firm, Ferrate Treatment Technologies (FTT), has developed technology to produce small quantities of sodium ferrate in solution. This sodium ferrate is unstable but proprietary equipment called a ferrator can generate ferrate from caustic soda and bleach *in situ* in a manner analogous to ozone generation. Using this technology, phosphorus in WW streams such as sewage and sewage sludges, ballast water and other municipal and industrial streams can be reduced to very low levels (<0.05 mg TP/L) by injecting a few milliliters of this nontoxic chemical that acts as an oxidant, a coagulant and a "green" disinfectant.

When injected into a WW, some of the ferrate (Fe^{6+} as a cation and FeO_4^{2-} as an oxyanion) will quickly react with dissolved phosphates in it precipitating them, while its simultaneous reduction into Fe (III) ions or ferric hydroxide will also lead to the precipitation of any other forms of phosphorus and many other CoCs in the water.

$$FeO_4^{2-} + 8H^+ + 3e^- \rightarrow Fe^{3+} + 4H_2O \quad (2.23)$$

$$FeO_4^{2-} + 4\,H_2O + 3\,e^- \rightarrow Fe(OH)_3\downarrow + 5\,OH^- \quad (2.24)$$

Ferrate will quickly oxidize a large number of organic and inorganic CoCs in aquatic environments and also kill viruses, bacteria, algae and other microbes as well as protozoa in the WW being treated.

* Equation 4.1 in Section 4.3 of Chapter 4 shows a more complex version of this equation.

2.6.3 The Question of Mercury

Mercury is a major pollutant that can be quite recalcitrant and difficult to manage in WWT facilities. Runoff and seepages from soils at former chlorine production and certain pulp and paper plants facilities that used mercury cells may be contaminated with Hg, as are those at the sites of *Artisanal and Small Gold Mining* (ASGM) facilities in a number of developing countries. Indeed, the use of EEB Systems at the latter sites may have great potential.

The problem with mercury is that, although it can be converted to very insoluble mercuric sulfide (HgS and cinnabar) under anaerobic conditions such as those in a BCR, it also likes to form much more soluble complexes with HS^-. Indeed, the chemistry of mercury and sulfur is very complex and there are major uncertainties about their interactions, especially under anaerobic conditions.

Further, the same SRB that lead to mercury–sulfide complexes may also produce extremely toxic methyl mercury if they have labile organic materials available to them under anoxic conditions.* Also, the active media of BCRs (see Chapters 4 and 5) are organic materials. The question then, in the context of sustainable mining, is: "Could such WWT be integrated into an artisanal gold mining flow sheet to manage the discharge of cyanide, mercury and other heavy metals into the environment at an ASGM site?" (Higgins & Anderson 2014).

Research will be required to develop a better understanding of the best mechanisms for mercury removal using EEB Systems. R&D and pilot-scale testing is suggested to address this question (see Section 9.3 in Chapter 9).

* Appendix E provides the background on ASGM and EEBs.

3

Aerobic Eco-Engineered Bioreactor Ecotechnologies

3.1 Types of Aerobic Eco-Engineered Bioreactors

It has long been recognized that if the DO content of a wastewater into or within a Constructed Treatment Wetland can be augmented, much better removals of its contaminants amenable to aerobic processes (e.g., ammonia oxidation) will occur. SSF CWs of any size can be aerated by injecting air into them (Davies & Hart 1990) or aerating their influents in various manners (Bowmer 1987). Indeed, a 1901 patent mentioned such a strategy for a natural WWT system (Liner 2013). Active aeration of a wastewater upstream of (i.e., influent to) a CW can be achieved by using cascades, trickle filters, rotary biological contactors, in-line venturis and other methods. Upstream Aerated Lagoons can also be used to oxygenate influents, or if they are not present but ponds or facultative lagoons are, these can be converted into them using various kinds of floating aerator–aspirators or by installing submerged air diffusers in them. Air can be bubbled into smaller CWs using aquarium aeration rocks (Jamieson et al. 2003). The DO content of a wastewater can be even increased by adding a chemical reagent such as hydrogen peroxide to it (Krohn 2007).

However, the most efficient way to aerate an SSF gravel bed CW is by placing aeration tubing connected to an air blower under or in its substrate. Such "intensified" aerated SSF wetlands have been referred to as *aerated gravel beds*, *biofilm reactors* and *biological aerated filters* (ACRP 2013; Liner 2013). However, modern EEB ecotechnology (that for which the term *Engineered Wetland* was first used to differentiate it from ordinary SSF CWs) dates from the development in Canada in the mid to late 1990s of fully aerated gravel bed EEB Systems during the *BREW Project* (see Section 1.4 in Chapter 1).

As was introduced in Chapter 1, there are currently two kinds of aerated SSF EEBs of this sort:

- BBRs
- SAGR bioreactors

By increasing the oxygen supply in an aerated gravel bed, the rates of transformation of oxidizable CoCs can be increased dramatically above those possible in unaerated beds (Higgins 2000a; Higgins 2003b; Nelson 2013). For example, the removals of ammonia by nitrification (see Section 2.4.3 in Chapter 2) that are 30%–70% in an SSF CW can be increased to over 98% (often 99%+, see Tables 6.1 and 6.2 in Section 6.2 of Chapter 6) in an aerated SSF EEB (Higgins 2006; Higgins et al. 2006a,b). Also, since the conversion of ammonia is often rate limiting in the treatment of many wastewaters, adding air can dramatically reduce the "size" (surface area) of a bioreactor treatment system.

BBRs (see Section 3.2) and *SAGR Bioreactors* (see Section 3.9) are types of in-ground, semi-passive, aerated Eco-Engineered Bioreactors that evolved from early EWs. With both, the air is introduced through perforated tubing located under or in their granular aggregate substrates.* There is also a kind of EEB System involving another kind of aerobic mixed HSSF/VSSF EEB: *Engineered Stormwater Wetlands* are EEB Systems for treating intermittent municipal stormwater flows (see Section 3.10).

The aerated EW ecotechnology was developed in the 1990s during the *BREW Project* the initial objective of which was to develop advanced natural, passive treatment, CW-based processes, primarily for the treatment of domestic wastewaters. However, the relatively low treatment rates of such natural WWT systems and their large land area requirements restrict CW use to smaller flows, and another goal of the *BREW Project* was to define systems that would allow the effective treatment of wastewaters at higher flows, up to those usually treated by active mechanical WWTPs (thousands of m^3 per day). In addition, a CW's need to contain wetland plants has restricted the adoption of these systems in Canada and the northern US due to the lack of plant growth in winter. Mediocre year-round performance of CW systems in cold climates has led to the development of EEB Systems that are not dependent on plant growth for treatment performance.

3.2 BREW Bioreactors

As was mentioned, *BBRs* are the basic aerated form of EEBs and their operating mode is semi-passive (i.e., air is added to them mechanically to maintain aerobic conditions in them). Their cells involve beds of gravel, underlain by

* While other kinds of aggregates and even other granular materials might be considered for the substrates of aerated EEBs, in most cases gravel is used.

or incorporating in them parallel rows (runs) of aeration tubing* into which air from nearby blowers (low psi air compressors) is injected. While in early EEBs the aeration tubing was simply porous irrigation tubing (often placed inside lengths of perforated plastic pipe to prevent the weight of the gravel from crushing the tubing), now much more efficient kinds of aeration tubing are available that can be laid directly in the substrate and will not crush under its weight.

BBRs come in both downflow† VSSF varieties in which the influent is distributed near the gravel surface and flows down through the substrate (countercurrent to rising air bubbles) and saturated bed HSSF varieties (in which wastewater flow is parallel to the substrate surface‡ and air bubbles follow tortuous ascending paths among the gravel particles "perpendicular" to the wastewater's bulk flow).

HSSF and VSSF BBRs can be used to treat wastewater flows between 10 and over 10,000 m^3/day, with VSSF versions usually being more suitable for flows in the higher end of this range. In the early EWs used for domestic WWT, flow was HSSF while flow in most recent larger BBRs treating specific kinds of CoCs (e.g., glycol-contaminated streams at airports) has been mostly VSSF.

In HSSF BBRs, the influent enters the bioreactor via a single perforated or valved§ inlet distributor oriented perpendicular to flow and located on or buried near the surface of the gravel at the upstream end of the usually rectilinear¶ cells. The wastewater being treated then percolates down into the gravel bed and flows more or less parallel to the substrate surface until collected in a single perforated outlet distributor (also oriented perpendicular to flow) located at the base of the gravel at the downstream end of the cell.

In VSSF BBRs, the wastewater being treated flows down from multiple inlet distributors (parallel rows of perforated pipes located under infiltration chambers) to multiple outlet distributors (usually runs of perforated pipe lying on the floor of cells) located right on top of the cell liners (usually HDPE plastic sheeting but sometimes other materials or compacted clay). Outlet

* Typically 1/2″ diameter tubing is used with runs placed 6–12″ apart. Several of the pictures in this chapter show the placement of such aeration tubing in BBR cells.

† Upflow VSSF CWs are not recommended as they represent a potentially unstable flow regime that will result in short-circuiting (Kadlec 2001). R&D is proposed (see Section 9.3 in Chapter 9) to determine whether such would also be the case with EEBs, but for prudent design upflow operation should be avoided until the matter is resolved.

‡ While the bottoms of HSSF cells are sometimes sloped 1%–5% in the direction of flow depending on terrain and other aspects, they need not be, and the bottoms of VSSF BBR cells are usually flat.

§ In smaller HSSF BBRs where the inlet distributor is located at the gravel surface, a series of ball valves on it is a highly efficient and economic way of ensuring that the influent is distributed evenly across the cell width.

¶ But not always. In the case of the two HSSF BBR-based EEB System arrays at Casper (see Section 3.4), the cells were "pie" shaped.

distribution piping is often buried in a layer of gravel coarser than that of the bulk of the substrate.

Inlet distributors are usually buried in the gravel just below its surface (see Figure 3.9). In some cases, a second set of distributors may be located below the first, part way down in the gravel bed, and these can be used to backwash the beds, back up the higher up inlet distributors, and in autumn to "prime" the bed with large amounts of bacteria before cold weather. Both inlet and outlet distributors are connected at one end or the other (or both) to risers (usually plastic pipe) that allow their backwashing and occasional cleanout.*

Effluents from the outlet distributors of BBR cells usually flow into downstream outlet control structures (OCSs, usually separate, buried, vessels) that are used to control water levels in the upstream bioreactor cells† using adjustable weirs in them set to control water levels in the upstream EEB cells at or just below their gravel surfaces. Flow rates from the cells are often measured at the same places. OCSs may also contain a provision to allow the recycle some of the effluents to feed. Effluent sampling often takes place in OCSs. In cold weather areas, these vessels may contain stock heaters and/or air spargers to prevent freezing.

In some EEB Systems, inlet control structures are also placed in advance of the BBR cells to allow the levelling, directing and measuring of flows, the injection of nutrients and influent sampling. Where there is a potential of solids in feedstocks to BBRs, filters may be placed upstream of the inlet control structures for final grit, sediment and/or oil and grease removals.

Gravel thickness in BBR cells ranges from 0.7- to 3-m thick, with 1.5 or 2.0 m being common selections.

Most BBRs are designed with vertical sides. However, the sides of VSSF BBR cells need not always be vertical; in some cases, the cells are surrounded by earthen berms with slopes ranging from 2:1 to 4:1, although system morphologies of this sort require more surface area than those where vertically sided cells are used.

Figure 3.1 shows a cross section of one corner of a bermed VSSF BBR cell with a single set of inlet distributors. As may be seen in it, the infiltration chambers (shown as upside down horseshoes, see Figure 3.8 for a picture) are buried in the gravel just below its upper (often insulated) surface and the perforated influent lines are located under them.

* Both hydrogen peroxide solutions (for organic contaminants) and muriatic (hydrochloric) acid solutions (for inorganic contaminants) are available to flush tubing and pipes in BBR cells.

† In smaller BBR cells, standpipes alongside the cell can be used to control water levels.

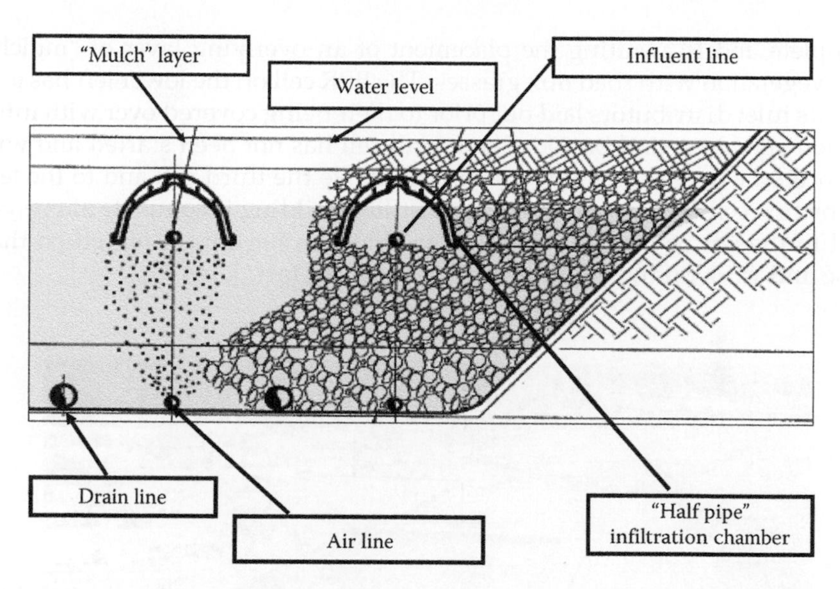

FIGURE 3.1
Sketch of a corner of a bermed VSSF BBR cell.

Figure 3.2 illustrates an in-ground vegetated HSSF BBR cell.

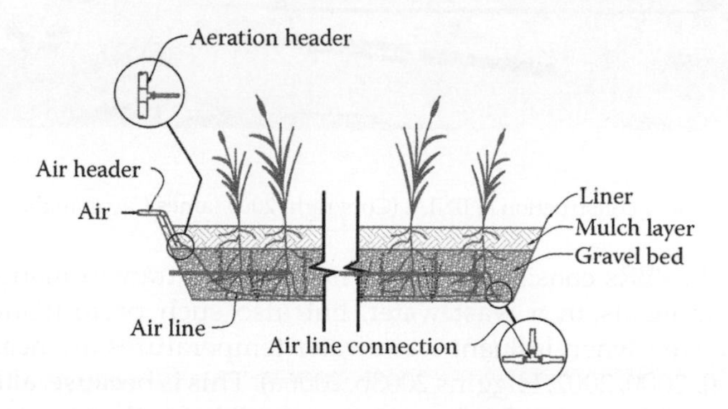

FIGURE 3.2
HSSF BBR cell.

BBR cells can be as small as those in 1 m^3 chemical totes or larger than football fields. They can contain up to tens of thousands of cubic meters of gravel that have thousands of kilograms of biomass in their biofilms.

Figure 3.3 (Copyright 2008 James Cavanaugh) shows three of the very large VSSF BBR cells under construction at BNIA in 2006. Figures 3.3 and 3.20 and the picture on this book cover show three of the original four BBR cells at BNIA under construction. The one on the upper left has just been excavated, lined and had its aeration tubing and outlet distribution piping installed prior to gravel substrate addition. The one on the lower right is complete and is awaiting the placement of an overlying layer of "mulch" and vegetation with road mix grasses. The BBR cell on the lower left has just had its inlet distributors laid out prior to their being covered over with infiltration chambers and buried. The fourth cell has not been started and will be located where the staging area is just above the third cell and to the left of the first cell. The base of the future Utility Building is complete alongside and to the left of the third cell. Its four blowers have been placed on that base and work on the building shell is about to start.

FIGURE 3.3
VSSF BBR cells under construction at BNIA. (Copyright 2008 James Cavanaugh.)

Not only do BBRs consistently allow much better transformations of oxidizable components in a wastewater, but also such performance can be maintained even when influent wastewater temperatures are near freezing (Kinsley et al. 2000, 2002; Higgins 2003b, 2006a). This is because, although the autotrophic and heterotrophic bacteria responsible for the CoC transformations in the biofilms of BBRs do not grow much in very cold waters, those already present under the highly aerobic conditions in them will continue to metabolize the CoCs they are exposed to[*,†] and the microbial biodegradation reactions in BBRs are exothermic in any case.

In addition to carbonaceous materials, the bacteria in BBR cell biofilms need a balanced diet of nitrogen, phosphorus and potassium (along with some

* In addition to biodegrading wastewater contaminants, BBRs will also remove some dissolved metals and other CoCs in a wastewater by redox reactions, biosorption and other processes.
† The design of BBRs is usually based on the rates for these aerobic transformations as determined during a treatability test at the coldest anticipated operating temperature with feedstock spiked with CoCs to "worse possible" conditions expected (see Chapter 6).

micronutrients) to ensure their steady growth. Nutrient requirements for a BBR are based on the estimated composition of bacterial biomass (Grady et al. 2011). To determine them, influent COD concentration is multiplied by an assumed bacterial yield (mg-biomass/mg COD). The biomass is multiplied by an empirically derived elemental fraction to calculate nutrient requirements.

Table 3.1 compares the amount of surface area for BBRs and other kinds of WWT methods need.

TABLE 3.1
Typical Areas Needed for Various Treatment Technologies

WWT System	Area Required (m^2/m^3 of Wastewater Treated)
Facultative lagoon-based WWT systems	100–200
FWS CWs	75–125
HSSF CWs	20–40
Intermittent sand filters	15–25
Active (mechanical) WWTPs	1.5–2.5
BBRs	2.0–12.5

While BBRs still take up more space than alternative active (mechanical) WWTPs, the difference is very much less than might be expected and perceptions that they require extremely large areas as do CWs are mistaken.*

Table 3.2 compares the aeration energy requirements for BBRs and other kinds of WWT systems (Brix 1999).

TABLE 3.2
Aeration Energy Requirements

Treatment Type	Energy Input (kWh/m^3)
Ordinary activated sludge-based WWTPs	0.51–2.30
Extended aeration-based WWTPs	1.06–2.39
BBRs	0.14–0.18

Aeration is very efficient in BBRs probably because of the tortuous paths that air bubbles have to take while rising among the biofilm-coated/permeated particles of gravel substrate, and these the impact of the sizes of the bubbles and the diffusion of oxygen out of them. The bubbles moving through the interstitial spaces are smaller and/or longer, and this results in more surface area, allowing much better contact with the biofilms than is the case with other fixed film WWT processes.

The effect of aeration is illustrated by comparing the rate constant for BOD removal determined in pilot-scale testing during the BNIA treatability test

* Again, it is emphasized that, during comparative evaluations of WWT method efficiencies, BBRs (and other kinds of EEBs) should not be grouped with CWs or classified with or as being wetlands.

(see Section 6.7 in Chapter 6) which averaged 5.4 per day for aerated SSF cells compared to that for parallel, otherwise similar but unaerated cells where it was <0.5 per day. Indeed, with aeration, the rate constants for the transformations (largely oxidation) of specific CoCs may be 5–30× those of CWs (Higgins 2000b; Wallace et al. 2006; Higgins et al. 2007a; Liner 2013). The relatively long residence times of WWs in BBRs – days versus only hours in many kinds of active WWTPs such as the suspended growth tanks of many conventional active WWT facilities – contribute to aeration efficiency as well.

The impact of aeration on a gravel bed is further shown by considering the mass flux rate (see Section 8.8 in Chapter 8) of oxygen-limited, passive treatment, Reed-vegetated HSSF CW cells at London Heathrow Airport (prior to that system's upgrade to an EEB mode, see Section 3.8) which was then 4–13 g COD/m^2.day compared with the loading of the oxygen-saturated, semi-passive treatment VSSF BBR cells at BNIA which was (and still is) up to 160 g COD.m^2.day, more than 20 times the efficiency of the CW (Kadlec & Wallace 2008).

BBRs then are attached growth, aerobic bioreactors and, since biofilm microbes (mostly bacteria) in their substrates allow greatly enhanced oxidation rates, have conditions in them in some cases approaching those of complete mix reactors. As such, they largely rely on metabolization by aerobic autotrophic and heterotrophic bacteria for the removal of organics and other CoCs.

It is normal to vary the wastewater feed rate to BBRs to maintain loadings within acceptable ranges. Influents have to be controlled in such a manner that target daily treatment loadings (anywhere from 1000 to 1500 kg BOD per day) are not exceeded. This necessitates impounding feedstocks upstream of the BBR cells in balancing tanks/ponds/vaults, and then metering the amounts of influents from them entering each train of BBR cells (i.e., controlling flow rates), as well as the concentrations of organics in the water flowing into them. This allows the control and optimization of performance. The superiority of the BBR ecotechnology in removing oxidizable contaminants from wastewaters of all sorts has been demonstrated in a variety of pilot-, demonstration- and full-scale projects, and may now be regarded as fully proven (Wallace 2001b).

Online analysers are available that can measure BOD, COD or TOC, but the latter have proved most reliable and can be calibrated to provide correlations between the meter reading with COD, BOD and laboratory-derived TOC values. Flows can be measured using v-notch weirs in control structures or flumes, or by using in-line flow meters. Signals from TOC and flow meters can be utilized by programmable logic controllers (PLCs) forming parts of SCADA systems to determine and control the appropriate loading to the BBR cells.

Over 100 aerated HSSF and VSSF BBRs are now in operation year-round treating a wide variety of wastewaters (Wallace 2001b, 2004; Higgins 2003; Redmond 2012) at a number of locations, including those in Table 3.3.

TABLE 3.3

Examples of Locations with Operating BBRs

Buffalo Niagara International Airport, Buffalo, NY	McArthur Airport, Islip, NY
Edmonton International Airport, Leduc, AB	London Heathrow Airport, UK
Former Amoco Refinery, Casper, WY	Fields of St. Croix, Lake Elmo, MN
Williams Pipeline Terminal, Watertown, SD	Jackson Meadows, St. Croix, MN
ARCO Refinery, Wellsville, NY	Lake Allie, Renville Co, MN
The Dunes, Abu Dubai, UAE	Lutsen Sea Villas, Lutsen, MN
Yargon River Authority, Tel Aviv, Israel	Anoka County Landfill, Ramsey, MN

By most measures, for WWT they are the *BATEA* (Higgins et al. 2007a–d).

BBRs are being used to treat a wide variety of wastewaters but their most important uses are for treating municipal sewage, MIWs of all sorts and the glycol-contaminated runoff resulting from de-icing activities at northern airports in cold weather. In the latter service, many are still referred to as EWs, although the BBR designation is becoming more common instead, even for them.*

Up to the present (2017), all operating BBRs have used the same kind of aeration tubing, and it has served well. However, there is growing concern about the build-up of organic and/or inorganic materials in BBR aeration tubing over longer terms. If this occurs (and it is not always the case), it can result in back-pressure on the air blowers, and in some cases this can lead to them tripping off. This results in the cessation of aeration in the BBR cells, and although they have great "inertia" (resistance to flow and concentration upsets), running the BBRs without air for will eventually lead to reduced CoC transformations and off-spec effluent. As was mentioned, the build-up of deposits in aeration tubing can be resolved by their periodic flushing with solvents and/or acids. Although doing so can easily be carried out, it can be inconvenient and costly, especially for BBRs whose aeration tubing is harder to access. While for future BBRs, aeration system design will be such as to facilitate such flushing, research is also planned (see Section 9.3 of Chapter 9) to see if more plug-resistant and/or easier to flush out aeration tubing can be identified. For existing BBRs, especially the smaller ones treating municipal WWs, when the gravel substrate in BBR cells is removed and washed before reinstallation every decade or so (a recommended practice), it is also recommended that the aeration tubing be replaced at the same time.

The process flow diagram (PFD) in Figure 3.4 (next page) illustrates flows in a comprehensive BBR-based EEB System, in this case a two-cell facility for treating complex, glycol-contaminated streams at an airport. This system is piped so that the BBR cells can be operated (under the control of a SCADA system) in

* In fact, the use of BBRs in treating glycol-contaminated wastewater streams from airports is one of the most important uses of EEB ecotechnology and, as such, merits an appendix (Appendix D) to describe them as well as to address the relevant chemistries and properties of the glycol and related de-icing chemicals involved.

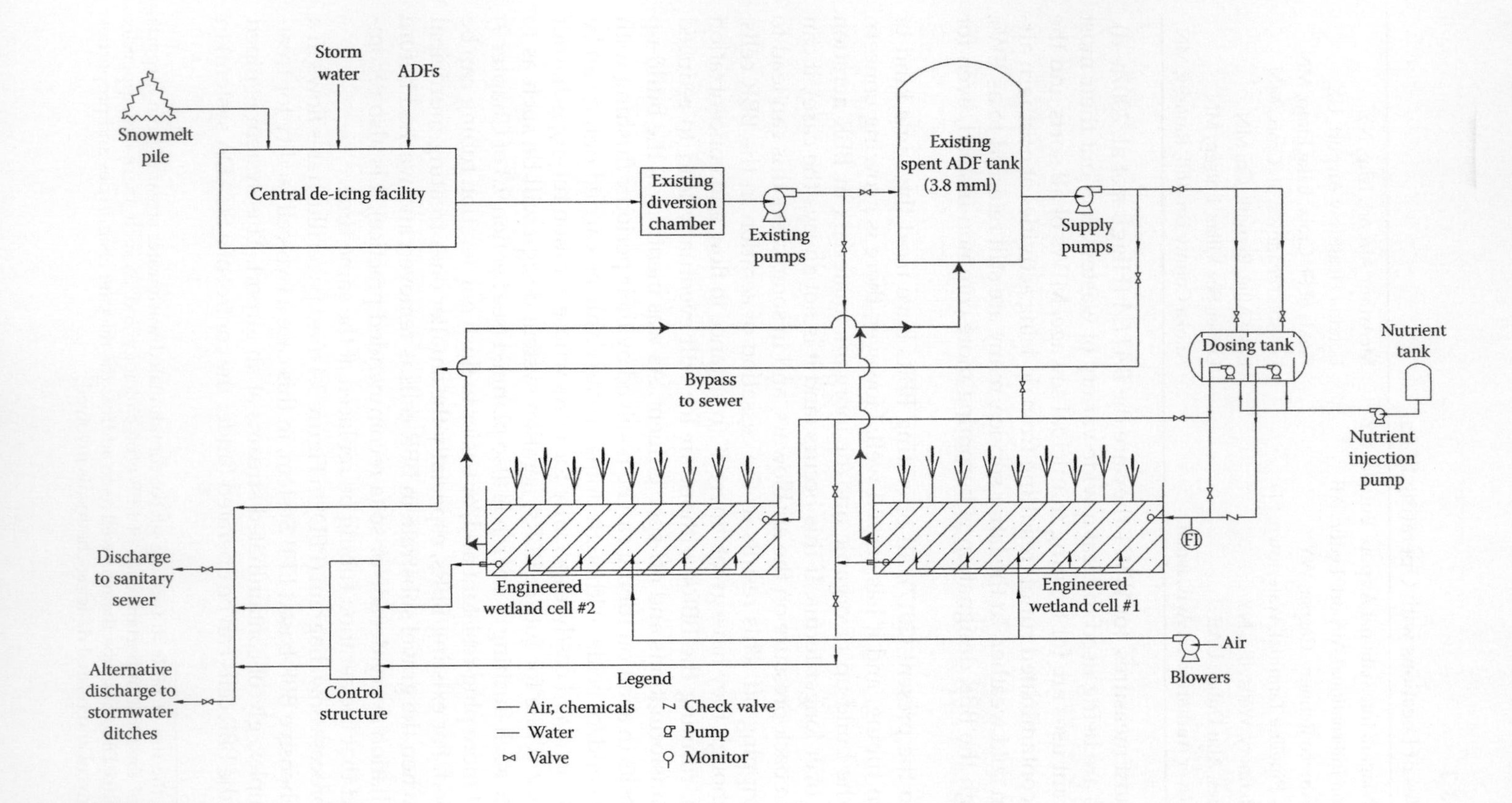

FIGURE 3.4
PFD for a two BBR cell EEB System.

parallel or in series, and has provision to recycle some treated effluent to feed for situations where influent feedstock is excessively strong (i.e., high in BOD).

BBRs and the EEB Systems in which they form the secondary treatment component are also proven and demonstrated technology; highly efficient at removing CoCs; are modular; can operate successfully at very much higher contaminant loading and mass flux rates than can ordinary CWs (Kinsley et al. 2000; Wallace et al. 2006); often have CAPEXs about half those of conventional mechanical WWTPs (even less if the upgrading of a lagoon-based WWT facility is involved); have very much lower OPEXs and life cycle costs than mechanical WWTPs (Higgins et al. 2010b); and can operate even under the coldest winter conditions (Higgins 2003; Nivala 2005; Higgins et al. 2010a–c).

BBR-based EEB Systems treating up to 6000 m^3/day of wastewater are successfully operating, and their nature (multiple trains) means that wastewater of several times this rate are feasible (just add trains). BBRs can and do operate at high efficiencies treating wastewaters in the tropics where water temperatures are in the 30°C+ range, and in cold weather areas with water temperatures as low as 0.1°C. BBRs represent proven and demonstrated ecotechnology.*

3.3 The Construction of a BBR Cell

As was mentioned, the gravel substrates of BBRs can be up to 3-m thick and this can be compared to the maximum 0.6 to 0.7-m thick substrates possible in ordinary SSF CWs. The thicker substrates are possible as the oxygen for aerobic microbial contaminant transformation processes in BBRs is supplied (in excess) from the introduced air rather than having to be sourced from DO in influents and diffusion of it from the roots of wetland plants as is the case in SSF CWs.

As a result, BBRs need not be vegetated with wetland plants at all (although some still are), and instead can be planted with herbaceous terrestrial vegetation (e.g., terrestrial grasses, see Figure 1.2 in Chapter 1), or even left unvegetated. In northern areas, a layer of insulating material ("mulch") usually placed on the open gravel surfaces to facilitate winter operation (see Figure 3.1 as well as Figure 3.10). In addition, the aeration itself helps to inhibit freezing in cold weather. BBRs are designed to operate at high efficiencies even when treating wastewaters with water temperatures near freezing.

BBR cells are often constructed with vertical sides and with their insulated surfaces flush with the ground level of the surrounding terrain.

Figures 3.5 through 3.11 illustrate the construction of a small VSSF BBR cell of this sort.

Figure 3.5 illustrates how supported (on the outside) plywood sheeting is constructed to form the sides of an excavated basin. Usually, a layer of sand or similar material is placed in the bottom of the excavation. Once the

* BBRs are summarized in Section 9.1.1 of Chapter 9.

plywood sides are in place, they are insulated with styrofoam and usually an impermeable plastic liner (HDPE or similar) is placed over them that covers the bottom of the cell and the sides.

FIGURE 3.5
Construction of a BBR cell.

Figure 3.6 shows a lined cell with effluent lines (perforated plastic pipes, six in this case) connected to a header outside the cell (not shown).

FIGURE 3.6
Cell liner and effluent collection piping in place in a BBR cell.

Once the outlet distributors are in place, a layer of gravel is often placed over them, as is shown in Figure 3.7.

FIGURE 3.7
Gravel substrate covers the BBR cell effluent collection system.

The aeration lines are then laid out on the raised gravel surface and connected to an aeration header which in turn has a connection to air blowers located nearby, as is illustrated in Figure 3.8.

FIGURE 3.8
The installation of aeration lines in the BBR cell.

Gravel is then placed over the aeration lines and the BBR cell filled nearly to the top, after which runs of perforated inlet distribution piping are laid near the gravel surface, and each is covered over with a plastic infiltration chamber as is shown in Figure 3.9.

FIGURE 3.9
BBR cell inlet distribution piping covered by infiltration chambers.

The infiltration chambers are then covered over with the gravel substrate to achieve the desired total thickness.

FIGURE 3.10
BBR cell infiltration chambers are buried in gravel.

The gravel substrate is then often covered over by some sort of insulating "mulch" material (e.g., compost preparation rejects) and growth medium (if the cell is to be vegetated) as is illustrated in Figure 3.11.

FIGURE 3.11
Completed BBR cell.

In Figure 3.11, the tops of the capped cleanout risers for the effluent collection runs are on the right, as are hatches for access to the aeration header.* Such BBR cells may be planted with wetland or terrestrial vegetation, or left un-vegetated to be colonized over time by local plants.†

The construction of HSSF BBR cells is somewhat similar to the above, although gravel placement is much more critical with the HSSF morphology (to avoid areas of uneven compaction that would cause channelization) and much more attention to gravel particle sizing is required. VSSF BBRs can, if desired, have varying gravel sieve sizes for various layers, while such can occur in HSSF cells only in bands perpendicular to flow. For example, the original CW cells of the Heathrow Airport EEB System had (and still have) gabion baskets filled with coarse rock both upstream and downstream of the gravel. (These cells which were HSSF CW cells have now been upgraded to HSSF BBR cells.)

3.4 The Casper Project

In the early 2000s, a 6000 m^3/day (~1.5 MM USG per day) BBR-based EEB System was constructed at the remediated site of a former BP Refinery in Casper, WY,

* It is noted that once the cell is filled with substrate, the plywood sides are no longer needed and it does not matter if they deteriorate.

† Figure 1.4 in Section 1.5 shows a completed BBR cell vegetated with terrestrial vegetation.

USA bordering the North Platte River in order to clean up groundwater under it contaminated with BTEX-based dissolved hydrocarbons and light, nonaqueous phase hydrocarbons (LNAPLs) (Wallace & Kadlec 2005; Wallace & Austen 2008). Figure 3.12 illustrates the site with the extraction wells used.

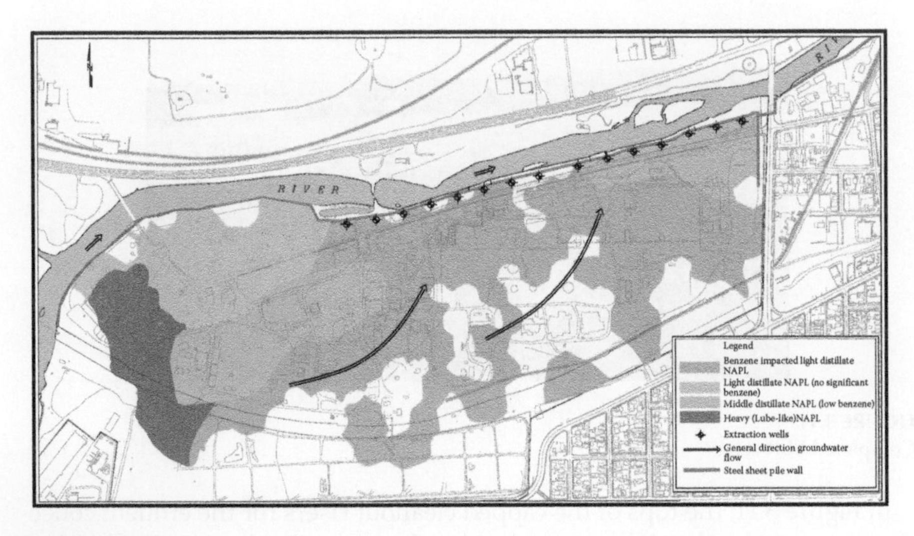

FIGURE 3.12
LNAPL distribution at the Casper site.

Figure 3.13 shows plans for remediating the site into a golf course, a white-water rafting park, commercial office space and light industrial park space.

FIGURE 3.13
Reuse plan for the Casper site.

As shown in Figure 3.14, the EEB System at Casper involves preliminary treatment consisting of an oil–water separator, a cascade aerator located near the groundwater extraction wells and, integrating into the golf course, two 0.6 ha surface area FWS CW cells (primary treatment), followed by secondary treatment in two, circular HSSF BBR secondary treatment cells with a combined area of 1.3 ha (0.9 and 0.4 ha) containing gravel substrate 1-m thick.

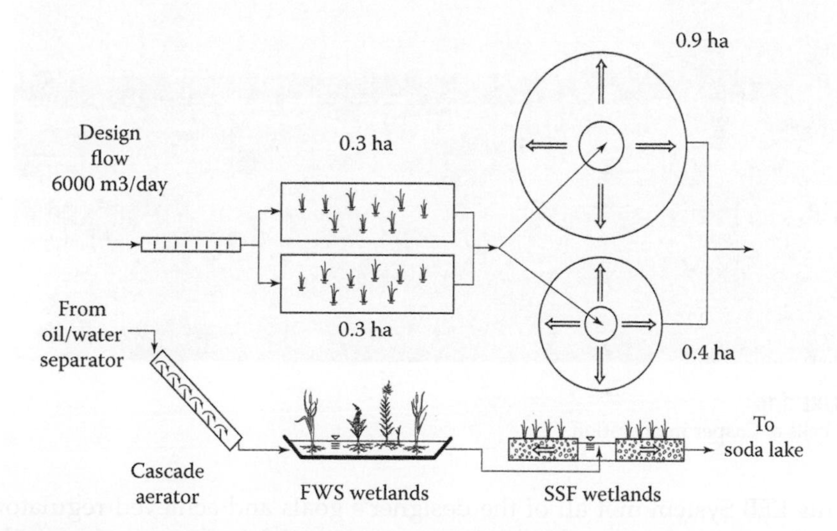

FIGURE 3.14
Process flow at Casper.

The Casper EEB System was designed to reduce influent benzene concentrations that average 300 μg/L to effluent concentrations of 0.05 μg/L or less.

Figure 3.15 shows the two BBR cells under construction.

FIGURE 3.15
Construction of the Casper BBR cells.

Figure 7.1 in Chapter 7 shows gravel substrate being placed in one of the BBR cells at Casper.

Figure 3.16 shows the two HSSF BBR cells in operation on the golf course after construction; the Reed-vegetated cells function as "rough" areas for the course.

FIGURE 3.16
BBR cells at Casper in operation.

This EEB System met all of the designer's goals and achieved regulatory compliance within 1 week of start-up. (Section 6.8 in Chapter 6 provides details on the pilot-scale treatability test that preceded its final design.)

3.5 CW and EEB Systems at Airports

There are only a few options for the off-site disposal of glycol-contaminated wastewaters from airports. In some cases at gates, de-icing pads or CDFs, Spent Glycols that are vacuumed up by GRVs and concentrated materials collected in dedicated sewers and drains can be sent to recycling facilities or municipal WWTPs (if such are available locally and if their operators will accept such high BOD streams*). However, even if all of the Spent Glycols are diverted or recycled, much of the ADAFs used will still enter the huge volumes of stormwater runoff inevitable due to the large areas occupied by airports, and for GCSW complete off-site treatment is usually not an option.

This then dictates that on-site management be considered (Switzenbaum et al. 2001). There are various methods, all involving assembling the GCSW

* The EEB System project at BNIA was partly motivated by the operators of local municipal WWT facilities asking the airport to find other methods to deal with its Spent Glycols.

into ditches and sewers and directing it into one or more detention ponds, basins or flow balancing tanks, to even out its periodic nature, skim off any floating oil and grease and debris, and allow grit to settle out. However GCSW is collected, it must eventually be dealt with.

At some airports, GCSW is still discharged untreated to receiving waters, but the reality is that, more and more, its on-site treatment will be required before disposal. Although there are several options for managing Spent Glycols, there are a very limited number of feasible, on-site treatment options for GCSW at airports (Higgins & MacLean 1999; Higgins et al. 2001), and simple impoundment and natural attenuation is often no longer an option. Many kinds of active WWT technologies (e.g., activated sludge-based WWTPs, anaerobic bioreactors, carbon filters, membrane systems) might be considered for treating high-glycol-concentration Spent Glycols (tens of thousands of mg glycol per L), but mostly they are not suitable for the high volumes, often cold temperatures and intermittent natures of much lower-glycol-concentration GCSW (usually hundreds to a few thousand mg glycol per L). Also, the maintenance requirements of such processes would be undesirably high anyway. If there is sufficient land available at an airport, large FWS CWs could handle airport runoff streams but these might attract waterfowl, creating birdstrike hazards.

With SSF CWs, the risk of their attracting large waterfowl and other birds is minimized, and there are several examples of airports where they are (or were) used. These include facilities at Pearson International Airport (YYZ) in Toronto, ON, Canada (Flindall & Basan 2001); at the DHL facilities at the Wilmington Airpark, OH, USA (ILN); EIA (YEG) near Edmonton, AB, Canada (Higgins & MacLean 1999); Heathrow International Airport (LHA) near London, UK (Richter et al. 2004); the Westover Air Reserve Base (CEF) at Chicopee, MA, USA (ACRP 2013) and small pilot facilities (no longer operating) in Zurich, Switzerland, Berlin, Germany and Kalmar, Sweden (Thoren et al. 2003). However, until one at EIA was built, only the ones at ILN and LHA represented practical, operating, full-scale facilities.

Initially, the YEG and LHA facilities for treating glycol-contaminated waters involved ordinary HSSF CWs.* That at ILN still involves pulse flow VSSF CW cells (fill and drain, see Section B.4 of Appendix B), and that at Westover involves a single 0.6 acre Reed-vegetated SSF wetland containing gravel 3 ′ thick vegetated with 2000 reed rhizomes planted on 3″ centres in a 3″ layer of 3/4″ gravel. It treats an average of 100,000 US GPD of stormwater runoff from a 162 acre watershed and has had effluent BOD reductions of from 11% to 78% from influents containing BOD ranging from 165 to 2655 mg/L (2002 data).

The CWs at YYZ, EIA, LHA and ILN are associated with Sedimentation Ponds in front of the wetlands. The CW at YEG treats/treated GCSW contaminated with EG; that at YYZ: EG; that at LHA: EG, PG and DEG and that in ILN: PG. The HSSF CW at YEG operated only part of the year, and was

* They have subsequently been upgraded to BBR-based EEB Systems.

(and still is as a combined CW/EEB facility) frozen in the coldest weather. The LHA CW system was (and now is as an EEB System) able to operate year round. The ILN HSSF CW attempts to operate most of the year but tends to impound water in the very coldest periods. The CW system at YYZ was designed to manage stormwater from a 382 ha area including the paved areas around the airport's main terminal and is sized to treat runoff from a 25-mm storm event. It involves a 0.42 ha, Reed-vegetated, multi-gravel-size layered VSSF CW cell and a 1.38 ha FWS CW cell. Its operability has been poor and the AOA, the Greater Toronto Airport Authority, is considering decommissioning it or upgrading it to an EEB System.

The 12 CW cells at YEG were vegetated with transplanted Cattails (two CW cells have now been converted to un-vegetated BBR cells). The 12 CW cells at LHA were (and still are following upgrading to HSSF BBR cells) vegetated with Reeds, as are those at YYZ. The Wilmington VSSF CW cells are not vegetated. At YEG and ILN, influent flow rates are keyed to water temperatures, with lower throughputs occurring when the water is colder. None of the CWs is insulated. All use gravel as their substrates: it is 0.7-m thick in YEG, 0.8 m at YYZ (VSSF cell), 0.6-m thick at LHA and 2.1-m thick at ILN (Higgins & MacLean 1999; Higgins et al. 2007a–d). In the past, none of these three SSF CWs was mechanically aerated, although there was (and still is) an aerated pond upstream at LHA treating some runoff. (This resulted in some plugging of the downstream HSSF cells with biosolids sludge formed in the aerated pond.)

In summary, the treatment of GCSW and snowmelt (and even Spent Glycols) at airports using CWs has been, and is being, practised, although very large areas are required to do so (and many airports do not have such space). Owing to birdstrike concerns, this usually dictates the use of SSF CWs rather than FWS ones. (e.g., waterfowl and other birds have been a problem with the FWS CW cell at YYZ.)

SSF CWs at airports can be very expensive because of the amounts of gravel substrate needed. Two of the major airports where large-scale CWs were used to treat glycol-contaminated streams (YEG and LHA) have already now partially or fully upgraded to BBR-based EEB Systems and, given the availability of BBRs now, it appears unlikely that CWs will be used at airports in future.

3.6 The VSSF BBR-Based EEB System at BNIA (BUF)

The seminal application of very large-scale EEB Systems ecotechnology is at *Buffalo Niagara International Airport* in upstate NY, USA where there are four very large rectilinear VSSF BBR cells (51 m × 91 m each) located in an open area near the airport's main runway (see Figure 3.3 for a picture of three of them under construction). Each cell contains gravel substrate 1.5 m thick and the system has been successfully treating up to 4600 m^3/day of glycol-contaminated wastewaters during de-icing seasons for several years now.

The airlines using BNIA use PG-based ADAFs during the de-icing season, and de-icing is carried out at terminal gates. Glycol-rich Spent Glycols from around the de-icing pads and from a snowmelt pad are collected in dedicated sewers and these and Spent Glycols collected by GRVs on the pads are stored in underground "glycol" storage tanks (~1100 m^3 [30,000 US gallons] in total) and are piped to allow disposal at a local municipal WWTP. A separate sewer system tied to a 2 MM USG mostly underground vault is used to collect stormwater runoff from around most of the airport property, and this discharges into a small local stream.

Despite Spent Glycols collection, a large fraction of the ADAFs used at BUF (50%–60%) still was lost, most of it ending up in the airport's stormwater system where it sometimes led to exceedances of the airport's mandated 30 mg BOD per L NPDES discharge limit. It also became increasingly difficult to deliver the Spent Glycols to the local WWTP, and the AOA, the NFTA, wished to find a new way to manage them, so after a feasibility study and treatability test carried out by Jacques Whitford, it was decided to install a BBR-based EEB System at the airport to treat both Spent Glycols and GCSW together.

After kinetic and other scale up parameters were determined during a treatability test (see Section 6.7 in Chapter 6), the design and construction of a full-scale BBR-based EEB System for BNIA proceeded. (See Figure 1.2 in Chapter 1 for a picture of a completed and operating BBR cell there.) Calculations at the end of the feasibility study indicated that about 3.4 ha of 1.5-m thick BBR cells would be needed at BNIA to treat all of the Spent Glycols and GCSW all of the time under worse case conditions. Owing to funding constraints, the NFTA decided to only proceed with a first phase involving four BBR cells of 1.9 ha of surface area, while retaining the option of impounding stormwater and still diverting some Spent Glycols to the local WWTP if extreme conditions merited doing so. (Planning is now underway [early 2017] for adding two to four further BBR cells.)

In addition to the four BBR cells, the first phase of the EEB System at BNIA included a utility building; four 250 HP blowers; aeration equipment; grit removal vessels; cell inlet chambers; pumps and their control chambers; piping systems; instrumentation and controls; cell OCSs and an electrical supply system. The EEB System has been integrated with BNIA's existing stormwater system involving stormwater and Spent Glycol sewers, the buried spent glycol storage tanks and the large stormwater vault. One of the BBR cells (the one in the background on the right in Figure 3.3) has been built slightly depressed with low berms around it, allowing it to act in an ESW mode (see Section 3.10) as a shallow emergency stormwater storage pond under extreme stormwater volume conditions. The adjacent utility building houses the blowers, some pumps, instruments and electrical equipment as well as chemical injection equipment and associated infrastructure.

The nominal de-icing season at BNIA is 190 days, and design was based on a total glycol usage of about 1150 m^3 of pure PG per year. Flows rates for design were up to 820 m^3/day for Spent Glycols plus 3800 m^3/day for GCSW.

The BBR cells of the EEB System at BNIA can accommodate large fluctuations in flow and influent BOD concentrations and the EEB System operates easily during the coldest weather and idles in summer. It is designed to ensure compliance with the airport's NPDES permit (and achieves this goal) and not only removes the glycol but also any other organics, grit, suspended solids, dissolved metals and any ammonia present.

Construction of the EEB System at BNIA was completed in 2009, and measured performance has been close to that predicted during the treatability test. Figures 3.17 through 3.19 show various stages during the construction process.

FIGURE 3.17
Preparing a BBR cell at BNIA.

FIGURE 3.18
Liner installation in a BBR cell at BNIA.

Figure 3.17 shows the vertical plywood sides of one of the four BBR cells at BNIA being built, while Figure 3.18 shows the HDPE liner being installed.

Notice that the liner overlaps the tops of the styrofoam-insulated plywood side walls.

Figure 3.19 shows the aeration tubing and outlet distributor piping installed on the floor of one of the BBR cells* as the gravel substrate is being added.

FIGURE 3.19
Gravel being added to a BBR cell at BNIA.

The four below-grade aerated VSSF BBR cells at BNIA contain 52,000 m^3 of screened and washed dolomite gravel (10–15 cm dia.), and are designed to sustain on the gravel particles a resident, attached community of bacterial as biofilm with a mass of about 4500 kg acclimatized for the specific task of PG removal.

Figure 3.20 shows three of the four BBR cells under construction, illustrating the very large size of each of the cells. The concrete pad on the left with the four white plastic covered blowers already in place on it is where the utility building was later built. One of the cells (top left) has just had its liner installed, the one in the bottom right has just had its aeration tubing laid out and the cell at the top right is ready for gravel installation.

Figure 3.21 shows the near surface inlet distribution piping for another cell covered over by half-pipe plastic infiltration chambers.

* It is noted that at BNIA, the aeration lines and the outlet distributors are installed at the same level directly on the liner. In other BBR cells, the outlet distributors on the floors of the cell are covered over with a layer of gravel or crushed rock, and the aeration tubing installed above them a few centimeters in a separate layer. Either option is OK and will depend on specific project considerations and the design agreed on.

FIGURE 3.20
Three of four BBR cells under construction at BNIA. (Copyright 2008 James Cavanaugh.)

FIGURE 3.21
Infiltration chambers over inlet distribution piping.

Figure 7.2 in Section 7.2 of Chapter 7 shows the dolomite gravel used at BNIA. Figure 3.22 shows a completed and operating BBR cell at BNIA.

As may be seen, it appears to be a grassed area and does not look anything like a "wetland," reinforcing the recommendation to preferentially refer to such cells as BBRs, not EWs.

The currently operating BBR-based EEB System at BNIA successfully manages Spent Glycols, snowmelt and GCSW containing BOD up to that equivalent to that which would be generated by 50,000 homes.

Figure 3.19 illustrates one arrangement of aeration tubing (a header connected to runs of aeration tubing) placed in a vertically sided BBR cell under construction at BNIA. As may be seen, in this case the outlet distributor piping and the aeration tubing are placed on the same level at the

FIGURE 3.22
One of the four operating BBR cells at BNIA.

base of the gravel substrate. In other cases, the aeration tubing is located above the outlet distribution piping in a separate layer of gravel, sand and/or crushed rock.

The BNIA EEB System continues to operate well through the harshest winter conditions. It was designed to meet the following criteria:

- Siting near an operating runway
- Simplicity of operation with minimum attention
- The ability to treat cold and variable strength wastewaters (both Spent Glycols and GCSW) year round
- Integration into the airport's existing stormwater management system
- The ability to handle large seasonal variations in influent flows and to operate steadily even under the harshest winter conditions
- To produce effluent to stormwater system meeting discharge criteria
- Not to be any sort of birdstrike or other airside hazard
- To produce little or no odours
- Not to produce organic sludge requiring additional management
- To have low operating and maintenance costs

The BNIA EEB System has been able to meet all of these objectives. The savings obtained by the NFTA by treating the Spent Glycols and GCSW (e.g., minimizing the need to send Spent Glycols to the local WWTP, and by eliminating most of the GRVs at the de-icing pads) much more than compensate for the additional costs incurred in operating the EEB System (e.g., electricity for blowers, nutrient chemicals).

The EEB System at BNIA had some early "teething" problems (e.g., the aforementioned sliming incident, and the plugging of aeration lines resulting in blower operation problems), but these have been identified and successfully resolved. An example of an early result (March, 2010) indicated the reduction of average mixed spent glycol/GCSW cBOD concentration from 2400 mg/L in the influent to 30 mg/L in the effluent.

3.7 The Aerated VSSF EW Cells at EIA

For over a decade, there has been a large SSF constructed wetland successfully operating at *EIA* (EIA or YEG) treating ethylene GCSW runoff from aircraft de-icing. The catchment area at EIA is very large, and this, coupled with the airport's tight clay soil, results in very large amounts of stormwater runoff. About 10%–20% of the ADAFs used annually in colder weather at EIA for the glycol de-icing of aircraft were, and still are, collected by GRVs and pumped to a local WWT plant, but the balance is allowed to freeze in piles of contaminated snow beside the airport's de-icing pads, eventually ending up in stormwater ditches during warmer periods. The airport's AOA, the ERAA, decided to install an on-site treatment system to handle the GCSW at the airport prior to its disposal (Higgins & MacLean 1999; Higgins et al. 2001).

The multiphase project first involved a systems analysis carried out by the Lead Author as project manager for Jacques Whitford* to define options for the management of the highly glycol-contaminated part of the stormwater runoff at the airport. This systems analysis evaluated not just the impact of de-icing glycol on stormwater, but the impacts of pavement de-icing chemicals as well, and whether or not new CDFs were economical in the light of other glycol management options.

Various treatment options were considered including pond/lagoon treatment, various kinds of bioremediation and phytoremediation, and on-site conventional (mechanical) and natural WWT processes of several types. The systems analysis concluded that an HSSF CW system would be the most practical and economic option for EIA at the time (late 1990s). Since at that time accurate values for the biological degradation of de-icing glycols under field conditions were not available in the literature, the systems analysis included off-site pilot-scale treatability testing on a simulated ethylene glycol-contaminated runoff indicative of that under worse case contaminant loading conditions that might be encountered at EIA.†

* Another of the other Co-Authors (Stiebel) was Jacques Whitford's Project Executive for the EIA CW system project.

† Although the design and construction of a CW system proceeded after the treatability test, Jacques Whitford used the opportunity to also carry out early aerated gravel bed testing as well, and the results supplemented those from the BREW Project, although they did not at that time result in the installation of BBR cells instead of SSF CW cells at EIA. (This upgrading occurred later.)

The implementation of the EIA CW project was authorized to proceed as part of a larger stormwater management project in 2000, and start-up occurred in August of 2001 (Higgins 2002a,b). The HSSF CW system at EIA consisted then of 12 square, very large gravel-filled cells with sides of 47.5 m each.

Figure 3.23 provides an aerial view of the northwest corner of the 740 ha airport showing the locations of the terminal, runways and taxiways, de-icing pads, the CW system (top, left-hand corner, its inlet flow equalization pond, [the Gun Club Surge Pond, located with a private skeet club alongside, a factor which discourages waterfowl use]) and the receiving waters, a cold water trout stream.

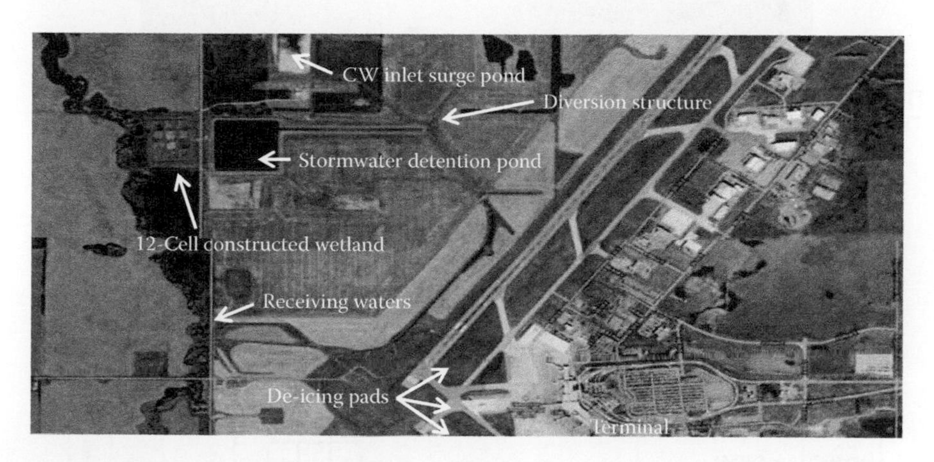

FIGURE 3.23
Facilities locations for the CW system at EIA.

The cells at YLG are arranged in six trains of two cells each and influent is pumped into each train from the 91,000 m^3 Gun Club Surge Pond via a lift station located on the wetland's central berm.

A system of open ditches transports stormwater and snowmelt from the airport facilities northwest to a large diversion structure, a series of large manually operated gates, which allows water to be directed either to a high flow condition, normally dry, stormwater detention pond (for collection and bypassing of relatively uncontaminated stormwater, outside the de-icing season) or to the CW system inlet equalization pond, the Gun Club Surge Pond (when there is glycol in the water requiring treatment).

Figure 3.24 shows the diversion structure and its manually operated gates.

Figure 3.25 shows the flow scheme for the CW system, while Figure 3.26 shows a layout diagram for the CW system.

In winter, during the de-icing season at EIA, glycol de-icing is carried out at de-icing pads near the terminal and spent ADFs and AAFs not vacuumed up by GRVs collect as contaminated snow in piles near the pads. When the weather warms in spring, the snow melts and the glycol-contaminated

FIGURE 3.24
The diversion structure at EIA.

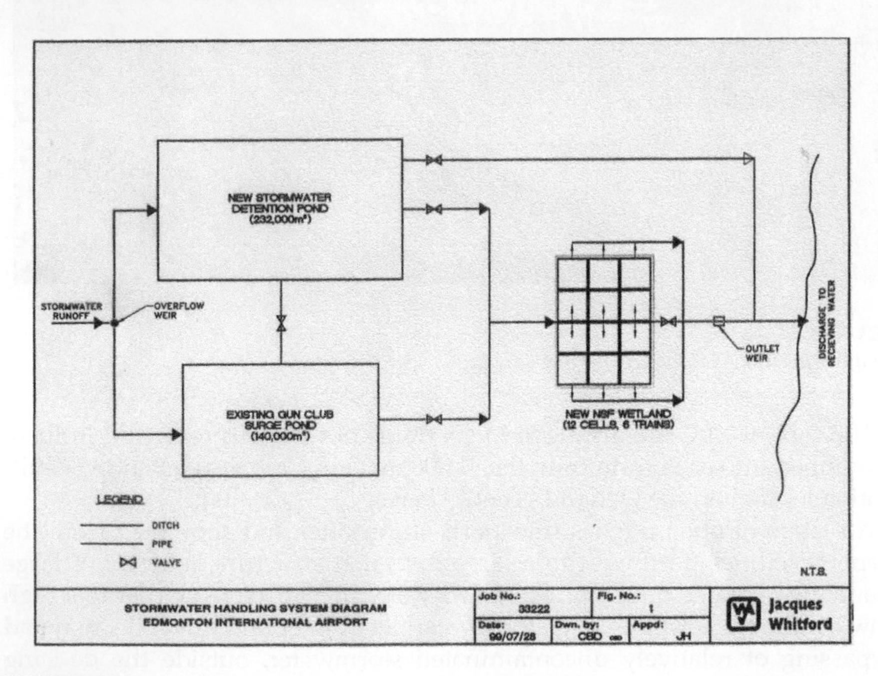

FIGURE 3.25
Process flow at EIA.

runoff flows through the stormwater ditches to the Gun Club Surge Pond via the diversion structure.

Each of the 12 cells of the EIA CW was vegetated with Cattails transplanted from airport stormwater ditches. The gravel substrate in the 12 CW cells is/was 0.7-m thick and its surface area is 2.7 ha, with a "footprint" (including cell separator berms, etc.) of 5 ha.

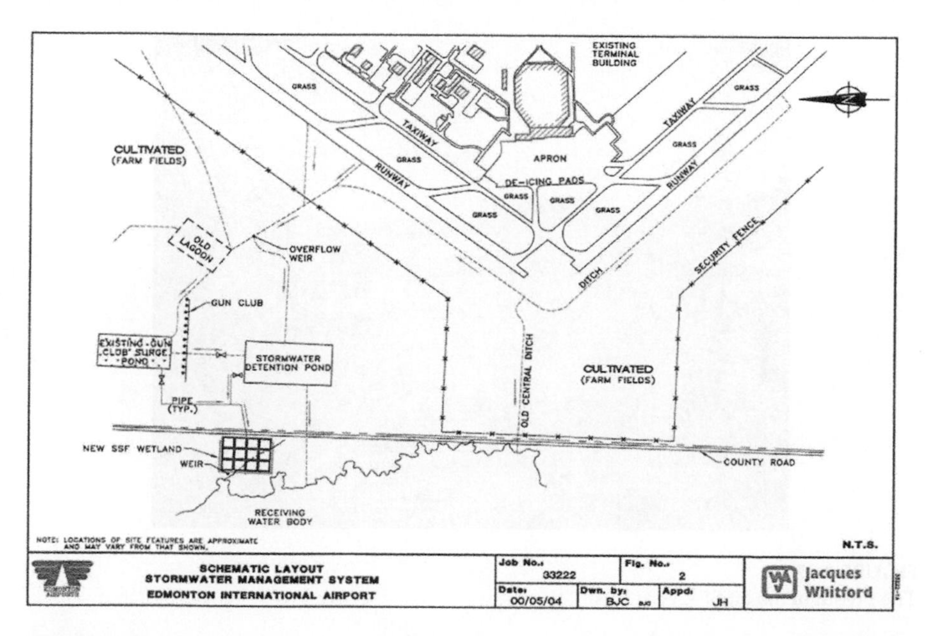

FIGURE 3.26
CW system layout diagram.

Design conditions for the CW were for the treatment of stormwater runoff contaminated with up to 1350 mg EG per L at flows of up to 1250 m^3/day. The CW system included the wetland itself, the lift station, associated ponds and ditching, piping between existing and new ponds and the CW, an outlet weir system, mandated continuous sampling facilities, fencing and diversion facilities allowing less contaminated water to bypass the wetland via the normally dry, very large stormwater detention pond.

At EIA, influent glycol-contaminated runoff flow rates (8–28 L/s) were keyed to water temperatures, with lower throughputs occurring when the water was colder. The CW did not operate in the coldest parts of winter and was (and still is) left frozen during these periods when no runoff occurs in any case.

OPEX costs for the wetland system were very low, and it was (and still is) designed for unattended operation other than periodic monitoring. Figure 3.27 shows an aerial photo of the EIA wetland following construction of the cells, just before the liners and gravel substrate were emplaced.

The 12 HSSF cells at EIA (six trains) operated successfully in the CW mode, meeting effluent targets for discharge into the receiving waters, a nearby cold water trout stream. However, owing to the proximity of the Athabasca Oil Sands, traffic at the airport has increased steadily during the period between 2001 and 2010, and in 2011 two of the six trains (the four cells on the right in Figure 3.27) were upgraded (Liner 2011, 2013).

The upgrade is designed to provide over 711 kg BOD per day of treatment, and to allow increased flow rates as the influent BOD drops during

FIGURE 3.27
The layout of the HSSF CW at EIA.

the freshet. The first cell of each of the upgraded trains was modified from an HSSF CW cell to a VSSF BBR cell by installing an aeration system under its gravel and taking gravel from the second cell of the train to increase the first's gravel thickness to 1 m.

The second cell of each of the two upgraded trains was reconfigured as a polishing FWS CW cell* with a remaining gravel depth of 0.3 m overlain by 0.3 m of water. Nutrient addition capability was added as were two 1500 SCFM blowers to supply air to the two upgraded primary cells. The incorporation of a recycle system within the new BBR cells permits the expedited start-up of the EEB System in springtime.

In essence, a six-train, 12-cell HSSF CW with hydraulic constraints was converted to a two-cell BBR-based EEB System with each of the two EEB cells followed by tertiary treatment FWS CW cells, the remaining five CW trains becoming idle. (Plans are now in place to upgrade a third train.)

3.8 The BBR-Based EEB System at Heathrow (LHA)

An advanced natural treatment system for glycol-contaminated from Heathrow Airport (IATA code LHA) near London, UK is located in an area

* Normally, FWS CWs are not recommended at airports because of birdstrike concerns (their open water may attract waterfowl and other birds) but at YEG, the mixed EEB System/CW system is far from the runways, and the AoA, the ERAA, has allowed their construction in this case.

known as Mayfield Farm which is just south of the airport. With this system, GCSW and Spent Glycols are collected from two areas at LHA, a Southern Catchment and an Eastern Catchment (290 ha in total area), and delivered to Mayfield Farm via peripheral firewater lines at the airport that are already in place for emergency purposes.

Depending on BOD content, wastewater from the Eastern Catchment enters either the "clean" side (<40 mg BOD per L, measured using on-line BOD meter) of what is called the Middle Pond (a large impoundment with a floating plastic barrier dividing it roughly in two) from which it is discharged to the River Crane, or the "dirty" side from which it can be pumped at up to 100 L/s into what is called the main reservoir (an Aerated Lagoon) covered with floating reed bed rafts.*

Runoff from the Southern Catchment at LHA at 80 L/s is pumped into a diversion chamber from which, if it has a BOD of <40 mg/L, it is discharged untreated to the River Thames via a Southern Reservoir. Otherwise, it enters the main reservoir. Pretreated water from the main reservoir is pumped into a Reed-vegetated wetland treatment system at 40 L/s via a balancing pond.

This SSF CW originally involved 12 pairs of Reed-vegetated HSSF cells of varying sizes operating in parallel. The original CW cells consisted of two discrete beds of six cells each that had cells of differing sizes, as well as differing aspect ratios and treated flow rates (Richter 2003; Murphy et al. 2014).

Figure 3.28 shows one of the system's unplanted HSSF CW cells and Figure 3.29 shows a view of the wetland after planting.

FIGURE 3.28
Completed HSSF CW cell at Mayfield Farm prior to planting.

* Figure C.5 in Appendix C shows a picture of these rafts.

FIGURE 3.29
View over the CW cells after planting with Reeds in the autumn of 2001.

As may be seen from Figures 3.28 and 3.29, there are vertical concrete walls around the treatment cells. Each consists of a 2 m wide open water area on the inlet side followed by a 0.6 m wide gabion of coarse rock, then an approximately 20 m × 20 m vegetated gravel bed treatment area. Further, there is a similar gabion after each gravel bed. When it was a CW system, gravel (porosity, 0.45) filled the bed part of each of the cells to a depth of 0.6 m. Flow was HSSF (and still is). The total cell area (all 12 cells) is 2.1 ha for the gravel bed parts, 1.1 ha for the gabions and 2.3 ha for the open water areas.

There were problems with the operation of the CW system at Mayfield Farm and, since the development of BBR ecotechnology during the *BREW Project* had indicated that the performance of CWs might be greatly enhanced by aerating them, an on-site test project involving two of the existing 12 CW cells was approved and proceeded in 2005. One cell (the control) was to continue to operate as a CW while a second, otherwise identical, one was converted to an aerated gravel bed cell (a BBR cell, although the designers did not use that name) by "ploughing" aeration tubing into its gravel substrate (Murphy et al. 2014).

Figure 3.30 shows the addition of the aeration tubing to one of the two test cells in 2005.

The two test cells were then operated in parallel and their respective performances monitored. It was found that the cBOD removals in the aerated (BBR) cell were 14 times that found for the unaerated (control) cell. This, and the example from the construction and subsequent start-up and successful operation of the BBR-based EEB System at BNIA in the United States, demonstrated the superiority of the semi-passive BBR ecotechnology over the passive CW method.

FIGURE 3.30
Converting an HSSF CW cell at LHA into a BBR cell.

Accordingly, a decision was made to upgrade the entire system (i.e., the remaining 11 CW cells) into BBR cells as well and this conversion was carried out in 2010/2011. Aeration tubing was "ploughed" into each of the cells, and gravel depth was increased to 1.1 m. All of the 12 upgraded cells operate in parallel and provision is made to recycle effluents to upstream parts of the system if its cBOD exceeds 40 mg/L. As with BUF, a SCADA system controls pumps and flows, and at LHA six PD Blowers (240 KW max.) provide aeration air. These operate during the de-icing season on a duty/assist configuration, with all six in operation during peak load conditions due to precipitation events and/or high de-icing periods. (The system continues to operate during the summer months but usually no aeration is provided during this period.) Nutrient dosing occurs at various points. Design basis is to remove 3500 kg $cBOD_5$ at 40 L/s with a total system residence time of 21 days.

Table 3.4 (adapted from Murphy et al. 2014) compares some design information for the original HSSF CW system at LHA constructed in 2001 with that for the upgraded BBR-based EEB System constructed in 2010/2011.

TABLE 3.4
Comparison of the Former CW System at LHA with Its Upgraded EEB System

	CW System	EEB System
Design $cBOD_5$ loading (kg/day)	374	2073
Design flow rate (m^3/day)	3456	6912
BBR substrate gravel depth (m)	0.6	1.1
Implied O_2 transfer (g/m^2.day)	2.4–7.7	75.6
$cBOD_5$ removal (g/m^2.day)	4–13	75.6

The EEB System at Mayfield Farm now successfully treats glycol-contaminated runoff from Heathrow Airport and has handled peak loads of up to 9000 kg cBOD per day.

3.9 SAGR Bioreactors

Another kind of aerated SSF EEB ecotechnology evolved from early HSSF EWs and this is that of Submerged Attached Growth Reactors (SAGR™ Bioreactors*). These specialty EEBs for nitrification are provided by Nexom (formerly Nelson) and are used in the upgrading of lagoon-based WWT systems in northern locations (Liner & Kroeker 2010).

The concept of SAGR Bioreactors was developed during the Alexandria Project (see Section 6.5 in Chapter 6), which featured (among other innovations) converting a facultative municipal WWT lagoon to an Aerated Lagoon and following it by three HSSF "EW" [BBR] cells in parallel located in another drained lagoon).

As with BBRs, aeration tubing in SAGR Bioreactors is placed at or near the floor of the EEB cells, and aeration air is supplied from a nearby blower or blowers (Higgins et al. 2009d; Nelson 2013).

For SAGR Bioreactor-based EEB Systems, the bulk of suspended solids removal, ammonification of organic nitrogen in the influent municipal wastewaters, and BOD removal occurs in an upstream Aerated Lagoon, leaving the downstream HSSF aerated EEB cell or cells (which may be also be located in a converted facultative lagoon, or alongside of it) to be dedicated to nitrification and final polishing.

In 2007, building on their earlier work for Jacques Whitford at Alexandria, BNIA and elsewhere, Nelson and NAWE constructed a two-train SAGR Bioreactor demonstration facility near Winnipeg (Steinbach), MB, Canada to test the nitrification concept of the new ecotechnology. At the time, regulators across Canada and the United States were tightening ammonia effluent regulations on small- and medium-size municipal sewage treatment systems in response to ammonia toxicity issues. The Steinbach demonstration project used aerated EEBs that had significantly deeper gravel beds than had been previously used for BBRs (2-m thick). For this SAGR Bioreactor demonstration-scale test, the average winter effluent ammonia nitrogen concentration was about 0.1 mg/L in wastewater that could be colder than 1°C.

The testing at Steinbach went on for over 3 years, with weekly testing. On the basis of the data collected, design criteria were developed which discounted the influence of oxygen diffusion from plant roots or water surface transfer. In fact, when the test SAGR Bioreactor beds at Steinbach

* The redundancy is deliberate for clarity.

were decommissioned in 2010, it was found that few, if any, plant roots penetrated from the insulating mulch layer on top of the SAGR Bioreactor cells into the aggregate bed. *In situ* hydraulic conductivity testing of the Steinbach demonstration site after three years of operation showed no significant bed fouling or deterioration of hydraulic performance. Since the SAGR Bioreactor process does not rely on plant uptake of nutrients, no biomass harvesting or removal is required to maintain treatment performance.

Figure 3.31 shows the successfully operating Steinbach SAGR Bioreactor cells on a day when the air temperature was 40 below.

FIGURE 3.31
Operating SAGR Bioreactors at Steinbach in winter.

In addition to the nitrification performance of SAGR Bioreactors, polishing treatment has also been observed with them (e.g., disinfection, further organics removals). Effluent cBOD and TSS at the demonstration site were typically below detection (<2.0 mg/L). Subsequently, Nelson (now Nexom) has advanced the SAGR Bioreactor with a number of innovations.

On the basis of the results from the Steinbach test unit (Liner and Kroeker 2010), the performance of SAGR Bioreactors as measured by BOD removals is little affected by season and cold water. Also, on the basis of the performance of the Steinbach site, a second demonstration-scale SAGR Bioreactor test site was set up at Lloydminster, SK, Canada, and the first full-scale installation was built at Glencoe, ON, Canada.

The Lloydminster SAGR Bioreactor consists of a single horizontal HSSF aerated gravel bed, divided up lengthwise into several zones, operated in series. The zones provide the ability to have multiple water quality sampling points throughout the SAGR Bioreactor. Wastewater flow is collected at the end of each zone and redistributed across the beginning of the ensuing zone. Water sampling ports are connected to the redistribution piping at the end of each zone to ensure a constant flow of water for representative sampling. Each zone is fitted with a coarse bubble aeration system to supply the required oxygen for nitrification. SAGR Bioreactors achieve excellent nitrification rates even in very cold water conditions (down to 0.1°C) using a process that allows the growth of two or more discreet biomass colonies in the SAGR Bioreactor cell when the water temperature is moderate in late summer or fall, providing sufficient biomass to provide full nitrification in winter with cold influent conditions.

The Aerated Lagoon/SAGR Bioreactor process has made the transition from the variable treatment rates observed in CW systems to consistent cold weather nitrification performance. A large number of SAGR Bioreactor-based EEB Systems have now been designed and built by Nexom and are very successfully operating across northern US and southern Canada (Higgins et al. 2009b; Liner & Kroeker 2010), as is illustrated in Table 3.5.

TABLE 3.5

Locations for Some SAGR Bioreactors

Guthrie School, ON	Walker, IA
Lloydminster, AB[a]	Sylvan Lake, SD
Perth, ON[a]	Shellbrook, SK
Blumenort, MB[a]	Kingsley, IA
Doaktown, NB	Balcarres, SK
Dawson Creek, BC	Colesburg, IA
Mentone, IN	Kennard, IN
Glencoe, ON	Grand Rapids FN
Mountain View, WY	New London, IA
Lamar, MO	Long Plain FN, MB
Mentone, IN	

[a] Demonstration scale.

More SAGR Bioreactors have been built, are under construction and are planned (Nelson 2013). SAGR Bioreactors can be installed following a wide variety of primary and secondary treatment systems including facultative lagoons.*

In addition to the nitrification performance of SAGR Bioreactors, polishing treatment has also been observed. Full-scale SAGR Bioreactor-based EEB Systems in operation have demonstrated consistent cBOD/TSS of <10/10 mg/L. Even systems following facultative lagoons instead of aerated

* Full-scale SAGR Bioreactors are in place following facultative lagoons as well as aerated ones.

ones have demonstrated consistent cBOD/TSS levels below 20 mg/L, despite large algae blooms in the lagoons in summer.

SAGR Bioreactors have also been found to provide a significant level of disinfection.* At several sites where SAGR Bioreactors are now operating, effluent fecal coliform levels have been consistently below 200 CFU/100 mL, and some full-scale systems have now been approved without further disinfection being needed.

SAGR Bioreactors used for nitrification in EEB Systems for northern municipal lagoon WWT upgrades can be regarded as fully proven technology.

3.10 Engineered Stormwater Wetlands

Influent wastewater flows to most EEBs (and CWs) are maintained at relatively steady and continuous rates using upstream balancing ponds, tanks or vaults, while Stormwater Wetlands (see Section B.1 in Appendix B) are designed to accommodate intermittent precipitation events yielding varying influent runoff flows (but, however, they only provide minimal contaminant removals).

There is, however, a kind of EEB System which can clean up municipal stormwater runoff without resorting to upstream impoundment and influent flow control, and which can provide good contaminant removals despite this. Engineered Stormwater Wetlands (ESWs) are stormwater treatment systems that involve special kinds of mixed HSSF-VSSF, high freeboard, unaerated EEB cells that operate in a "Pulse Flow" (fill and drain, see Section B.4 in Appendix B) mode and are used to treat variable flows (Higgins et al. 2009d, 2010c).

ESWs are usually designed to upgrade wet ponds and dry ponds, kinds of stormwater management facilities (see Section B.1 in Appendix B) that provide stormwater volume control but little treatment. ESWs are kinds of EEB Systems that involve innovative types of oil–grit separators (OGS units) upstream of the aforementioned kinds of EEB cells to provide primary treatment by removing sediment and suspended solids from the runoff, and also often include downstream pond or free water surface CW cell (see Section B.3.3) to polish effluent from the EEB cell (tertiary treatment).

Figure 3.32 shows the site plan and hydraulic profile for the operating ESW in the Town of Aurora, ON, Canada.

During minor storm events, grit-removed OGS unit effluent enters an ESW's EEB cell via an upstream permeable berm, and fills all or part of the EEB cell's gravel substrate, before percolating out VSSF in a fill and drain

* The same is true of BBRs.

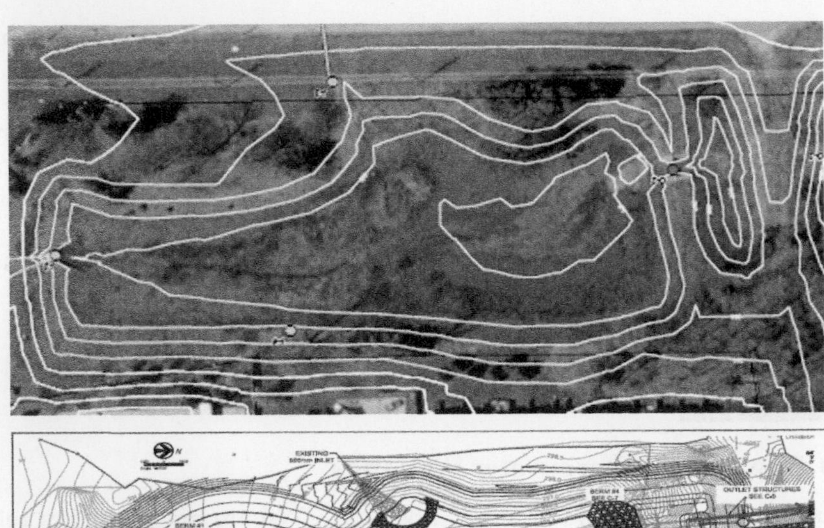

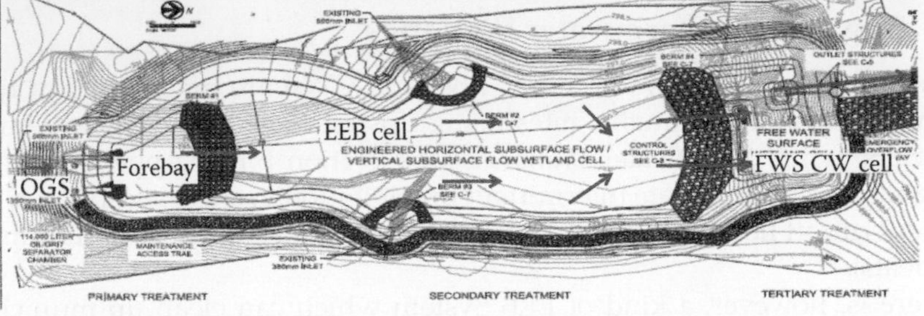

FIGURE 3.32
Upgrade of a Stormwater Wet Pond in Aurora, ON to an ESW system.

flow mode. Any base flow through the EEB cell (and there is usually some in any municipal stormwater system) is treated in the EEB cell in an HSSF mode. During a major storm event, water fills the EEB cell and impounds above its surface in the cell's relatively high (~2 m) headspace. After the storm is over, the impounded water percolates down into the substrate and out of it again in a normal VSSF fill and drain flow mode. Currently (2016), two full-scale ESW systems are operating successfully in ON, Canada and several more are planned.

3.11 Standard Oxygen Transfer Efficiency Testing

With aerated WWT facilities, clean water oxygen transfer tests (referred to as Standard Oxygen Transfer Efficiency [SOTE] tests) are often carried out to determine the efficiency of the aeration processes (Redmon et al. 1983). As part of the BNIA EEB System project, Jacques Whitford subcontracted Nelson to carry out SOTE testing at an indoor test facility in Winnipeg, MB, Canada from January 24 to 26, 2007 on an aerated gravel bed EEB test cell using various aeration tubing configurations. Nelson, in turn, engaged Redmon Engineering Company of Racine, WI, USA to assist with the test.

The SOTE test involved deaerating the water into an EEB cell using sodium sulfite promoted by cobalt chloride hexahydrate followed by re-aeration to near saturation levels with DO concentrations being measured at several points in the bioreactor. The data obtained were analysed using a simplified mass transfer model to estimate the apparent mass transfer coefficient steady state DO saturation concentration.

The clean water oxygen transfer efficiencies for an aerated gravel bed EEB found during the BNIA SOTE Test ranged from 1.2% to 1.7% per foot of water depth, averaging about 1.5% per ft. (For comparison, values to 0.8% per ft. and ~2% per ft. are typically found for Aerated Lagoons with submerged coarse and fine bubble diffusers, respectively.) Accordingly, aerated gravel bed EEBs can be expected to have roughly similar SOTEs to those of Aerated Lagoons and the SOTE Test data can be used in the design of aeration systems for them.

Figure 3.33 (Wallace & Knight 2006) illustrates the relative aeration efficiencies of BBRs and other kinds of aerated technologies.

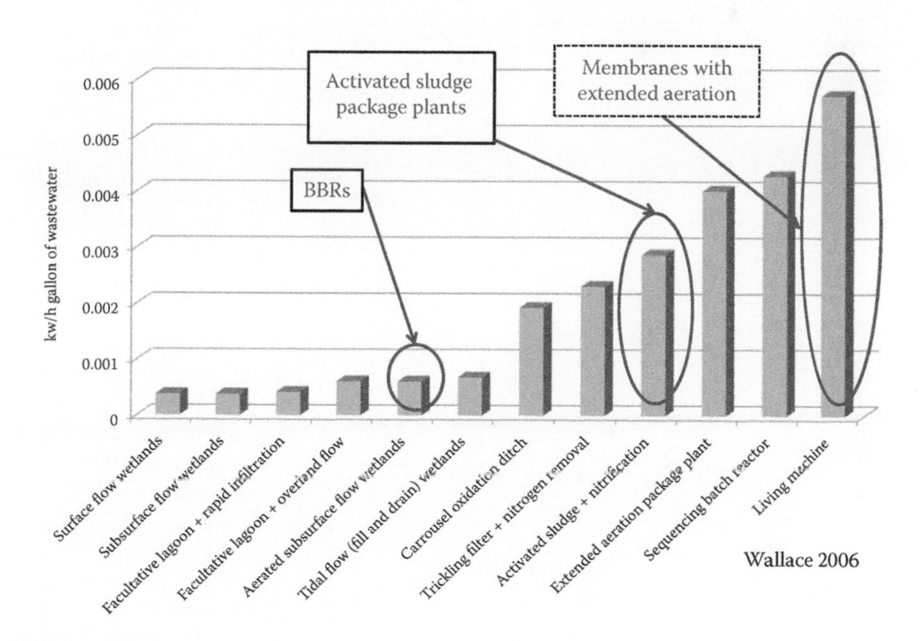

FIGURE 3.33
Relative aeration efficiencies.

4

Anaerobic Eco-Engineered Bioreactors

4.1 Kinds of Anaerobic Eco-Engineered Bioreactors

As was introduced in Chapter 1, in addition to the aerobic EEBs discussed in Chapter 3, there are various kinds of other anaerobic EEBs, particularly certain kinds of anaerobic bioreactors (ABRs).*

The main kinds of EEB ABRs are

1. Denitrification bioreactors (DNBRs)
2. SAPS Bioreactors
3. Passive treatment BCRs
4. Semi-passive treatment BCRs

As the name suggests, DNBRs remove nitrates. SAPS Bioreactors are intended to neutralize ARD while inhibiting ferric oxyhydroxide formation. As was introduced in Section 1.3 of Chapter 1, there are two basic kinds of BCRs and these are discussed in some detail in Chapter 5.

There are also several kinds of specialty BCRs (which are usually passive treatment BCRs, see Section 4.4) and these are designed to remove specific metal(loid)s (e.g., As, Se, Cr), and there is a rapidly developing kind of BCR called an electro-biochemical reactor (see Section 4.4.2).

While in the past many Stand-Alone ABRs were designed for HSSF† flow, such a mode is no longer recommended for EEB ABRs, and they now should be designed so that WWs being treated in them flow VSSF. All EEB ABRs are either buried or located under water covers (e.g., at the bottoms of ponds) to maintain anaerobic conditions in them (Higgins & Mattes 2009).

* A number of Stand-Alone ABRs have been built, but hereafter, unless specifically indicated otherwise, all mention of ABRs herein will refer to EEB types.

† The Interstate Technology and Regulatory Council in the United States has published a Guidance Document titled Biochemical Reactors for Mining-Influenced Waters (the ITRC Report) which summarizes information on projects mostly for passive treatment, Stand-Alone, HSSF BCRs (although some other kinds of ABRs are also addressed in it as well, ITRC 2013).

The reduction reactions carried out by microbes in the biofilms of anaerobic bioreactors require a source of carbon and this is supplied by organic active media material. In passive treatment BCRs, the active media is a solid material forming part of a more comprehensive permeable substrate. Semi-passive treatment BCRs* contain aggregate substrates similar to those of BBRs (albeit without aeration systems) and the active media (carbon source) needed to facilitate their reduction reactions are pumped into them as carbonaceous liquids along with the wastewaters being treated.

Most ABRs are passive in nature and these kinds contain permeable substrates that incorporate solid organic active media, the degradation of which by various kinds of microbes (largely fermenters and anaerobic heterotrophs such as cellulose degrading bacteria [CDB] and acid-producing bacteria [APB]) provide breakdown products that other kinds of microbes such as characterizing bacteria (e.g., DNB, SRB) use in their metabolisms (Brown et al. 2002; Higgins et al. 2003c; Gusek 2008).

In addition to the solid active media, most passive treatment ABR substrate beds usually contain inert support materials (e.g., quartz gravel, sand, crushed rock) to provide permeability to the matrix and further attachment sites for biofilms. ABRs can also contain reactive support media (e.g., limestone, dolomite, fly ash) either mixed in their main substrate beds, or before or after them as separate layers (Watzlaf et al. 2000).

Passive treatment ABRs may also have relatively thin layers of separator and bedding materials such as sand, and have covering layers of water (i.e., they are located at the bottom of impoundments) or are covered over with soil, clay, gravel and/or geomembranes to ensure anaerobic conditions in them. Table 7.1 in Chapter 7 lists a few of the kinds of materials that have been used as solid active media in ABRs (mostly passive treatment BCRs).

Since the carbonaceous active media in passive treatment ABRs may (and probably will) leach various materials (e.g., suspended solids, organics, ammonia), it is common practice in EEB Systems to include one or more aerobic cells (e.g., BBRs, Aerobic Wetlands) downstream of anaerobic EEB cells to remove these contaminants before discharge (Mattes et al. 2007, 2010; Higgins & Mattes 2009).†

Periodically (every 5–10 years), the substrates in passive treatment ABRs should be examined to see if they may have to be removed and cleaned out and, if necessary, replaced. However, doing so is not particularly onerous, and the cost will be relatively small compared to the cost of operating an

* BCRs in general as well as a Specialty BCR Project for removing arsenic are addressed in the next chapter.

† The liquid active media added to semi-passive BCRs are less likely to result in residuals and degradation products in bioreactor effluents (their flow rates can be adjusted to minimize these) but usually semi-passive treatment BCR-based EEB Systems also include downstream (tertiary treatment) cells to ensure that such do not cause problems.

alternative mechanical WWTP over the same period. The gravel substrates of most existing BBRs have not required replacement after operating periods that for many are now well into their third decades (largely for HSSF ones treating domestic wastewaters) and it not anticipated that the situation will be much different for the VSSF BBRs (which have not yet been operating as long). The same will probably be true for gravel-substrate, semi-passive treatment BCRs. The situation with passive treatment BCRs may be different as their substrates contain organic active media that will get "used up," hence the recommendation to periodically evaluate whether their substrates may need replacement. Nevertheless, for prudent financial planning, the operators of EEB Systems should budget for potential EEB substrate replacement or cleaning every decade, even though in most cases the funds may not be needed.

The detailed designs of EEB ABRs (as is also the case with aerobic EEBs) should be based on Reaction Kinetic models determined during treatability tests carried out using the same substrate materials to be used full scale, and with the wastewater under consideration spiked with the main CoCs to worse case possible conditions (see Chapter 6).

Owing to their natures and the more complex hydraulics and hydrology, throughput in EEB Systems involving passive treatment BCRs* has to be lower than that possible with ones involving semi-passive treatment BCRs. In addition, passive treatment BCRs have much more complicated morphology including substrates the components of which may be much more difficult to access, store, blend and use.

Generally, Reaction Kinetic models (see Section 8.3 in Chapter 8) for passive treatment EEB ABRs assume that plug flows are found with ABRs (see Figure 1.5 in Chapter 1 for an example of the results of a tracer test that include those for a BCR).

4.2 Denitrification Bioreactors

Many wastewaters contain particulate and dissolved nitrogen compounds in the forms of ammonia, organic nitrogen compounds and nitrates. In aqueous environments, the organic nitrogen compounds can be mineralized (ammonified) by facultative bacteria to ammonia, both aerobically and anaerobically (see Section 2.4.2 in Chapter 2). Thereafter, under aerobic conditions, the ammonia is nitrified to nitrates (see Equations 1.1 and 1.2 in Chapter 1), often leading to wastewaters contaminated with significant amounts of nitrates (see Section 2.4.3). Providing sufficient bioavailable

* Section 5.4 in Chapter 5 addresses the relative advantages of the various kinds of BCRs.

organic matter is present, nitrates can be microbially denitrified to nitrogen gas by DNB in anaerobic EEBs under anoxic conditions (see Equation 2.14 in Section 2.4.3).

Denitrification bioreactors (DNBRs) are ABRs that are used to remove nitrates from wastewaters. Alternative names for DNBRs are denitrifying bioreactors, denitrification beds, denitrification biofilters, tile drain biofilters and (recognizing the most common active medium used in many of them) woodchip bioreactors. DNBRs are usually passive treatment ABRs, but could be designed with gravel substrates and injected active media to operate in a semi-passive mode.

DNBRs to date have mostly been Stand-Alone passive treatment HSSF bioreactors used for the treatment of tile drainage and runoff from agricultural fields that are contaminated by nitrogen-containing fertilizers and their degradation products. Christianson (2011) has provided a review of these bioreactors.

In the substrates of passive treatment DNBRs, the carbonaceous active media may be supplemented with gravel or other support media to improve permeability. The slow degradation of the fixed carbon in them by fungi, CDB, APB and other microbes provides the simpler chemicals that are used by the DNB as their energy and carbon sources.

In the past, the amounts of nitrates in treated effluents from municipal and industrial WWTPs and natural WWT systems were usually not regulated, and streams containing elevated amounts of them could be discharged to receiving waters or groundwaters without regard to their nitrogen concentrations. However, regulatory authorities are increasingly mandating that these effluents meet either strict total nitrogen criteria (e.g., 5–10 mg TN per L) or nitrate limits (e.g., 10 mg NO_3-N per L).

One way to meet these new guidelines would be to treat the wastewaters in passive or semi-passive treatment VSSF DNBR that form parts of EEB Systems. While these kinds of DNBRs have not yet been built, there is no reason to believe that such bioreactors (which would have similar morphologies to BCRs, see the next section) would not work, but until one is, the DNBR EEB ecotechnology cannot be said to be proven,* although there are a number of successfully operating Stand-Alone HSSF DNBRs.

* Jacques Whitford once tested a pilot-scale EEB System consisting of a BBR cell and a "BCR" cell to treat leachate from a municipal landfill. This leachate contained both nitrates and dissolved metals. The system gave excellent results in removing oxidizable contaminants in the BBR cell, and also removed nitrates and reducible aromatic and chlorinated organics in the following passive treatment "BCR" cell, but no removals of dissolved metals occurred at all. What appeared to happen was that DNB in the second cell of the pilot unit (which was operating as a DNBR cell instead of a BCR) outcompeted any SRB present and suppressed sulfate reduction (see Table 1.1 in Section 1.5 of Chapter 1).

4.3 SAPS Bioreactors

One semi-passive way to deal with acidic wastewaters such as ARDs is to neutralize them by contacting them with limestone in systems such as oxic and anoxic limestone drains (see Section C.3 in Appendix C). However, this can result in the armoring of the limestone particles (coating with ferric oxy-hydroxides, inhibiting further neutralization) and even eventually with the plugging of their interstitial spaces with sludge. To address this problem, a passive treatment bioreactor that combined the functions of an anaerobic SSF CW with a limestone drain in a single downflow basin was conceived. This combination is called a Successive Alkalinity Producing System (a SAPS Bioreactor, sometimes also referred to as a Reducing and Alkalinity Producing System [RAPS], a vertical flow wetland, a vertical flow pond or just an alkalinity producing system).

As may be seen from the following equation,* the hydrolysis of ferric iron adds acidity to the ARD and lowers its pH while forming a complex oxyhydroxide sludge:

$$Fe^{3+} + (3+x)H_2O \rightarrow FeO(OH)(H_2O)_{(1+x)}\downarrow + 3H^+ \tag{4.1}$$

The iron oxidation and hydrolysis reactions are strongly controlled by ARD pH and DO concentrations and can be buffered by alkalinity (Kepler & McCleary 1995; Gould & Kapoor 2003). The relative solubilities of relevant (hydr)oxides are

$$\underset{\text{Insoluble}}{Fe(OH)_3 = MnO_2} > Al(OH)_3 \gg \underset{\text{Soluble}}{Fe(OH)_2 > Mn(OH)_2} \tag{4.2}$$

Accordingly, it was felt that if the iron in an ARD could be kept in its ferrous form, it would stay in solution rather than precipitating out when neutralized.

The concept of a SAPS Bioreactor then is to first convert any ferric iron in an ARD into ferrous iron in a carbonaceous substrate bed without raising the pH, then to raise the pH to circum-neutral levels in a following, non-substrate limestone bed. (As is discussed in Section 4.5, reality is quite a bit different from this conception.†)

In most of the Stand-Alone SAPS Bioreactors that have been built in the past, design was such that ARD flowed into the headspaces of basins where a covering layer of water 0.3–3-m deep ensured anaerobic conditions below (Younger 2000; Brown et al. 2002; Wildeman & Schmiermund 2004). It then

* Equation 2.20 in Section 2.6 of Chapter 2 shows a simplified version of this equation.

† A major pilot-scale test project, the first part of which has been already carried out at the CAWT facilities, the EW Test Project (see Section 4.5), has commenced to better define the kinetics and genomics of SAPS Bioreactors (and BCRs), and the results of this R&D are beginning to better elucidate mechanisms and microbiology.

was supposed to percolate down through the substrates (which were 0.1–1+ m thick) where ferric iron would be reduced to ferrous iron. Finally, the ARD would enter underlying limestone gravel layers (which were 0.5–1.5+ m thick) where the ferrous iron solution was supposed to be neutralized.

In summary, a SAPS Bioreactor is intended to

1. Microbially reduce through the actions of IRB any ferric iron in an influent ARD to ferrous iron

$$Fe^{3+} + e^{-} \rightarrow Fe^{2+} \tag{4.3}$$

thereby creating conditions under which ferric iron hydrolysis in the substrate (Equation 4.1) is prevented.

2. In an underlying non-substrate limestone bed, raise its pH to circum-neutral values without armoring its particles or plugging the bed with ochre.

The above is a great simplification. In the bioreactor treatment of sulfate-contaminated ARDs and other highly acidic waters, acidophilic prokaryotes (both bacteria and archaea) will catalyse dissimilatory iron oxidation or reduction (or both), and the dissolved ferric iron will not be present as Fe^{3+} but rather as $Fe(SO_4)_2^-$ and $Fe(SO_4)^+$ complexes (Barrie Johnson et al. 2012). Accordingly, the microbial processes will be much more complex than those imagined. In addition to the targeted iron reduction in a SAPS Bioreactor, sulfate reduction and other microbial reactions that can undesirably raise pH may also occur. As is the case with all EEBs, whatever the bulk situation (anaerobic in this case), there will always be aerobic microenvironments in them, so aerobic processes also need to be considered.

The performance record of Stand-Alone SAPS Bioreactors has not been good; they have not achieved anticipated operating lifetimes, and often have not operated as anticipated (Neculita et al. 2007). Failures have been common. This has negatively affected the perceptions of many potential users of this method as well as other ABR ecotechnologies. Some of the reasons for their poor performances include:

1. Making poor choices in active media selection (e.g., using alkalinity-raising materials as all or parts of the active media in their bioreactors' substrates as it may cause undesirable ferric iron hydrolysis)
2. Attempting to carry out two kinds of processes in the same bioreactor (see below)
3. Using empirical design criteria
4. Using information from units developed for one situation when designing for an entirely different one (e.g., the correlations presented in the literature for SAPS Bioreactors are for ones treating

coal mine drainage and may be inapplicable for treating other kinds of wastewaters)

5. Not allowing for the fact that as active media are metabolized, they will leach materials (e.g., ammonia, organic degradation products) into effluents

For example, some constructing Stand-Alone SAPS Bioreactors have used spent mushroom compost (which contains lime from its preparation) as part of their active media, and expected that, along with the desired ferric iron reduction by IRB, ferric iron hydrolysis would not occur. Unfortunately, including such alkalinity-generating materials in the active media results in the creation of dissolution/precipitation "fronts" (reaction zones) in the bioreactor substrates and these move downwards during operations, leading to the formation of areas of hardpan and preferred flow channels. These situations and hydrolysis due to uncontrolled pH rises in substrates have led to poor performances that have hampered the widespread use of the SAPS Bioreactor ecotechnology (Jage et al. 2001).

The EW Test Project is already indicating that it is preferable to use (for the active media part of a SAPS Bioreactor's substrate) carbonaceous materials which do not generate alkalinity (either chemically or biologically), leaving the pH of the ARD passing through the substrate to remain as close as possible to (or below) the influent pH (i.e., <3) until it reaches the limestone. This will inhibit any significant amounts of ferric iron hydrolysis, allowing the neutralization of the ARD to only occur chemically in an underlying limestone bed. Neutralized ARD can then be released into Sedimentation Ponds or CW cells where the ferrous iron will oxidize to ferric iron on exposure to air, and where the resulting Fe^{3+} will quickly hydrolyse to produce amorphous sludges that will then settle out. Alternatively, and more appropriately, a SAPS Bioreactor can be followed by a BCR which will allow the iron and other dissolved metals in the neutralized wastewater to be precipitated under anaerobic conditions in it as less voluminous (and much less potentially later mobile) sulfides, carbonates and hydroxides.

Influent WWs to SAPS Bioreactors should not contain any appreciable amounts of dissolved aluminum (>1 mg/L). However, if it does, when the pH of the ARD being treated is raised above 5 in their limestone beds, gelatinous aluminum hydroxide will precipitate out

$$3CaCO_3 + 2Al^{3+} + 6H_2O \rightarrow 3Ca^{2+} + 2Al(OH)_3\downarrow + 3H_2CO_3 \qquad (4.4)$$

While aluminum hydroxide does not cause armoring, it can build up in the void spaces between the limestone particles, reducing throughput and possibly plugging the beds. A passive way to remove aluminum hydroxide sludge from an EEB SAPS Bioreactor is to attach a dosing siphon to its outlet (see Section C.5 in Appendix C), and allow it to periodically flush itself to prevent the build-up of the floc (Vinci & Schmidt 2002).

There are a number of pilot, demonstration and full-scale SAPS Bioreactors that are or have been in operation, including those shown in Table 4.1.

TABLE 4.1

Some Locations for SAPS Bioreactors

Augusta Lake, IN	Oven Run, PA
Bowden Close, UK	Palena, Wales
Brandy Camp, PA	REM, PA
Buckeye, OH	Schepp Rd., PA
Filson, PA	Simmons Run, UK
Hanchan, Korea	Wheal Jane Mine, UK
Hohe Warte, Germany	Whitworth, Wales
Howe Bridge, PA	#40 Gowen, OK
Jennings, PA	

Most of the above bioreactors were designed as Stand-Alone units and are/were not EEBs; all except for Hohe Warte (fluorspar mine) are (or were) used for treating coal mine ARD, Jage et al. (2001); none are in far northern areas; and none of them so far are designed to treat ARD under extreme weather conditions.

The amount (mass M) of limestone required for the limestone layer of a SAPS Bioreactor can be calculated using the same formula as that given for ALDs in Equation C.7 in Section C.3 of Appendix C.

As was mentioned, a common practice so far has been to include the limestone in a SAPS Bioreactor inside of it as a separate layer below (downstream of) the substrate, although there is no reason in an EEB System involving such cells that the limestone cannot be placed outside of the bioreactor in a separate cell (a buried one to ensure anaerobic conditions).

The limestone layer of a SAPS Bioreactor may also remove some of the ferrous iron (and any manganese present) in the influent ARD by the formation of carbonates. Some of the influent iron entering a SAPS Bioreactor will also be sorbed in the substrate.

SAPS Bioreactors cannot as yet be considered proven and demonstrated technology, although the work of the continuing EW Test Project (see Section 4.5) should soon rectify this.

4.4 Other Kinds of Eco-Engineered Bioreactors

4.4.1 Types of Specialty Eco-Engineered Bioreactors

There are two main kinds of specialty EEBs: specialty BCRs and electro-biochemical reactors (EBRs).

Most specialty BCRs are passive treatment BCRs that are used:

1. Where it is desirable to target the removal of a specific dissolved metal(loid) (e.g., As, Cr, Mo, Sb, Se), and managing the subsequent products will be more complex than those in an ordinary BCR
2. Where bacterial consortia dominated by appropriate bacteria are available to inoculate a BCR using bioaugmentation
3. Where particular conditions can be created to enhance the removal of certain species

Most specialty BCRs have the same morphologies (i.e., substrates, designs) as other kinds of passive treatment BCRs, and are usually used as parts of EEB Systems in association with other kinds of aerobic and anaerobic EEBs whose cells may be placed upstream and downstream of them.

Particular kinds of microbes are found in many specialty BCRs such as those that remove chromium (Mattes et al. 2010) or arsenic* (Duncan et al. 2004). These microbes can be accessed, in some cases grown, and inoculated (bioaugmented) into bioreactor substrates. Where chromium is the metal being removed (and it does not form a sulfide under aqueous conditions), the substrate has to be designed so that it will strongly bind the otherwise potentially mobile chromium hydroxide formed in the bioreactor (Higgins & Mattes 2009).

Another kind of speciality BCR is one used to treat a metal-contaminated wastewater that does not contain enough sulfates to normally allow sulfate reduction to occur so that they can be removed as metal(loid) sulfides, etc. In this case, either a soluble sulfate can be added with the influent wastewater or a solid sulfate-generating material may be incorporated in the substrate.

Examples of these kinds of wastewater are the stormwater runoffs from electrical utility pole storage yards that become contaminated with copper chrome arsenate (CCA) and other wood preservatives that are used to protect the wooden poles after they are installed. Figure 4.1 shows an indoor pilot-scale EEB System test unit that was used to evaluate the removal of CCA-contaminated leachate products from utility poles. It involved a water-covered, upflow VSSF passive treatment specialty BCR cell, which in this case was a sulfate-reducing BCR (foreground). The BCR cell was followed by a vegetated downflow VSSF BBR cell with gravel substrate. The active medium in the BCR cell's substrate was pulp and paper mill biosolids and, in addition to sand and gravel support material, its substrate contained gypsum to supply sulfate to a feedstock from a soak tank (not shown) containing chipped CCA-preserved utility poles.

The testing was carried out at ambient (~20°C) temperatures and (after enclosing the pilot's main components in a refrigeration unit) at about 10°C. Copper, chromium and arsenic in the test unit influent (concentrations up

* See Section 5.4 in Chapter 5 for an example of such a system.

FIGURE 4.1
Pilot-scale EEB System for treating CCA-contaminated water.

to 20 mg/L) were removed to levels below regulatory requirements. The successful pilot testing led to outdoor demonstration-scale test involving the same sort of EEB cells (a BBR cell following a passive treatment BCR cell, the latter of which was buried in this case) at an electrical utility pole yard. At this pole yard, there are wooden utility poles stored that are preserved with pentachlorophenol (PCP) and creosote as well as CCA. As a result, the yard stormwater runoff contained phenols, chlorinated hydrocarbons, PAHs and dioxins & furans, as well as particulate and dissolved copper, chromium and arsenic. The EEB System proved excellent in removing all of the CoCs.

As was mentioned, not all specialty BCRs involve sulfate reduction. Selenium is a particularly recalcitrant contaminant that can involve soluble bioavailable, toxic forms that can bioaccumulate. Selenium is often found in coal mine MIWs and certain other wastewaters. This metalloid may occur in aqueous systems in its insoluble and nontoxic elemental form (Se_o); as soluble and bioavailable selenate oxyanion (SeO_4^{2-}) which involves the sometimes toxic Se^{6+} ion; as the also soluble and bioavailable selenite oxyanion (SeO_3^{2-}) which involves the always toxic Se^{4+} ion; and as selenide (Se^{2+}) which occurs in organic and inorganic forms (Oremand et al. 1989; MT 2005; Walker 2010). The reduction of SeO_4^{2-} should occur at a slightly lower ORP than that required for nitrate reduction but at a slightly higher ORP than that for sulfate reduction (Sonstegard et al. 2010, also see Table 1.2 in Section 1.2 of Chapter 1). Both selenate and selenite can be reduced to Se_o by selenium-reducing bacteria (SeRB) in a specialty BCR, as is illustrated in the following equation for selenate reduction to selenite using acetate as an electron donor:

$$CH_3COO^- + 4SeO_4^{2-} + H^+ \rightarrow 4SeO_3^{2-} + 2CO_2 + 2H_2O \quad (4.5)$$

Similar processes occur for other electron donors such as lactates, and reduction may continue until elemental selenium is formed. Since the optimal ORP range for selenium reduction by SeRB falls between those for nitrate and sulfate reduction, any wastewater containing significant amounts of Se (more than a few tens of micrograms per liter) would have to be treated semi-passively using an EEB System consisting of an upstream DNBR (if nitrates were present as well) and a downstream sulfate-reduction BCR (if the removal of other dissolved metal[loid]s was also required).

Thus far, passive treatment specialty BCRs other than EBRs that have been targeted for selenium removal have only been used at the pilot scale, although one such test (Walker 2010) resulted in up to 98% removals, reducing selenium concentrations from 90 ppb in a bioreactor influent to 0.5 ppb, and removals remained effective (>90%) even under winter conditions. Selenium removals were, however, affected by the presence of alternate electron acceptors such as nitrates and sulfates. Nitrates are more favorable electron acceptors than selenates and it was demonstrated that when they were present as well, they had to be completely removed before selenate reduction occurred in the bioreactor. (DO, MnO_2, tungstate and chromate will also inhibit selenium reduction, and precipitated elemental selenium can be oxidized by nitrates and ferric oxyhydroxide (Oremand et al. 1989), illustrating the need for careful EEB System design if successful selenium removal is to be achieved.)

In addition, although sulfates are less favorable electron acceptors than selenates and their reduction should have to be minimized, some degree of sulfate reduction will occur as well when sulfates are present along with selenates and selenites.

4.4.2 Electro-Biochemical Reactors

An advanced new kind of specialty BCR is the patented electro-biochemical reactor (EBR) (Adams & Peoples 2010; Adams & Miller 2010; INOTEC 2011; Adams et al. 2012; Franks 2012; Lovley 2012; El-Naggar & Finkel 2013; Opara et al. 2013a,b,c). An EBR stimulates biological processes using conventional equipment.

In an EBR, a low electrical potential (1–3 V) and a mA current are applied across its microbial biofilm on high surface area substrates. (1 mA of current can provide 6.24×10^{15} electrons per second to the microbes and bioreactor system.) The directly added electrons supplement/replace electrons from the metabolisms of organic electron donors and are usable by the microbes, assisting them in more quickly and efficiently removing CoCs. The electrons can be supplied from a battery, a solar panel or a power supply. This "free" energy – from the microbial perspective – is used for growth, maintenance and CoC conversion, and provides a controllable adjustment of the reactor's oxidation–reduction potential (ORP) environment. Since microbial metabolism is not required to acquire these provided electrons, CoC removals are especially improved at low temperatures.

In addition, the amounts of nutrients required are greatly reduced (by as much as half) and these lower nutrient requirements result in significantly lower effluent suspended solids concentrations, often eliminating the need for the treatment of solid active media components such as biosolids. Also, required reactor residence times to obtain good results are shortened and removal efficiencies are improved, significantly reducing CAPEX and OPEX. Most importantly, by modifying a BCR for operation in an EBR mode, the often wide fluctuations in ORP in the substrate of a bioreactor are smoothed out as is illustrated in Figure 4.2 which presents the results from pilot-scale testing treating a wastewater containing multiple dissolved metal(loid)s.

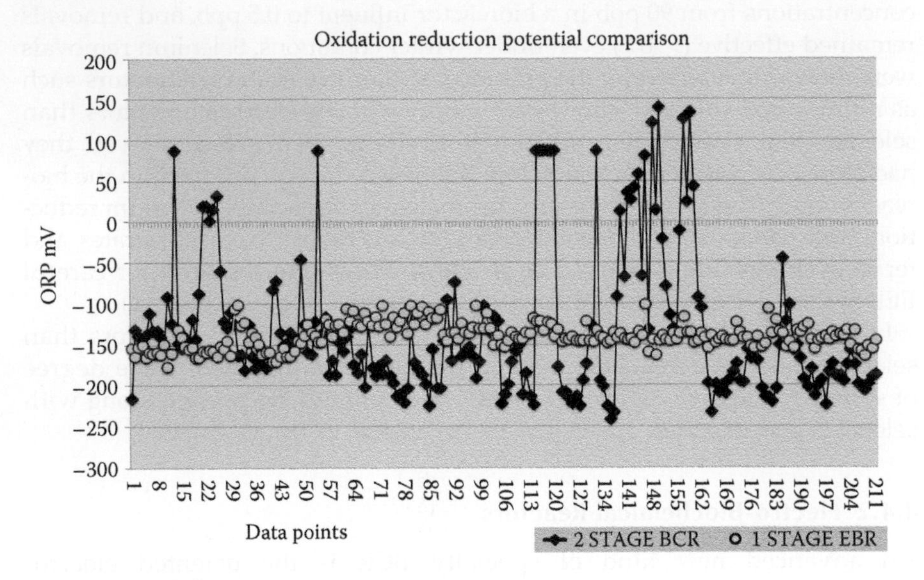

FIGURE 4.2
Comparison of ORP in an EBR and an ordinary BCR.

The EBR process has been tested with excellent results in numerous bench- and on-site pilot-scale tests. Pilot-scale EBR systems have been tested for the removal of multiple contaminants from various mining-influenced, coal-fired power plant and other wastewaters, and have demonstrated significant improvement over conventional (active treatment) bioreactors in the removals of both primary and secondary contaminants in waters containing multiple CoCs and alternate electron acceptors such as nitrates, sulfate, arsenic, selenium, antimony, cadmium, copper, lead, molybdenum, nickel, silver, uranium, zinc and mercury. A full-scale unit for removing nitrate and selenium is now operating (Adams 2010).

Section 6.9 in Chapter 6 presents the results of pilot-scale testing of an EBR-based system to remove high levels of nitrates and selenium from a closed gold mine MIW.

4.5 The EW Test Project

A seminal project was the EW Test Project and it illustrated the potential of genomics to enhance the treatment of ARD, especially MIWs generated at mine sites after mine closure. That study's test system was intended to remove iron from a sulfate-rich synthetic ARD, treating the water to a point where it had a quality that would have been suitable for release to the environment. The EW Test Project involved a pilot-scale EEB System consisting of a SAPS Bioreactor test cell and a BCR cell in series. It was carried out at the CAWT test facilities (see Section 6.3 in Chapter 6). Its purpose was to better define the kinetics, chemistry and microbiology of the two ABR cells (Higgins & Mattes 2014).

For the EW Test Project, the two pilot cells were constructed inside a walk-in cooling unit at CAWT's greenhouse, and testing was carried out successively at 20°C, 15°C, 10°C, and 5°C. Anaerobic samples were taken from nine ports that sampled the ARD being treated from inside the two cells. The wastewater into, out of and within the two ABR cells was analysed for pH, ORP, DO, conductivity, SO_4^{2-}, S^{2-}, Fe_T, Fe^{2+}, and Fe^{3+}. The testing was carried out for two weeks at each temperature, with weekly analyses of the above parameters. In addition to determining the information normally determined during a treatability test (the kinetics of CoC removal and other scale-up aspects needed to design field-scale facilities, see Chapter 6), special samples were collected for genomic testing, frozen and shipped to the UoG where microbial analyses were carried out.

Figure 4.3 shows the process flow of the pilot unit.

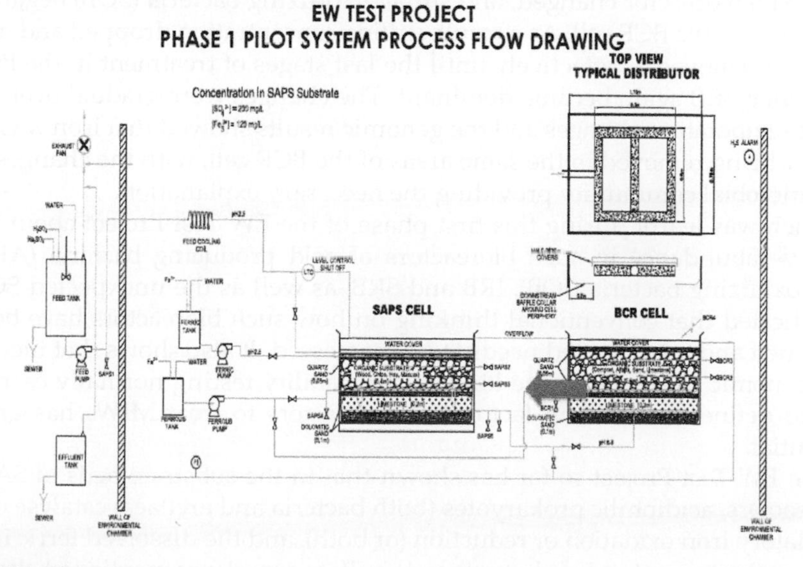

FIGURE 4.3
EW Test Project PFD.

During the EW Test Project, the microbial communities in the two EEB cells were allowed to stabilize and adjust at each temperature level before the temperature of the pilot unit was reduced by 5°. The iron concentration in the pilot unit's influent was maintained at a very high level (>100 mg Fe^{3+} per L). The expected result according to textbook descriptions of SAPS Bioreactors and BCRs was that, regardless of temperature, ferric iron would be reduced to ferrous iron by iron-reducing bacteria in the SAPS Bioreactor cell's substrate; that the cell's lower limestone layer would then increase the ARD's pH to circum-neutral levels; and the resulting still dissolved ferrous iron would be reduced to sulfide by sulfate-reducing bacteria SRB and precipitated in the "downstream" BCR cell's substrate. At the low pH and ORP of the SAPS Bioreactor's substrate, IRB were expected to dominate over SRB, and sulfate removal was not expected to occur until the ARD had been neutralized and passed into the BCR cell.

Such was not what was found.

Although the analysis of the EW Test Project's results has not yet been completed and further testing is planned, preliminary determinations of the changes in the populations of the most common bacteria and other microbes found in them at each temperature have revealed surprising facts about the operation of the two kinds of ABRs, ones that will have significant effects on the designs of full-scale EEBs of these sorts.

For example, sulfate was almost totally reduced to sulfide in the SAPS Bioreactor cell, together with a very large reduction in the concentration of the ferric iron it contained. At 20°C, this resulted in an almost total removal of iron. Since the temperature was reduced, the microbial community in the SAPS Bioreactor changed, and sulfate-oxidizing bacteria (SOB) began to dominate in the BCR cell. As a result, sulfide concentration dropped and iron was not removed as effectively until the last stages of treatment in the BCR cell when SRB again became dominant. The changes were gradual over the three temperature changes and the genomic results showed that iron was no longer being removed in the same areas of the BCR cell, with the changes in the microbial community providing the necessary explanation.

Much was learnt during this first phase of the EW Test Project about the relative abundance in such bioreactors of acid producing bacteria (APB), iron oxidizing bacteria (IOB), IRB and SRB, as well as the unexpected SOB. It indicated that conventional thinking on how such bioreactors have been designed and are operated needs to be reassessed. It also shows that including genomic testing with the kinds of treatability testing normally carried out to define the scale-up factors for bioreactors to treat MIWs has great potential.

The EW Test Project so far has shown that in the substrate beds of SAPS Bioreactors, acidiphilic prokaryotes (both bacteria and archaea) catalyse dissimilatory iron oxidation or reduction (or both), and the dissolved ferric iron may not be present as Fe^{3+} but rather as sulfate complexes mentioned above.

The EW Test Project further showed that, in this case, a SAPS Bioreactor can best be modelled using Reaction Kinetics by a TIS-with-delay model (see Section 8.7 in Chapter 8), as is indicated by the following normalized best fit, tracer test E-Curve (see Section 1.9 in Chapter 1) for the part of the phase 1 testing carried out at 20°C where C_{in} and C_{out} are the measured concentrations of KBr tracer into and out of the bioreactor (Figure 4.4).

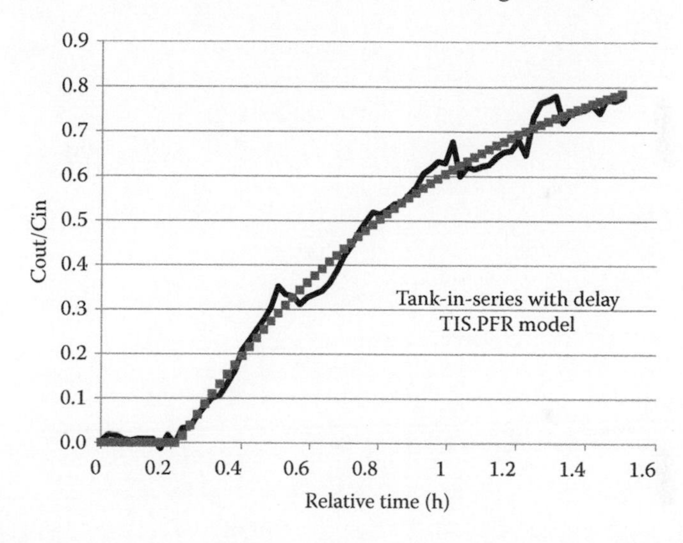

FIGURE 4.4
Best fit model for the SAPS bioreactor.

5

Biochemical Reactors

5.1 BCR Background

The use of "passive" systems to treat ARD was first suggested and studied in bench-scale applications as early as 1978.* Later, Anaerobic Wetland systems (see Section B.7 in Appendix B) were designed whose intent was to maximize SRB populations using organic materials in them to promote bacterial growth, mostly in services to treat ARD. These then evolved into passive treatment BCRs.

The main purpose of biochemical reactors (BCRs) (sometimes also referred to as *sulfate-reducing bioreactors, sulfate-reducing wetlands or, earlier and erroneously, as the only kind of ABR*) is to carry out biological sulfate reduction, while converting many dissolved metal(loid)s in wastewaters into insoluble precipitates, thereby removing them from the wastewater passing through them.

While physical and chemical transformations can and do also occur in BCRs, soluble metal(loid)s in wastewater are "bioremediated" by SRB and other microbes in them in a number of ways (Gusek 2008):

- Biomineralization
- Biosorption
- Bioleaching
- Biodegradation of chelating agents
- Enzyme-catalysed reduction

Biomineralization is the transformation of a soluble metal(loid) into an insoluble form (Gould et al. 1997). While this includes the processes that typify BCRs: the precipitation as sulfides of many divalent metal(loid)s (see the next section on BCR Chemistry), biomineralization in a BCR can also include

* Government of Pennsylvania website: Science of Acid Mine Drainage. The West Fork EEB System (see Section 5.5) was the first large-scale anaerobic system built that was not just a kind of Anaerobic Wetland. The first large-scale non-wetland **aerobic** system (2000 US gpm) for treating MIWs was built by the TVA in 1992 (Gusek et al. 1998).

processes such as those that produce certain phosphates, carbonates and hydroxides (again, see the next section).

Biosorption involves the physical adsorption of a metal(loid) onto microbial biomass. While in some anaerobic systems such as Anaerobic Wetlands this can include phytoremediation mechanisms such as rhizofiltration (see Section C.6 in Appendix C), in BCRs this involves sorption onto the SRB and other microbes in the bioreactor. Biosorption may also involve microbially enhanced chemisorption which leads to the absorption of the CoC into the biomass.

Bioleaching is the process of using bacteria to mobilize (dissolve) metal(loid)s such as As, Au, Co, Cu, Ni, and Zn. In a BCR, it can also involve the formation of soluble chelates due to the action of organic acids.

Biodegradation of chelating agents occurs in BCRs when these chelates are further biodegraded into the soluble metal(loid) ions and CO_2.

Enzyme-catalysed reduction (sometimes called biotransformation) involves the reduction of certain oxidized, soluble metal(loid)s (e.g., Fe^{3+}, Mn^{4+}, Cr^{6+}, Se^{5+}) in a WW being treated in a BCR into insoluble forms in reactions catalysed by bacterial enzymes.

The net result of the biological and related reactions in a BCR is to convert the mobile (soluble, chelated, etc.) metal(oid)s in an MIW or other WW into insoluble forms that are stable over a wide range of pHs and other conditions. Indeed, once formed, these insoluble forms are generally so difficult to remobilize that doing so would have to involve draining a BCR, dismantling it and leaving its substrate bed material exposed to the elements over a long period.

As was introduced in Section 1.1 of Chapter 1 and the previous chapter, there are two kinds of BCRs, differentiated by the nature of their substrates: *passive treatment BCRs* (solid organic active media) and *semi-passive treatment BCRs* (liquid active media). These are discussed in more detail in Section 5.3.

Unlike SAPS Bioreactors (which, as was mentioned in the previous chapter, have been used mostly to treat coal mine ARDs), BCRs (mostly passive treatment ones) have found applications in treating waters from both coal and metal mines (Higgins et al. 2007b; ITRC 2013), and for other services such as the treatment of landfill leachates. Many have operated and continue to operate in locations where they are exposed to extreme winter conditions. They include ones for removing As, Cd, Cr, Cu, Mo, Ni, Pb, Sb, Se and Zn at concentrations ranging from thousands of mg per L (ppm) down to low mg per L levels, and in some cases for certain CoCs and situations even to μg per L levels.

As with any EEB, the Detailed Design and Engineering of BCRs of either sort should be based on Reaction Kinetics, and scale-up information should be determined during a treatability test.

It should be noted that for passive treatment BCRs, it can be easily calculated (and experience has shown it to be true) that the space taken up during operations by precipitated and sorbed sulfides and other compounds is very much less than the space in a substrate by the organic active media material biodegraded, and there is usually little chance that these products will plug the substrate beds. The same is true for semi-passive BCRs where the amounts of pore space in their gravel substrate beds are very much larger than the volumes of any precipitates.

Influent to a BCR should be circum-neutral in pH, although the provision of a non-substrate limestone layer before the active media-containing substrate in a passive treatment BCR or an upstream ALD before a semi-passive treatment BCR can allow wastewaters with pHs in the 4–6 range to be treated, provided aluminum and ferric iron have been removed from the wastewater well upstream in an EEB System.

The BCR ecotechnology is mostly not patented. Designing BCRs (and the other ABRs mentioned) depends as much on "know-what" as on "know-how."

The BCR ecotechnology may be considered to be demonstrated and proven.

5.2 BCR Chemistry

There are two purposes of biochemical reactors:

1. To raise the pH of an MIW or other WW being treated
2. To remove metal(loid)s largely as:
    a. Sulfides (e.g., CdS, CuS, FeS, PbS, ZnS)
    b. Hydroxides (e.g., $Fe(OH)_3$, $Al(OH)_3$, MnO)
    c. Carbonates (e.g., $FeCO_3$, $MnCO_3$, $ZnCO_3$)

In the gravel substrate beds of semi-passive BCRs, SRB and other biofilm microbes metabolize the relatively low molecular weight added (injected with the feedstock) active medium (e.g., methanol) or mixed organic media resulting in the ultimate biodegradation products: carbon dioxide, methane and hydrogen.

In the solid active media-containing substrate beds of passive treatment BCRs, populations of CDB, APB and other microbes break down the high molecular weight organics into simpler compounds that the SRB, etc. can metabolize. As is shown in various sketches presented in the literature and at conferences and short courses (Wildeman & Schmiermund 2004; Higgins et al. 2006a; Wildeman et al. 2006, 2007), heterotrophic bacteria and other microbes (e.g., methanogens and even in some cases eukaryotic fungi) will progressively break down complex organics in the active media of a passive treatment BCR's substrate, producing progressively simpler compounds by various pathways, ultimately resulting in simple compounds such as hydrogen, ethanol and acetate that can be metabolized by the characterizing microbes (e.g., SRB, see Table 1.2 in Chapter 1 and Equations 2.3 through 2.5 in Chapter 2).

The sulfate-reducing bacteria may be either autotrophic or heterotrophic anaerobes, function best above pH 5.5, and are also capable in part of volatilizing the sulfate in a wastewater to hydrogen sulfide gas, H_2S, and reducing bicarbonate (alkalinity) in it (Postgate 1984; Gusek et al. 1998; Wildeman et al. 2006; Mattes 2013).

$$2CH_2O + SO_4^{2-} + 2H^+ \rightarrow H_2S + 2H_2O + 2CO_2 \quad (5.1)$$

$$2CH_2O + SO_4^{2-} \rightarrow HS^- + 2HCO_3^- + H^+ \quad (5.2)$$

where CH_2O represents a simple carbonaceous material.

It is noted that two moles of alkalinity and one mole of acidity result from the reaction illustrated in Equation 5.2,* so the combined action of the SRB is to raise alkalinity and buffer the wastewater.

Depending on pH, in the anaerobic environment of a BCR substrate bed, the HS^- may dissociate as follows:

$$HS^- \Leftrightarrow H^+ + S^{2-} \quad (5.3)$$

The sulfide ion has a high affinity for many metals, and for these heavy metals (and some metalloids) this affinity results in the creation and precipitation of metal sulfide solid phases that are retained in the substrate matrix as follows (illustrated for divalent cations):

$$Me^{2+} + HS^- \rightarrow MeS\downarrow + H^+ \quad (5.4)$$

where Me represents a typical divalent metal cation such as Ag^{2+}, Cd^{2+}, Co^{2+}, Cu^{2+}, Fe^{2+}, Ni^{2+}, Pb^{2+} or Zn^{2+}. Chacophile ("sulfur-loving") metals (e.g., Cd, Cu, Pb, Zn) are easier to precipitate than siderophile ("iron-loving") ones (e.g., Co, Cr, Ni) and BCR design has to take this into account as the SRB in them have to produce enough HS^- to precipitate all of the Fe before any of the siderophiles will precipitate (Fortin 2000, see Equation 5.8).

* As is shown by the equations and discussion that follow, the chemistry of the removals of metal(loid)s in BCRs is more complex than just that illustrated by Equation 5.2, and many more mechanisms and processes occur and are involved, and, as was mentioned, many kinds of microbes other than SRB may be involved. This is fully recognized. However, to simplify terminology, unless there is a need to be more specific, in this book the biological removal of metal(loid)s in BCRs will be referred to as occurring by sulfate reduction.

In addition, bicarbonate ions in a BCR will precipitate some metals as carbonates and hydroxides, as indicated for zinc and aluminum, respectively.

$$Zn^{2+} + HCO_3^- \rightarrow ZnCO_3\downarrow + H^+ \tag{5.5}$$

$$Al^{3+} + 3HCO_3^- \rightarrow Al(OH)_3\downarrow + 3CO_2 \tag{5.6}$$

These reactions are facilitated by consortia of sulfate-reducing bacteria and other microbes. To be effective, the pH-raising reactions have to dominate over the metal precipitation ones.

The sulfide/carbonate/hydroxide precipitation reactions are not the only ones that occur in BCRs. Any iron II or manganese II hydroxides in the feedstocks to a BCR (and most MIWs will contain a lot of these) can also be precipitated as sulfides

$$Fe(OH)_2 \text{ or } Mn(OH)_2 + S^{2-} \rightarrow FeS_2\downarrow \text{ or } MnS_2\downarrow + 2OH^- \tag{5.7}$$

Ferrous sulfide ions will compete with those of other bivalent metal sulfide ions

$$FeS_2 + 2Me^{2+} \rightarrow Fe^{2+} + 2MeS_2\downarrow \tag{5.8}$$

Many metal(loid)s that usually form oxyanions in water (e.g., Mo, As) can also be removed as sulfides using sulfate reduction (Mattes et al. 2004, 2007, 2010). For the reduction of molybdate ions (MoO_2^{4-}) to Mo(IV) and the precipitation of molybdenite, $MoS_2(s)$, the reaction is given by

$$MoO_2^{4-} + 2HS^- + 6H^+ \rightarrow MoS_2\downarrow + 4H_2O \tag{5.9}$$

However, if there is an excess of iron in a WW being treated in a BCR, its reactivity when S^{2-} is present will result in the preferential formation of its sulfide rather than the metal(loid) sulfide (e.g., FeS rather than As_2O_3 when arsenic is one of the target CoCs, Wilkin et al. 2003; Mattes 2013).

Metal(loid)s that do not form sulfides in water but do form oxides and hydroxides (e.g., Cr^{3+}) or carbonates (e.g., Mn^{2+}) can also be made to precipitate like the sulfides (Mattes et al. 2007; Mattes 2013). Table 5.1 shows the theoretical solubilities (K_{sp}'s*) of divalent cation sulfides and hydroxides in pure water.

* Tabulations of sulfide and hydroxide solubilities in water are available under standard conditions are available from the literature (e.g., Postgate 1984), but such data are pH dependent and values will vary with this quality. K_{sp} data do not show such variability.

TABLE 5.1

Theoretical Solubilities of Divalent Cation Sulfides and Hydroxides (mg/L)

	Sulfides	Hydroxides
Cadmium	1×10^{-27}	7.2×10^{-15}
Chromium	No precipitate	2×10^{-16}
Cobalt	5×10^{-22}	5.92×10^{-15}
Copper	8×10^{-37}	2×10^{-15}
Iron	8×10^{-19}	2.79×10^{-39}
Lead	3×10^{-28}	7.40×10^{-14}
Manganese	3×10^{-11}	2×10^{-13}
Mercury	2×10^{-53}	3.6×10^{-26}
Nickel	4×10^{-20}	5.48×10^{-16}
Silver	8×10^{-51}	n/a
Tin	1.52×10^{-25}	5.45×10^{-27}
Zinc	2×10^{-25}	3×10^{-17}

As may be seen from the above table, assuming that adequate amounts of sulfate are available to SRB, Cd, Co, Cu, Ni, Pb and Zn ions should be relatively easy to remove by precipitation as sulfides in a well-designed BCR, and this is indeed what has been found. The presence of Fe and Mn, as has been discussed above, may result in the preferential precipitation of their sulfides until the supplies of their ions (e.g., Fe^{2+} in the case of iron) are used up.

Chromium does not form a sulfide but does form a hydroxide. However, this compound is potentially mobile and to design a Specialty BCR to remove it, the substrate needs to have a composition that will sorb and hold that ion's hydroxide. (This can be done, and doing so is part of the "know what" of BCR design.)

Indeed, a large number of metal(loid)s can be treated in BCRs. Gusek in ITRC 2013 has prepared a version of the periodic table that highlights the elements the dissolved ions of which can be treated in them. Many of the same processes/reactions that occur in BCRs also occur in the active media of PRBs (see Section C.7 of Appendix C) and in CWs (see Appendices A and B).

5.3 Kinds of BCRs

5.3.1 Semi-Passive and Passive BCRs

As was mentioned before, there are two basic kinds of BCRs*:

- Semi-passive treatment BCRs
- Passive treatment BCRs

* BCRs are summarized in Section 9.1.1 of Chapter 9.

Passive treatment BCRs* have solid organic active media that are part of *in situ* substrates while semi-passive treatment BCRs have aggregate substrates and their carbonaceous active media are added separately as liquids. The two kinds are addressed in more detail in the following sections.

5.3.2 Semi-Passive BCRs

Semi-passive treatment BCRs are similar to BBRs in that their substrates are beds of screened and washed gravel or gravel-like crushed rock. This gravel may be quartz gravel, or if the pH of the wastewater being treated is, or can be controlled to be, near neutral, limestone or dolomite gravel may be used as well, although it should be recognized that this will lead to having chemical sulfate reduction (see Equation C.5 in Section C.3 in Appendix C) complement biological sulfate reduction.

With a semi-passive treatment BCR, the active media carbon source for the SRB is not part of the substrate bed but is introduced as a liquid along with the influent wastewater being treated. Table 5.2 (ITRC 2013) lists some of the kinds of materials that have been/are being used as active media in semi-passive BCRs, either alone or as part of more complex mixtures.

TABLE 5.2
Some Liquid Active Media Used with Semi-Passive Treatment BCRs

Methanol	Beer making yeast by-products
Ethanol	Partially treated sewage
Glycols	Food processing wastewaters
Sodium acetate	Wine production wastewaters
Sucrose-rich streams	Liquid manures
Molasses	

Some of the injected carbonaceous materials (e.g., manures) may be used to introduce the microbial consortium as well. In most cases, however, the microbial consortium is provided by biostimulation or bioaugmentation using microbial inoculants. Examples of semi-passive treatment BCRs in operation are units at Naciamento and Leviathan, NV, USA and Tudsequeh, BC, Canada.

Care must be taken when selecting liquid active media. For example, pure glycols are excellent liquid active media for semi-passive BCRs, but

* Even though these kinds of BCR are largely passive, some active treatment may be involved from time to time. For example, some liquid active media may be added to influents of normally passive treatment BCRs during the coldest weather for short periods to ensure that treatment continues under extreme winter conditions.

using glycol-rich Spent Glycols (see Appendix D) from cold weather aircraft de-icing at airports as liquid active media should be considered carefully as, although it is reported (Johnson et al. 2001) that they do not inhibit aerobic biodegradation, they contain additives (the ADPAC chemicals), some of which can inhibit anaerobic biodegradation. Also, the carbonaceous by-products of local industries (see the above table) may be available quite cheaply, but if they are to be considered as liquid active media for a semi-passive treatment BCR, careful consideration of their constituents, homogeny and long-term availabilities must be undertaken before they are selected. The best liquid active medium for semi-passive treatment BCRs is methanol (CH_3OH). Commercial grades of this simple alcohol are available in bulk almost anywhere (including grades prepared to have the lowest freezing points, in the −30°C range), and quality is assured and consistent. Prices are relatively low (US $400–450 per tonne in 2017) and, in addition to its use with BCRs, CH_3OH has multiple uses including as an antifreeze, as a solvent, as a fuel, as a sorbent and as a heat transfer fluid, and most likely will already be available at any commercial or industrial site where EEB Systems might be considered for WWT.

5.3.3 Passive Treatment BCRs

In common with some other kinds of passive treatment EEB ABRs (e.g., DNBRs), *passive treatment BCRs* have substrate beds that involve solid, complex carbonaceous active media (e.g., wood chips, biosolids, composts) mixed with support materials. While for logistical reasons (see Chapter 7) the composition of the active media part of the substrate beds in passive BCRs should be kept as simple as possible, nevertheless, the mix has to include organic materials that take some time to break down (e.g., wood chips, composts) as well as organic materials that will degrade over shorter periods (e.g., manures, biosolids), and some "quick sugars" that will begin breaking down almost immediately (e.g., secondary WWTP biosolids, see below).

In addition to the organic active media (e.g., compost), most passive treatment BCR substrate beds contain support materials (e.g., gravel, sand) to provide permeability to the matrix. As was mentioned, these substrates can also contain reactive support media such as limestone (see Section C.3 in Appendix C) so that alkalinity addition in them can be both biological and chemical. The organic and other substrate materials in passive treatment biochemical reactors not only serve as a carbon source but also physically retain precipitated metal(loid) sulfides and other precipitates in available pore spaces that are unlikely to clog.

Table 5.3 lists where there are (or have been) a number of large pilot, demonstration, and full-scale passive treatment BCR-based systems in operation in the United States, Canada and elsewhere.

TABLE 5.3

Some Locations for Passive Treatment BCRs

Big Five Tunnel, CO	Laval Pole Yard, QC
Brewer Gold Mine, CO	Lutrell Repository, MT
Burleigh Tunnel, CO	Lilly Orphan Boy, MT
Calliope, MT	Park City, UT
Champagne Creek, ID	Penn Hills, PA
Coal Run, PA	Peerless Jenny King Mine, MT
Dixon Run No. 3 Mine, PA	Silver Box, MT
Elizabeth Copper Mine , VT	Smolnik Mine, Slovakia
Fabius Coal Mine, WV	Surething, MT
Forest Queen, CO	Teck Cominco Trail, BC
Fran Mine, PA	West Fork, MO
Golinsky Mine, CA	Wheal Jane Mine, UK
Haile Mine, SC	Wood Cadillac, QC
July 14, PA	Yankee Girl Silver, Ymir, BC
Lady Leith, Jefferson County, MT	

When designing a passive treatment BCR, it is important to ensure that several important requirements are met (Neculita et al. 2007; Gruyer 2012). These are

- A near neutral pH for the influent wastewater
- Available carbon from a suitable organic source
- A solid matrix (sand and/or gravel) onto which SRB can establish microenvironments
- An anaerobic environment
- A supply of sulfate in the water being treated sufficient that the SRB can outcompete methanogens for available carbon
- A way to physically retain the metal precipitates that will be produced

Most of these criteria apply to semi-passive treatment BCRs as well. To the list should be added the requirement that influents to BCRs of either sort should not contain appreciable amounts of nitrates or ferric iron. (These should be removed in the upstream, primary treatment cells of an EEB System.)

The most desirable solid active media for passive treatment BCRs if they are readily and economically available are wood chips* and the by-product biosolids from WWT plants at pulp and paper plants and paper recycling

* However, beware as not all wood chips are equal and, depending on the trees from which they are sourced, suitability and quality may vary. In all cases before any source of wood chips is considered as the organic active medium in the substrate of a passive treatment BCR, detailed analyses of chemistry and quality should be carried out prior to using the wood chips in pilot-scale BCR cells during the mandatory treatability test that must precede their use in any full-scale EEB System.

plants. These are usually available as mixtures of primary biosolids (largely short wood fibres from process waters being treated) and secondary biosolids (largely dead "bugs" from the WWTP's bioreactors).

5.4 BCR Morphology and Operations

BCRs can be (and have been) configured as: anaerobic EEB cells (ABRs), Stand-Alone bioreactors and backfill reactors in abandoned/disused mineshafts. In addition, there are active WWTP technologies involving anaerobic bioreactors that carry out all of the same functions as those carried out in BCRs. Examples are the *Thiopaq* and related *Biosulphide* technologies.

In BCRs, the microbial (largely bacterial) communities in/on their substrates usually take some time to get fully established, especially for passive treatment ones, even where the substrate beds have been inoculated with biostimulation or bioaugmention microbial solutions. Operators of BCRs should make allowance for such situations and plan for start-up periods where the bioreactors are dosed with nutrients and some feedstock, then left to develop their biofilms over several weeks.

Semi-passive BCRs only operate in a downflow VSSF mode but passive treatment BCRs can operate either downflow or upflow VSSF.* However, while it is felt that operation of a passive treatment BCR in an upflow VSSF mode may minimize the effects of channelling and compaction, and there are internal design methods that can minimize hydraulic instabilities, for full-scale passive treatment BCRs only design for operation in a downflow mode can be recommended at present until further R&D on hydraulics (see Section 9.3 in Chapter 9) clarifies aforementioned concerns about instabilities in larger bioreactors of this sort.† While Stand-Alone HSSF BCRs have been designed and built, this morphology is not as effective as VSSF modes, and HSSF EEB BCRs are not recommended.

The use of semi-passive treatment BCRs will be dependent on:

1. The long-term availability of liquid active media at low cost from sources relatively close to where BCR-based EEB System projects are being carried out (often as the by-products of local area manufacturing processes)

* These and all other kinds of EEBs, including BCRs, should be piped up so that they can be backwashed if desired to clear their beds. This will allow upflow VSSF operation in future if the R&D clarifies matters.

† It is noted that the Teck Demo (see Section 5.6) and the Arsenic Demo (see Section 5.7) operated for many years in upflow VSSF modes without problems, but the four passive treatment BCRs involved had internal mechanisms to prevent flow instability problems. (Such involved "know what.")

2. Where these kinds are being operated at remote sites, an assessment of the costs/benefits of importing the liquid active medium to the site over the bioreactor's operating lifetime

Table 5.4 compares some aspects of passive and semi-passive treatment BCRs.

TABLE 5.4

Comparison of Passive and Semi-Passive BCRs

Kind of BCR	Passive Treatment	Semi-Passive Treatment
Complexity	Higher	Lower
Flow mode	Upflow or downflow VSSF	Downflow VSSF
Throughput	A few hundred m^3/day max	Up to 15,000 m^3/day+
Substrates	Amended solid active media	Gravel
Active media	Solid	Liquid
Substrate complexity	High	Low
Substrate costs	Can be high	Lower
Substrate replacement	Possibly after 5–20 years	May not be required
Limestone	Can be internal layer	Upstream ALD
Primary treatment?	Balancing/Sedimentation Pond	Diversion pond
Tertiary treatment?	Always, as some organic active media degradation products will be found in effluents	Usually desirable but not always needed

The comparisons in Table 5.4 are all relative. For example, the organic solid media for a passive treatment BCR might be available locally for little or no cost, but so might a suitable liquid by-product from a local industry.*

Similarly, the cost of screened and washed gravel from local sources might be high. However, one of the largest potential markets for passive treatment BCRs is mining companies, and at many mine sites such firms will already have gravel and crushed rock preparation equipment and their gravel costs often are quite modest.

In any case, semi-passive BCRs are not very complex internally; like BBRs, they are simply beds of gravel with some internal piping.† In contrast, passive treatment BCRs may have different layers of materials in them and their substrates may be complicated to prepare.‡

The biggest problem with using passive treatment BCRs is that they are not practical for treating very large volumes of WW. If you question this, take a look at Figure 5.8 in Section 5.6 to see the scale of a passive treatment BCR

* See Table 5.2 and its related discussion about availability.

† Also, the semi-passive BCRs do not have the aeration systems of BBRs!

‡ See Section 7.4 in Chapter 7 to appreciate the logistics of preparing and installing passive treatment substrates.

cell that treats a maximum of 25 m^3/day of leachate! Many opportunities to use a BCR-based EEB System will be for situations where very much larger volumes of wastewater need to be treated.

It is recommended that semi-passive BCRs should always be preferred as their design, construction, operation, control and maintenance are very much better than those of passive treatment BCRs, and much higher flow rates can be sustained with them.* This does not mean that there will be no markets for passive treatment BCR-based EEB Systems; there are many places where relatively modest flows of WWs need to be treated, especially ones containing specific CoCs such as arsenic.

5.5 The BCR-Based EEB System at West Fork

One of the first full-scale BCR-based (i.e., anaerobic) EEB Systems that was not based on Anaerobic Wetlands was constructed in central Missouri, USA in mid-1996 (Gusek et al. 1998†; ITRC 2013). The purpose of the *West Fork Project* was to treat NMD at about approximately 6500 m^3/day from an underground lead mine that was being discharged to the Black River in that state. The pH 8 MIW contained 0.4–0.6 mg/L of lead and about 0.2 mg/L of zinc. The EEB System was designed and constructed to treat the NMD and remove these metals, especially lead. Its design was based on a phased program of earlier bench- and pilot-scale BCR testing. As part of this, an approximately 100 m^3/day outdoor pilot unit was constructed at the mine site in 1994 to facilitate the development of the design.

The resulting approximately 2 ha, gravity flow, full-scale EEB System at West Fork consisted of five cells: an initial 1° *Settling Pond*, two 2° *passive treatment BCRs* in parallel, a 3° *Rock Filter* and a 3° aerated *Polishing Pond*, the latter two in series with the BCR cells. All cells were HSSF and had sloped sides (mostly 2:1). All cells were HDPE-lined, except the Rock Filter which was clay-lined. The system was integrated with the mine's then existing WWT facilities.

Figure 5.1 (Gusek et al. 1998) shows the facilities at West Fork.

The purpose of the initial 0.3 ha, 3-m deep Settling Pond (the primary treatment part of the EEB System) was to remove grit and suspended solids from the NMD by gravity settling, and to act as a balancing pond for the rest of the EEB System.

* Section 9.2.2 of Chapter 9 presents state-of-the-art concepts for a semi-passive BCR-based EEB System treating large volumes of acidic MIWs at a northern mine.

† The authors of this paper did not use the EEB terminology used in this book, but in the interests of consistency, their terminology has been updated for the text in this section and units are presented in metric.

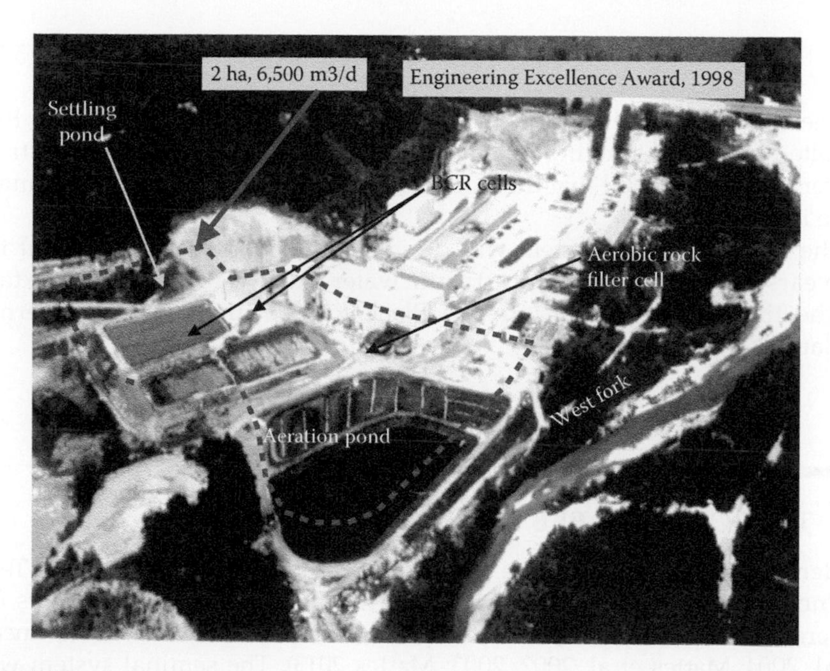

FIGURE 5.1
EEB System at West Fork.

The purpose of the two, 0.2 ha each, BCR cells (secondary treatment) was to remove most of the lead and zinc as sulfides via sulfate reduction, precipitating galena and sphalerite, respectively

$$Pb^{2+} + HS^- \rightarrow PbS\downarrow + H^+ \tag{5.10}$$

$$Zn^{2+} + HS^- \rightarrow ZnS\downarrow + H^+ \tag{5.11}$$

The two BCR cells used sawdust, limestone, composted cow manure and alfalfa hay in their 2-m thick substrates. The BCR cells reduced lead from a maximum of 0.6 ppm in the influent MIW down to meet NPDES effluent criteria of 0.027–0.05 ppm (Gusek et al. 1998). The BCR cells were not buried or located under water but their surfaces were covered by layers of limestone gravel.

Since the pH of the NMD was above 7, it was felt that any excess sulfide in the wastewater exiting the BCRs would not form H_2S there, but there was concern that lower pHs might develop downstream and it might be formed then, so an aerobic step (tertiary treatment) was needed after them to preclude its formation. The 0.6 ha, 0.3-m thick, limestone cobble-filled aerobic Rock Filter and the final 0.8 ha aerated Polishing Pond were included in the EEB System to re-oxygenate the BCR effluent so as to convert any excess sulfide in it back to sulfate:

$$S^{2-} + 2O_2 \rightarrow SO_4^{2-} \quad (5.12)$$

The two aerobic cells also controlled any BOD in the BCR effluents that resulted from any leaching from or degradation of their organic substrate materials. It was stated that the Rock Filter also removed any excess manganese in the NMD.

The West Fork EEB System was designed to last about 12 years but for 19 years met the stringent in-stream water quality requirements extant at the time for discharge into the Black River without a single permit violation.*

5.6 The BCR-Based EEB System at Trail BC

A demonstration-scale, passive treatment BCR-based EEB System (the Teck Demo) operated summer and winter for 6 years (and during summers for several years before that) at Teck Cominco's lead-zinc smelter (Duncan et al. 2004; Mattes et al. 2002, 2003; Mattes 2013). The seminal system was designed by Co-Author, Al Mattes, to treat wastewaters from a closed, highly metal(loid)-contaminated landfill and an old arsenic scrubber pond (collectively, Leachate) at Teck Cominco's site. Leachate was collected by French drains intercepting seepage from the two sources before it entered a creek that flows towards the smelter facilities site (Stony Creek—a tributary of the Columbia River).

Figure 5.2 shows the French drains constructed at the bottom of the landfill above the creek.

Leachate collected from the French drains below the two sources flowed into a central sump from which it was pumped both to the smelter's WWTP (an active lime treatment plant) and to the Teck Demo where it was treated. Figure 5.3 shows the control apparatus and housing at the central sump.

Feedstock for the Teck Demo, a slipstream of the Leachate, was taken from the central sump and pumped up a hill above it to the site of the Teck Demo.

The Teck Demo† consisted of two passive treatment, upflow VSSF BCR cells in series containing solid substrates consisting of pulp and paper mill

* This was without any need for substrate replacement/refurbishment. In early 2015, an underground mine shaft near the location of the Rock Filter collapsed, unintentionally decommissioning the EEB System (source: personal communication from J. Gusek). The West Fork NMD is now being treated at another WWT facility at another mine.

† Figure 2.2 in Section 2.2.5 of Chapter 2 shows a simplified version of the BBR-based EEB System of Figure 5.7.

FIGURE 5.2
French drain by creek at base of the landfill.

biosolids (the active media), limestone and sand (56.25%, 25% and 18.75%, respectively), followed by two HSSF CW cells, a FWS CW cell and a Pond Wetland cell. Treated Leachate from the Pond Wetland cell of the Teck Demo EEB System was disposed of by phytoirrigation onto a hybrid poplar plantation (see Section C.6 of Appendix C for a review of this phytoremediation method).

FIGURE 5.3
The central sump.

Figure 5.4 illustrates the configuration of the secondary and tertiary treatment EEB System cells of the Teck Demo, and provides typical ranges of CoC and other parameter concentrations and flow rates.

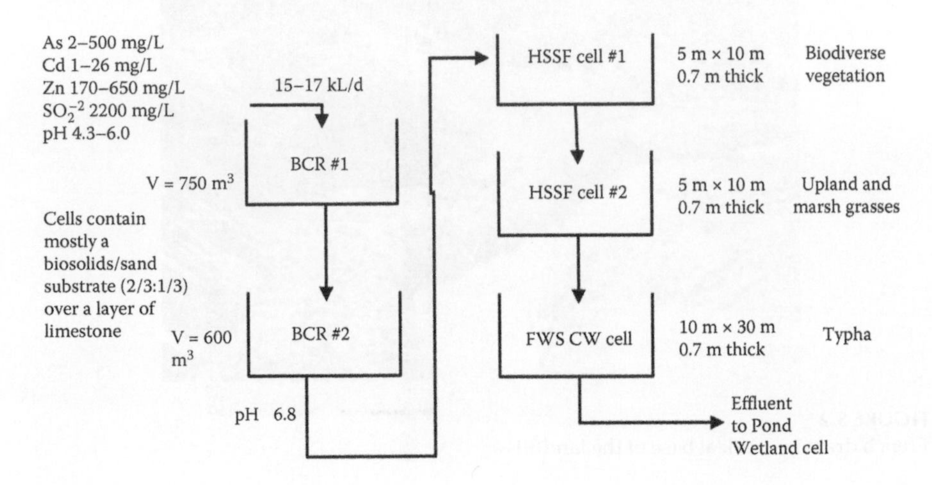

FIGURE 5.4
Teck Demo EEB System, average flows and concentrations.

Figure 5.5 shows an aerial view of the test site for the Teck Demo.

FIGURE 5.5
Teck Demo site (1) BCR #1, (2) BCR#2, (3) HSSF CW #1, (4) HSSFCW #2, (5) FWS CW, (6) Pond Wetland, and (7) hybrid poplar plantation.

Figure 5.6 shows the installation into BCR cell #2 of the Teck Demo of the premixed organic substrate used in both BCR cell (pulp & paper biosolids + limestone + sand, see Section 8.4 for a review of the logistics of preparing and installing passive treatment EEB cell substrates). Figure 5.6 provides a good illustration of the scale of such operations.

FIGURE 5.6
Substrate Installation in BCR cell #2 at the Teck Demo.

Figure 5.7 shows a view of the surface of completed BCR cell #1.

FIGURE 5.7
Completed BCR cell at the Teck Demo.

Treated Leachate from the second BCR cell then flowed to tertiary treatment in the two HSSF CW cells (1-m thick gravel-bed cells vegetated with terrestrial grasses) followed by the polishing FWS cell (one with 1-m thick organic substrate vegetated with wetland plants) and the Pond Wetland cell. The two HSSF CW cells were planted with select C4 grasses that had dense fibrous root systems and that are known to carry out most of their evapotranspiration at night, allowing them to grow in the often very hot daytime conditions of the Teck Demo site. Since the organic active media in the upstream BCR cell leached both its breakdown products and some of the sulfides and other products precipitated in it, the main purpose of the two downstream HSSF CW cells was to filter out such material (Fitch & Schoenbacher 2009).

Figure 5.8 shows the surface of HSSF CW cell #1.

FIGURE 5.8
Tertiary treatment HSSF CW cell at the Teck Demo.

Treated Leachate from the FWS CW cell then was collected in the Pond Wetland cell which acted as a final surge and settling pond for the EEB System and allowed the storage of treated Leachate prior to its disposal.

Figure 5.9 shows this final cell.

FIGURE 5.9
Pond Wetland cell at the Teck Demo.

Treated Leachate collected in the EEB System's Pond Wetland cell was disposed of by spraying it onto the nearby hybrid polar tree farm (see Figure 5.5). Figure 5.10 shows the tree farm.

FIGURE 5.10
Tree farm by the Teck Demo site.

The Teck Demo was designed to treat between 15 and 17 m^3/day of Leachate, with adjustments made seasonally and actually treated as much as 25 m^3/day at times. (Higher flow rates were possible in summer than in winter.) Over a 9-year period, more than 21% of the Leachate collected in the French drains around the landfill and scrubber pond (~53,000 m^3) was treated in the Teck Demo, with approximately 99% of the metal(oid)s in the influent (10,722 kg of As, Cd, and Zn) being removed.

Over the 9-year period of the Teck Demo's operations, suspended and dissolved zinc (average 205 mg/L), suspended and dissolved arsenic (average 120 mg/L), and dissolved cadmium (average 5 mg/L) from the Leachate and concentrations were reduced to a few mg per L (Zn) and much less than 1 mg/L (As and Cd). Table 5.5 lists average flow rates and cell outlet CoC concentrations over the 9-year operating period for the Teck Demo showing influent concentrations into the EEB System and the CoC concentrations out of each of the cells.

As may be seen from the results, over nine summers and winters of operation, the passive treatment BCR-based EEB System near the Teck Cominco smelter achieved excellent results, removing 97.7%–99.6% of total CoCs (As, Cd and Zn), and 97.7%–99.4% of the dissolved CoCs.

During the operations of the Teck Demo, a slipstream of the Leachate passing through the central sump was pumped uphill from it to the first cell in the Teck Demo, BCR #1, which was constructed at the high point of the site.

TABLE 5.5
Average CoC Concentrations for the Teck Demo over 9 Years

Outlet Concentration	TSS	Total As	Dissolved As	Total Cd	Dissolved Cd	Total Zn	Dissolved Zn
Influent	330	114	25	3.7	1.6	205	108
BCR cell #1	219	15.2	5.2	0.7	0.2	62	42
BCR cell #2	63	5.4	2.5	0.12	0.08	36	30
HSSF CW cell #1	12	2.4	1.5	0.02	0.01	35	32
HSSF CW cell #2	17	1.5	0.9	0.02	0.01	28	26
FWS CW cell	126	7.6	0.5	0.03	0.01	14	11
Pond Wetland cell	58	0.4	0.3	0.02	0.01	4.5	2.9
Percent removal		99.6	99.0	99.6	99.4	97.7	97.3

This allowed for gravity flow to feed the subsequent cells of the EEB System. BCR #1 (surface area: about 25 m × 18 m) contained approximately 700 m^3 of substrate and this and the subsequent #2 BCR cell were both configured as upflow VSSF ABRs.* The surface area of the more rectilinear BCR #2 was 25 m × 18 m (an aspect ratio of 1:4) and both BCRs were lined with 60 mill HDPE. From BCR #2, partially treated Leachate flowed through two planted HSSF CW cells (each with a surface area of 10 m × 5 m, see Appendix B.3 for descriptions of these kinds of CWs); the first vegetated with Eastern Gamagrass (*Tripsicum dactyloides*, a C4 grass species with a dense root system that was known to be tolerant of high concentrations of Zn), and the second with native grasses such as Bluejoint Grass (*Calamagrostis canadensis*).

Effluent from the HSSF CW #2 cell then flowed through the final polishing FWS CW cell (surface area: 50 m × 10 m) that was initially planted with Cattails (*Typha latifolia*) but when the still high zinc concentrations in the treated Leachate proved toxic to this species, the plants in the front half of this latter cell were replaced with the more Zn-tolerant Reeds (*Phragmites australis*). The long, narrow, open water FWS CW cell proved to be an excellent polishing cell, removing a high percentage of the arsenic (94%) that entered the cell and a lesser, but still significant, percentage of the zinc (64%). From the FWS CW cell, the treated Leachate was then delivered to the Pond Wetland cell (Figure 5.11) from which it was pumped and used to irrigate the tree farm (Figure 5.12). This effluent attained irrigation standards of the Province of British Columbia.

During the multiyear operations of the Teck Demo, weekly samples were taken and sent for analyses in the laboratory at the Teck Cominco smelter, not only of the CoCs but also of other important operating parameters such as pH, ORP, DO and the concentration of iron in the water as it flowed through

* The upflow operation did not seem to have any adverse impacts on the operations of the BCRs at Trail, but as mentioned elsewhere, until the matter of upflow VSSF EEB hydraulics and hydrology is resolved, any new full-scale BCRs should be designed VSSF downflow. HSSF BCRs are not recommended.

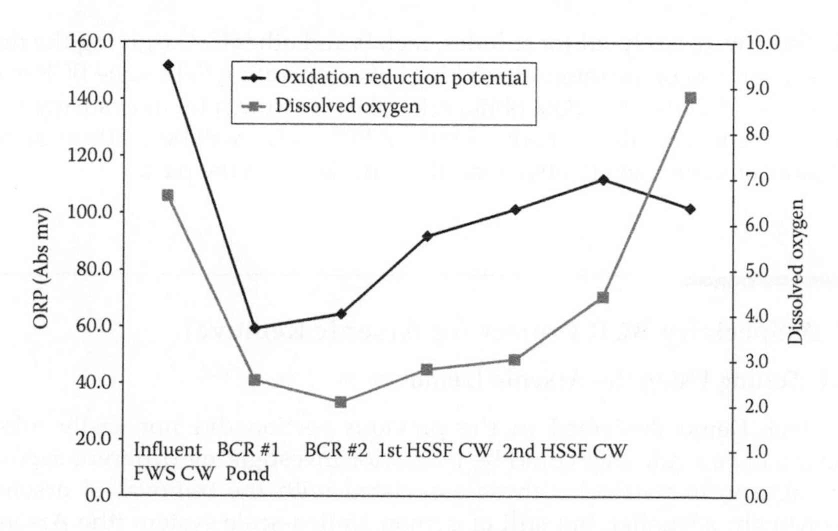

FIGURE 5.11
Changes in DO and ORP in the Teck Demo.

the EEB System. Flow rates were also regularly monitored. These data were then used to analyse the system to determine if anything required adjustment and if the system was operating according to the design specifications.

An example of how the data were used to provide operating information can be seen in Figure 5.11 which shows changes in DO and ORP. These data indicate a characteristic drop in the values across the two BCRs of the EEB System followed by a gradual increase as the treated Leachate moves through the tertiary treatment parts of the system (the CW cells).

Similar graphs and data analyses were carried out each year for each of the parameters so that a complete picture of the EEB System's operations and functioning could be described and comparisons then made to previous years. Originally designed solely as a demonstration-scale system, the Teck Demo proved to be robust enough to treat up to 125 m^3/day of leachate when required, thereby supplementing and backing up the smelter's WWTP.

The Teck Demo was very efficient in removing the CoCs, even when their concentrations temporarily increased dramatically. For example, a spike event took place in November and December of 2005 due to Teck Cominco making changes in WW drains and pipelines around the landfills. During the spike event, arsenic concentrations in the Leachate went up to 3600 mg/L, zinc concentrations rose to 3800 mg/L and sulfate concentrations fell to 500 mg/L, indicating that sulfate removal processes were taking place in the system up to the sump. In spite of the spike, the final EEB System arsenic and zinc concentrations in that fraction of the Leachate treated in the Teck Demo were still reduced by 99.0% and 99.8%, respectively, during this upset period.

Part way through the operation of the Teck Demo, the BCR #1 cell was opened up and samples of substrate were taken down through and across its substrate bed. These were analysed for sulfides, metals and other CoCs, giving the first complete picture of the internal workings of an operating field-scale BCR ever (Duncan et al. 2004). The data obtained will be instrumental in ensuring that future designs of BCRs (and other kinds of EEB ABRs) will be vastly superior and more efficient and effective than those designed in the past.

5.7 A Specialty BCR Project for Arsenic Removal

5.7.1 Testing Using the Arsenic Demo

The Teck Demo described in the previous section did not easily allow changes to be made that could be useful for investigating different aspects of treatment, in particular those associated with the removal of arsenic. Accordingly, a smaller, but still of demonstration-scale system (the Arsenic Demo), was constructed alongside it to test some particular aspects. This consisted of two parallel trains, each involving two smaller, aboveground passive treatment, upflow VSSF BCR cells (train AS 1 and train AS 2).

Two sets of tests were carried out using the Arsenic Demo:

1. Set 1: Tests with seepage collected directly from the French drains below the old arsenic scrubber pond (arsenic-rich seepage) and with the BCR cells of both trains filled with the same substrate as was used in the Teck Demo.
2. Set 2: Tests with the same Leachate feedstock as was used for the Teck Demo but with different substrates in the BCR cells than those of Set 1; the first BCR cell in each train containing iron slag and biosolids, and the second in each train containing biosolids and granular slag containing FeS_2. It also contained a very small amount of FeAsS.

For Set 1, each train (AS 1 and AS 2) had two cells: a downflow anoxic limestone drain cell followed by an upflow BCR cell.*

For Set 2, both of the cells of each train were configured as BCRs containing a largely biosolids substrate, the first mixed with iron slag to provide a source of iron and permeability and the second with ferrous sulfide and biosolids.

The effluent from these pairs of cells was then delivered to an optional downflow gravel-based VSSF CW cell that was planted with As-tolerant Ferns (*Equus sedum)* transplanted from a nearby area where they were growing in soil known to contain high levels of arsenic.

Flow for both sets was ALD–BCR–VSSF CW.

* See Section C.3 in Appendix C for a description of ALDs. Using an ALD cell followed by a BCR cell in a test unit is equivalent to having a limestone layer upstream of the substrate inside a field-scale BCR. Although these demonstration-scale test cells were upflow VSSF, it is suggested that only downflow VSSF cells be used for full-scale units.

Figure 5.12 shows the completed system showing the AS 1 and AS 2 trains with their waterproof covers.

FIGURE 5.12
Top view of the Arsenic Demo.

Figure 5.13 presents a view from the bottom of the installation showing the storage tank that was used during Set 1 to collect highly arsenic-rich seepage from the French drains below the old arsenic scrubber pond. (This was changed to a plastic tank for the Set 2 testing.)

FIGURE 5.13
Another view of the Arsenic Demo.

Soil was packed around the cells to hold them in place and insulate them.

5.7.2 Set 1 Testing Results from the Arsenic Demo

For the Set 1 testing, the top pair of vessels in each train (AS 1 and AS 2, see Figure 5.13) was configured as downflow ALDs (As 1-1 and As 2-1, respectively), while the lower vessels were set up as upflow BCR cells (As 2-1 and As 2-2, respectively).

To provide an inoculant to seed the Arsenic Demo's BCR cells, a small static bioreactor filled only with pulp and paper recycling plant biosolids was set up in early in 2001 and then charged with As-rich seepage. This was covered, then left for a period to allow for an appropriate bacterial population to develop. In early August, water samples were taken from this small, static (no-flow) bioreactor and analysed. The results of these analyses showed that a substantial removal of arsenic had occurred. (An initial concentration of 85 mg As/L had been reduced to 0.45 mg As/L.)

Construction of the Arsenic Demo's cells, piping and other infrastructure began in June of 2002 and was completed by mid-August. At that time, the BCR cells in the two trains of the Arsenic Demo were filled with As-rich seepage from the storage tank and "seeded" with an inoculant from the static bioreactor. The Arsenic Demo's BCR cells were then left undisturbed with no flow through them for 1 month to allow for the establishment of appropriate bacterial populations in their substrates. As-rich seepage from the Storage Tank was then pumped through each train. Flow into the first (BCR) cell of each train was approximately 125 L/day and sampling commenced. Weekly sampling continued until the end of October 2002 when both trains of the Arsenic Demo were drained and the EEB System winterized.

Table 5.6 shows the mean field and laboratory measurements of some important chemical parameters for the Arsenic Demo during 2003, 2004, and 2005. Values for pH and DO (mg per L) are for all 3 years, whereas those for ORP (mV), S^{2-} (mg per L) and total Fe (mg Fe_T per L) are for 2005 only.

TABLE 5.6
Set 1 Arsenic Demo Results

	pH	DO	ORP	S^{2-}	Fe_T
Influent	6.2	7.2	178	0.02	0.6
AS 1-1	5.8	2.5	−104	17.4	17.2
AS 1-2	6.3	2.8	−135	17.4	24.0
AS 2-1	5.8	2.3	−111	3.3	51.9
AS 2-2	6.3	2.0	−164	4.3	23.4

The results of some of the analyses of field-measured parameters were not what was expected. In both trains, pH was reduced from its input value

(6.2) by the output of the first pair of cells (limestone drains AS 1-1 and 2-1) and rose slightly above the input value at the output of the second pair of cells (BCR cells AS 1-2 and AS 2-2). Since the first cell in each train was an anoxic limestone drain, the pH was expected to rise. However, DO dropped sharply, as expected, and ORP values were negative in all cells in the system. Sulfide concentrations increased in both systems, more so in train AS 1 than in train AS 2, while iron levels showed high concentrations in all cells in both trains.

Figure 5.14 shows the differences between the removal results for the two trains.

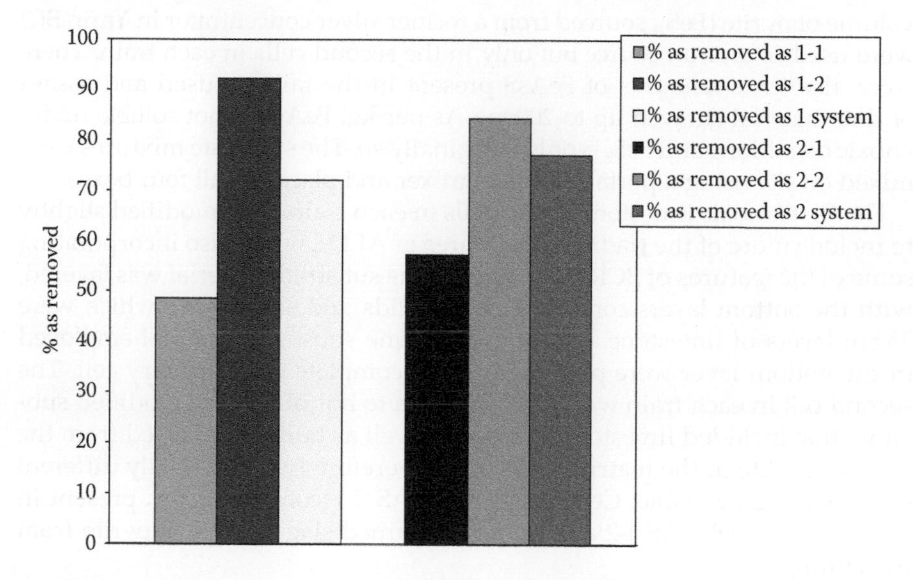

FIGURE 5.14
Comparison of Set 1 Arsenic Demo results.

The mean influent arsenic concentration for As-rich seepage into the two trains was 45.3 mg/L and the mean output concentration out of them was 7.5 mg/L. The mean percentage removed was approximately 77%. This was not as good as the results obtained in the Teck Demo during the same period, but the Arsenic Demo was treating a feedstock richer in arsenic, and in any case substantial arsenic removal still took place in the smaller system.

5.7.3 Set 2 Testing Results from the Arsenic Demo

After 3 years of operation, the Arsenic Demo was taken apart and rebuilt. At this time, each of the four main cells of the Arsenic Demo were emptied and refilled with different substrate mixtures and were all configured as BCRs, the first as downflow VSSF and the second as upflow VSSF. In addition, the tank that had contained the As-rich seepage used for the Set 1 testing was removed and replaced for the Set 2 testing with a 10,000 L Nalgene plastic tank and this was used during the Set 2 to contain the same feedstock that was being used with the larger Teck Demo.

To investigate the removal of arsenic using pulp and paper plant biosolids, these were mixed with limestone and slag from the Trail smelter in the same proportions as were biosolids, agricultural limestone and sand used in the BCR cells of the Teck Demo. The pelletized slag was an iron silicate material that was used as a nonreactive material in place of the sand to provide permeability to the mix. For the Set 2 testing, mine tailings containing a substantial volume of pyrite (FeS_2, sourced from a former silver concentrator in Ymir, BC) were used as an iron source but only in the second cells in each train. There were also trace amounts of FeAsS present in the tailings used and assays of that material showed up to 200 mg As per kg. FeAsS is not soluble under anoxic conditions and FeS_2 is only marginally so. The substrate mixtures were mixed on site using a portable cement mixer and placed in all four boxes.

The first (anoxic limestone drain) cells in each train were modified slightly to include more of the traditional features of ALDs, while also incorporating some of the features of BCR cells. In each, the substrate material was layered, with the bottom layers consisting of biosolids and slag, above which were 25 cm layers of limestone. Layers of the same substrate material employed in the bottom layer were placed on top to complete each primary cell. The second cell in each train was filled from top to bottom with a modified substrate that included limestone and slag as well as tailings collected from the Ymir mine site in the matrix. There were therefore two potentially different sources of Fe available: Cells AS 1-1 and AS 2-1 contained iron present in slag, whereas cells AS 1-2 and AS 2-2 contained slag as well as pyrite from the tailings.

Table 5.7 lists pH, DO (mg per L), ORP (mV) as well as the concentrations in mg per L of S^{2-}, total Fe, Fe^{2+} and Fe^{3+} in the two trains of the Arsenic Demo using different forms of Fe in the substrate and iron slag as a sand substitute.

TABLE 5.7
Set 2 Arsenic Demo Results

	pH	DO	ORP	S^{2-}	Fe_T	Fe^{2+}	Fe^{3+}
Input	6.2	6.3	112	0.01	0.1	0.05	0.08
As 1-1	6.4	1.5	−236	4.4	1.2	0.6	0.6
As 1-2	7.3	2.5	−136	0.1	3.6	0.2	3.5
As 2-1	6.4	2.1	−100	2.8	3.3	2.0	1.3
As 2-2	7.2	2.1	−109	0.5	4.4	1.5	3.0

Analysis of the operating parameters for the 1 year that data are available (2007) showed that values were as expected for the Arsenic Demo with only small differences between the first 3 years' operations with the same substrate as in the Teck Demo BCR cells in each cell and the current one in which the substrate had been modified. The pH was slightly higher at the input and rose in each subsequent cell in the train, whereas in initial operations using the Teck Demo substrate there was a drop in pH in the initial cell. Mean DO readings were lower at the input than in the previous test situation and they were lower in all cells except for train AS 2-2. ORP readings, which were all negative, were much lower in train AS 1-1, about the same in AS 1-2 and higher in both cells in the AS 2 train.

Sulfide ion concentration was lower in this phase of the study, with higher values in the first cells of both trains and lower values in the second cells. During Set 2, S^{2-} levels were very much lower than was noted for Set 1, particularly for the AS 1 train. Total iron was greatly different with average values of as low as 0.1 mg/L for cell AS 1-1 to as high as 4.4 mg/L for cell AS 2-2. Compared to Set 1 when the mean values were approximately 24 mg/L, this represents a very large difference. The presence of Fe^{3+} iron was tested for in this phase through calculations using the values for total Fe and the value for Fe^{2+} and its presence indicated the activity of iron-oxidizing bacteria. The smaller concentration of Fe can be attributed to the relative insolubility of the Fe present in the imported tailings as opposed to the reducible FeO present when sand was used.

Figure 5.15 shows arsenic removal percentages using iron silicate and mine tailings over a 2-year period.

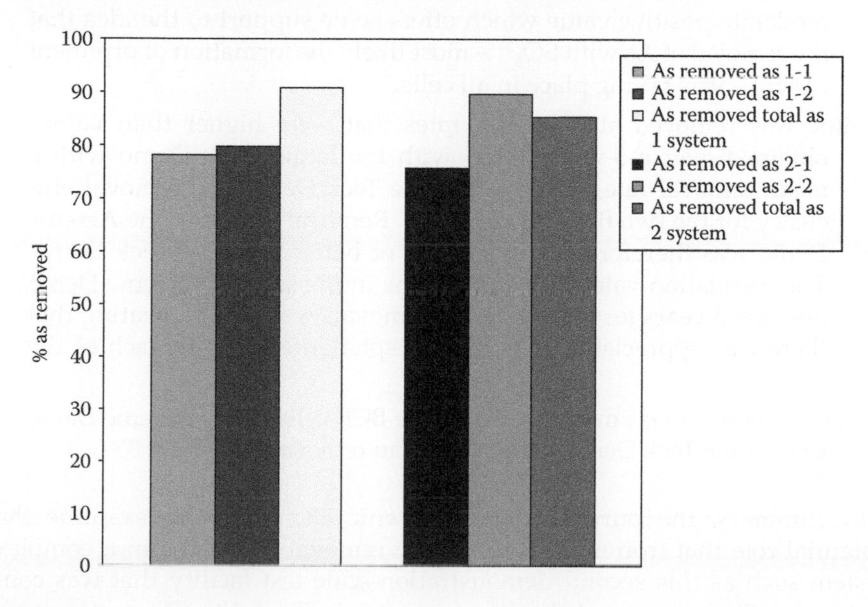

FIGURE 5.15
Comparison of Set 2 Arsenic Demo results.

The following conclusions can be made:

Each train of the Arsenic Demo removed approximately the same percentage of arsenic (~77%). The removal percentages were lower than the values found for the first two BCR cells in the Teck Demo (see Section 6.6) (equivalent to Arsenic Demo cells 1 and 2) where the two BCR cells removed 93.7% of the total arsenic. Since the Arsenic Demo operated only during the summer months, some of this could be due to the time it took for the bacterial community to become reestablished each year, but this cannot completely explain the differences. There was sufficient zinc removal, likely as ZnS, to allow for coprecipitation and/or adsorption of the arsenic on it. The high concentrations of iron noted in the effluent offer another explanation for the lower efficiency. Since both As (III) and As (V) are adsorbed to iron, the relatively higher levels of arsenic present in the effluent samples may reflect arsenic that was adsorbed to the colloidal iron that was also present. Since iron concentration increased in the second cells in both trains and the output of this cell went to the Teck Demo, it was not possible to complete a regression and correlation analysis on iron and arsenic.

Arsenic removal is fairly strongly correlated with the removal of zinc with a correlation value of 0.47 for both trains of the Arsenic Demo over the 3-year period. At the same time, the value for arsenic removal correlated with the removal of SO_4^{2-} was 0.31. This is a moderate positive value which offers some support to the idea that the removal of As with SO_4^{2-} – most likely the formation of orpiment (As_2S_3) – was taking place in all cells.

Zinc was removed at efficiency rates that were higher than values obtained over a 5-year period with the larger Teck Demo with a mean removal rate of 92.5%. For the Teck Demo, the removal efficiency for the two BCR cells is 84.8%. Removal of zinc in the Arsenic Demo was therefore comparable to or better than the Teck Demo. The correlation values for both trains in the smaller Arsenic Demo over the 3 years for Zn and SO_4^{2-} removals was 0.65, indicating that there was appreciable formation of sphalerite (ZnS) in each of the cells.

Sulfate removal was much higher in the BCR cells of the Arsenic Demo than in the Teck Demo where the mean removal was about 35%.

In summary, the purpose of the Arsenic Demo was to examine the potential role that iron might play in the removal of arsenic in a complex system such as this second demonstration-scale test facility that was constructed at Teck Cominco's lead-zinc smelter in Trail, BC. The testing was carried out over a period of years and involved many other analyses than those discussed above. For example, when the BCR cells were taken apart in preparation for the change in iron source (i.e., the beginning of Set 2), they were systematically sampled in three layers at nine points in each layer for a total of 27 samples per layer. The samples were dried and analysed using ICP-OES to determine the concentrations of the CoCs in each sample. These concentrations were then compared using statistical correlation techniques to determine the degree of association between, for example, zinc and sulfur, or arsenic and zinc. The same procedure was then carried out again when the Set 2 testing was finished and the cells of the Arsenic Demo were dismantled.

The correlation analyses appear to indicate the existence of two distinct mechanisms of arsenic removal during Set 2. The first was the formation of As_2S_3 and this appeared to be specific to the deeper, possibly more anaerobic, layers of the BCR cells. The second mechanism was coprecipitation with, or adsorption to, metal sulfides, in particular ZnS or FeS, or FeS_2 when it was present. A low concentration of iron present in the seepage from the old arsenic scrubber pond compared to zinc and arsenic indicated that little FeS was formed and the formation of ZnS and the subsequent adsorption of As or its coprecipitation dominated. Although the simple filtration of Kottigite (hydrated zinc arsenate, $Zn_3[AsO_4]_2{\cdot}8H_2O$) might have been a dominant feature of the removal system, when the BCR cells were dismantled, it was found that there was a smaller correlation between the two components (Zn and As) in the bottom layers of the BCR cell compared to when the Trail substrate was used in Set 1. This seemed to point to the formation of ZnS and the coprecipitation or adsorption of arsenic as being more important mechanisms in this latter system configuration. When FeS_2 was used as part of the substrate, the filtration of Kottigite appeared to be more important. This might have been due to the higher concentrations of arsenic and zinc present, as the source of the seepage had changed, or to the shorter period of operations – basically less than 2 years as opposed to the 3 years – for the Teck Demo testing.

The use of iron slag during Set 2 showed that it could be an ideal adsorbing surface for arsenic and perhaps a better material for filtering Kottigite from solution as well. Set 2 of the Arsenic Demo also showed that iron present as pyrite (FeS_2) could be an important factor in the removal of arsenic, but it did not appear to be as good a source of iron as the iron oxide coating on the sand used. If it were the only source of available iron, it might not play the same role as iron appears to have played in the Teck Demo testing.

Using statistical correlation procedures to determine mineralology was not ideal but it was the only tool available given the low concentrations of any one mineral present in the samples. The data suggest potential minerals that could be formed rather than absolutely pointing to their formations. The complexity of any BCR cell was highlighted in this study with variable removal profiles between different cells serving essentially different functions in a combined treatment system. When correlations were carried out on the removal of contaminants for the combined system in Set 2 using the slag and Yankee Girl substrate, they showed quite positive correlations amongst the removals of all three CoCs. Some of these positive findings were supported by analyses of the samples taken from the BCR cells, whereas others were not supported as completely as the removal data would have suggested. Overall, this was useful in pointing out the complexities of a BCR cell treating MIWs containing multiple metal(loid)s. The results also illustrated that no single explanation can be used to describe CoC removal mechanisms throughout all levels in a single BCR cell.

6

Treatability Testing

6.1 Reasons for Treatability Testing

Since every wastewater is different, as is every project/location/soil/climatic situation, to correctly design Eco-Engineered Bioreactors and their associated EEB Systems and determine the other scale-up factors needed to build them full scale, treatability testing is usually required in advance, and a separate test is required for each situation.

Why this is the case can be illustrated by comparing the situations at two medium-sized airports where BBR-based EEB Systems are being planned to treat glycol-contaminated residual streams generated during their de-icing seasons. Both airports use EG-based ADAFs in about the same proportions and one might imagine that the scale-up information from carrying out a treatability test at one would provide applicable information for the design of the other. However, this is not the case as both are quite different.

At the first airport, one in the west, most of the de-icing is carried out at-gate by its terminal building. At the western airport, there is a very sophisticated stormwater collection system consisting of collection piping alongside the airport's single runway and this system collects stormwater that during the de-icing season is impacted by the entire slate of contaminants listed in the first paragraph of Section 2.3.2 in Chapter 2, including surface (pavement) de-icing chemicals, runoff from grassed areas and Spent Glycols from the aprons and other areas where de-icing is carried out.

There is no snowmelt pad at the western airport (although one is planned) and ploughed snow from the aprons is allowed to melt into a trench connected to stormwater sewers.

At the second airport, one located in the east, de-icing is carried out on a modern CDF with an attached snowmelt pad for ploughed, Spent Glycol-contaminated snow from it. At this eastern airport, the water to be treated is almost all Spent Glycols, and is virtually un-impacted by GCSW, surface de-icing chemicals or runoff from grassed areas.

The bottom line is that an EEB System for the western airport will have to treat what is largely a stormwater stream contaminated by both Spent

Glycols and GCSW, as well as all manner of other chemicals including surface de-icing chemicals, rubber de-bonding chemicals, wash waters, fertilizers, oil and grease, etc., while an EEB System at the eastern airport will mostly treat Spent Glycols uncontaminated by anything else.

The required process configurations, BBR designs and sizing, and a host of other factors will be different for each airport, and only by carrying out separate treatability tests for each case can these be defined.*

Treatability testing can be carried out on- or off-site; indoors or outdoors and at bench, and pilot and/or demonstration scales. It may involve only the EEB(s) (the 2° treatment part of EEB Systems), or include the testing "upstream" and/or "downstream" 1° and 3° treatment components and cells as well as the 2° treatment of the EEB(s).

Treatability testing has several purposes, including:

1. Determining the kinetic, genomic and other scale-up parameters needed to design and engineer subsequent full-scale facilities
2. Showing clients considering EEBs that the wastewater that they wish treated can indeed be handled by such bioreactors
3. Conclusively demonstrating to regulators, bodies providing project financing, and other stakeholders that the EEB ecotechnology is indeed viable and will produce the desired levels of treatment

Indoor and outdoor treatability tests each have advantages and disadvantages. Outdoor test units installed on-site can treat an actual wastewater under ambient conditions that a later full-scale EEB System would encounter. However, varying temperatures, precipitation, CoC concentrations, lighting and other field conditions sometimes make it difficult and more costly with an outdoor test unit to accurately determine the kinetic data and other scale-up parameters needed. With indoor pilot-scale test units, a variety of confounding conditions (e.g., temperature changes, insolation, precipitation) can be fully controlled, allowing more accurate determination of the data needed for scale-up.

If necessary, indoor, off-site testing can be followed by outdoor, on-site testing to confirm that the feasibility and results found during in indoor testing will still stand up under field conditions. (Such was done at Rosebel, see Section 6.6.)

* At the Detailed Design and Engineering phase of a new EEB System project, some designs for BBRs determined using Reaction Kinetics and treatability tests may be cross-checked with those using aeration models (WEF 2010), but it is not recommended that such be used at the Conceptual Design phase of a project as such may lead to erroneous designs if used alone without first carrying out a treatability test and determining scale-up factors using Reaction Kinetics (see Section 8.3 of Chapter 8).

Off-site treatability testing is usually carried out at a *central test facility* and is often used to define the proof-of-concept of a new use of the EEB ecotechnology (i.e., will it actually work for this wastewater in this case?) using a relatively inexpensive methods, before later, often larger scale, more sophisticated (and costly) on-site testing is entertained.

Eco-Engineered Bioreactor central test facilities have on-site wet chemical and analytical test facilities associated with them, and these facilitate the testing and make it much more economical than what would be the case were all needed associated samples sent to outside commercial laboratories for analyses. By doing most of the testing in analytical and other labs at the off-site test unit location, rapid response to changing conditions in the test unit is facilitated.

6.2 Pilot-Scale Treatability Testing

While bench-scale EEB testing is often useful and is sometimes carried out before or in association with larger-scale testing, generally it is testing at a pilot scale that is practised in designing EEBs and EEB Systems. This is because it has been demonstrated that the results of such pilot-scale treatability testing with one cubic-meter test cells are usually scalable, while those from the bench scale often are not.

Generally, pilot-scale treatability testing is carried out indoors, although in some cases the test bioreactors (which are usually mounted on pallets) can be set up outdoors, or even temporarily moved outdoors if desired. Larger, demonstration-scale treatability testing of EEBs is carried out outdoors in excavated in-ground test basins and is more complicated; requires the import of much larger volumes of feedstock if carried out off-site; is more expensive; and is less flexible than pilot-scale testing. If desired, indoor pilot-scale EEB testing can be followed by outdoor pilot- or demonstration-scale testing, although such is not usually essential for EEB scale-ups.

Pilot-scale EEB testing is carried out by making used one-cubic-meter "totes" (commercial liquid chemical shipment vessels) into bioreactors by cleaning them out and removing the tops of their interior, semirigid plastic (usually Teflon or PVC) liners that are held inside outer steel cage assemblies.* Figure 6.1 shows a "topped" (lid cut off) chemical tote prior to its being used.

* Owing to insurance considerations, these vessels can only be used by chemical companies to ship liquid chemicals for certain periods before they are retired from service and, as a result, used ones are readily available for purchase in many locations.

FIGURE 6.1
A "topped" tote.

The topped totes (which already have bottom drains) then have inlet and outlet distribution and other piping as well as outside risers to control water levels, sampling points, etc. installed in them, and, as appropriate, they are filled with substrates and other materials to make them into EEB test units. Owing to their relatively small sizes, there is potential for some of the wastewater in them to flow down their walls, bypassing part of their substrates, especially if these substrates are granular in nature (e.g., gravel in BBR cells). This can be addressed by gluing plastic baffle "skirts" 10–20 cm wide around their peripheries at one or more locations down through the substrate. These will direct any water adhering to the cell walls back towards the centre of the cell, minimizing any wall effects.

During the pilot-scale treatability testing of EEB cells, the same kind of substrate that will be used in the later full-scale EEB System should be used to treat an imported or artificial wastewater spiked with enough of the main CoCs to simulate worse case conditions.

Generally, a pilot-scale EEB test unit involves one or more saturated-bed test cells (each having ~1 m^2 of surface area and containing 0.7–1 m^3 of substrate in one or more layers) with supporting tanks, mixers, pumps, piping, risers, an air blower (if a BBR cell is involved), valves and fittings, heaters and/or cooling coils (as required), instruments, grow lights (if vegetated cells are involved) and other equipment, as well as, if an EEB *System* is to be tested, appropriate associated primary and tertiary treatment test cells.

Figure 6.2 shows a typical morphology for the kind of test unit used during the *BREW Project* (see Section 1.4 in Chapter 1): a single Reed-vegetated HSSF BBR cell, a preliminary Mixing Tank beside it (a common feature of EEB test units that consists of a topped tote into which any required additives are mixed in and from which feed samples are taken), a large feedstock storage tank behind, a tipping bucket (in the foreground) for measuring flow rates and an automatic sampler (blue-bottomed barrel beside the tipping bucket).

FIGURE 6.2
Typical single EEB cell test unit.

Treatability testing was first carried out during the seminal *BREW Project* and during it indoor, 1 m^2 surface area indoor HSSF pilot-scale test units (such as that shown in Figure 6.2) and 25 m^2 outdoor, HSSF demonstration-scale test cells (not shown) configured as aerated HSSF BBRs (then called EWs) containing gravel substrates and the results with each kind were compared. It was determined that indoor pilot testing in the smaller cells was adequate to determine the needed scale-up factors (Higgins et al. 1999).

Pilot-scale EEB System test systems are designed so that they can be heated or cooled to control operating temperatures and determine scale-up parameters at the typical lowest and highest water temperatures that a full-scale unit might experience.

Figure 6.3 shows a test unit placed inside a walk-in refrigerator so testing can be carried out at low temperatures.

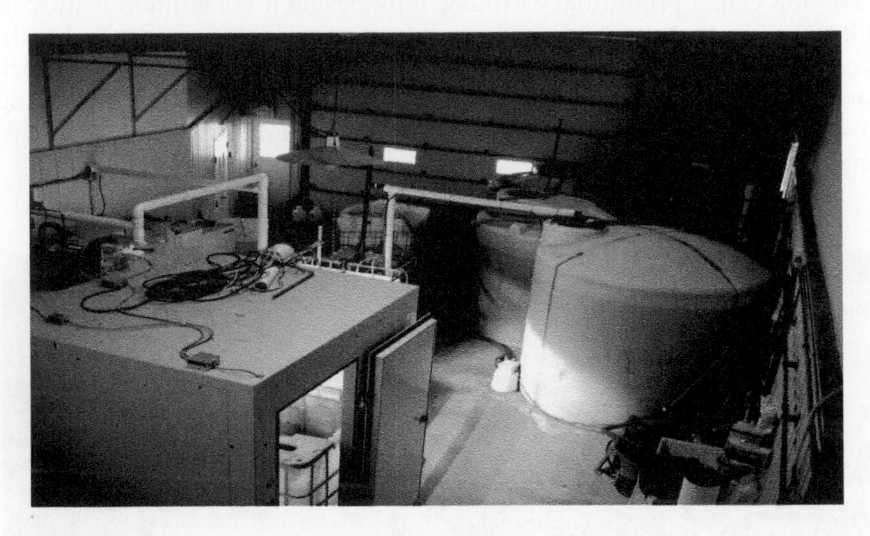

FIGURE 6.3
Treatability testing at low temperatures.

Section 8.9 in Chapter 8 details the concepts behind treatability testing at various temperatures.

A tracer test (see Section 1.8 in Chapter 1) is often carried out as part of the treatability testing to identify the Reaction Kinetic model (see Section 8.3 in Chapter 8) that will best simulate the operation of the Eco-Engineered Bioreactor cell(s). (Figure 1.5 in Chapter 1 shows the results of the tracer tests carried out during the treatability test with the test unit sketched below.)

Figure 6.4 shows a pilot unit that had two downflow EEB cells in series: a vegetated BBR cell in front (with grow lights over it)* and a vented ABR test cell behind it (in this case a BCR that might produce H_2S, hence the vent). This pilot-scale test unit also included a feed tank (in this case another tote, although larger vessels are often used as may be seen in Figure 6.2) and a Mixing Tank. It was used to test the treatment of a highly contaminated groundwater (aromatic and chlorinated aromatics, other hydrocarbons, ammonia and nitrates) from under a former industrial chemical site. The results of this treatability test indicated that the contaminated groundwater could be easily and efficiently treated in an EEB System, despite its varied and complex contaminants.

The rate constants of first-order Reaction Kinetic models (see Section 8.3 in Chapter 8) for any CoC may be expressed as either their area-based

* Figure 1.4 in Chapter 1 shows a sketch of this pilot unit.

FIGURE 6.4
A two EEB cell test unit.

(k_A, usually expressed in units of m per year) or volume-based equivalents* (k_V or simply k, usually expressed in units of day^{-1}).

An EEB can be modelled in several ways: (1) by assuming that hydraulic performance in it is indicated by a *plug flow reactor* (PFR models, see Section 8.4 in Chapter 8); (2) by a *continuously stirred reactor* (CSR models, see Section 8.5); (3) by *Tank-In-Series* (TIS) models (see Section 8.6); (4) by combinations of these models; or (5) by more complex models.

Given defined conditions (i.e., a defined wastewater flow rate [Q in m^3 per day]; a set of worse case influent CoC concentrations [C_i in mg per L]; a set of discounted target effluent concentrations [C_o in mg per L]; a measured substrate porosity [ε]; a known water depth in the test cell [h in m]; and the cell surface area [A in m^2]), the pilot unit results can be used to derive for the rate constant (k in d^{-1}] for any CoC at the particular test temperature. For example, the rate constant for the removal of any wastewater CoC (e.g., ammonia nitrogen, BOD, COD) in a CSR is given (Kadlec & Knight 1996) by

$$k_{CSR} = Q.(C_i/C_o - 1)/(\varepsilon.h.A) \quad (6.1)$$

TIS models assume that natural WWT systems such as EEBs can be simulated by a number of CSRs in series. For TIS models, the rate constant for any CoC treated in an EEB can be given by

* See Table 8.2 in Chapter 8 for a spreadsheet showing both. Volumetric rate constants are increasingly being used in design. The relationship between the areal and volumetric rate constants is shown in Equation 8.8 in Chapter 8.

$$k_{NTIS} = N.Q.\left((C_i/C_o)^{1/N} - 1\right)/(\varepsilon.h.A) \tag{6.2}$$

where N is an integer (usually between 1 and 5) representing the number of tanks in series needed to accurately model an EEB.* For example, larger FWS CWs can often be sized using the TIS model and assuming an N of 3 (i.e., the terminology used is that the wetland is able to be accurately simulated by a 3TIS model).

As was mentioned, exactly what model best defines the particular wastewater/situation can be determined using a tracer test. For example, for the BBRs of the EEB System project at BNIA, the tracer test carried out as part of the treatability test (see Section 6.7) indicated that a CSR model (Equation 6.1) had the best fit for the results, although a slightly more conservative 2TIS model was used for design. (Reaction Kinetic models are discussed in more detail in Section 8.3 in Chapter 8.)

The value of any rate constant will vary with temperature, loading and oxygen level, but these parameters can be controlled in a test cell. Once a rate constant is determined during a treatability test, it can be used to determine the minimum area of a full-scale EEB treating the same feedstock under worse case conditions using the same substrate.

Usually, for pilot-scale treatability testing, the feed rate (Q) of a quantity of the actual feedstock or of a made-up, synthetic feedstock is set at from 50 to 500 L/day, depending on the loading desired. Generally, pilot-scale treatability tests take from 3 to 6 months to complete.

Where multiple contaminants are involved, the equations have to be solved for each of the CoCs at each of the test temperatures, and those for the most extreme (worse-case) conditions are used for design. Generally, for reporting purposes, rate constants are adjusted to their values at 20°C using Arhennius coefficients (thetas) also determined during treatability tests carried out at different temperatures (see Section 8.9 in Chapter 8).

The kinds of results obtained during early BBR treatability testing (Higgins 1997, 2000b; Higgins et al. 1999, 2006c; Kinsley et al. 2000) can be illustrated by some of those obtained for the nitrification of ammonia (see Section 2.4.3 in Chapter 2). Two main feedstocks were tested during the *BREW Project:* septic tank overflow from facilities at the central test site, and an imported, very highly contaminated leachate from the then operating Keele Valley municipal landfill north of Toronto, ON, Canada. Table 6.1 shows average ammonia nitrogen removals (averages but not showing standard deviations) during the *BREW Project*, and compares them† to those

* N is not the number of cells in an EEB System, but instead is a mathematical parameter that allows the fitting of the model to actual data.

† These data are for early treatability tests when BBRs were still referred to as Engineered Wetlands, and when it was still felt necessary to vegetate the test cells with wetland plants. While the parameters determined during these treatability tests are still valid, it should be noted that the EEB ecotechnology has evolved a long way since they were carried out, and later treatability tests did not bother with or need vegetated cells.

from treatability tests for treating leachate from a closed Maidstone municipal–commercial landfill in Western Ontario (see Section 6.4); raw sewage from WWT lagoons in the Town of Alexandria (low and high temperature "runs," see Section 6.5); and ammonia- and cyanide-contaminated water from the tailings pond at the facilities of RGM in Suriname, South America (see Section 6.6).

TABLE 6.1

Early Pilot-Scale Treatability Tests Involving BBRs – Ammonia Nitrogen Removals

Project	Average Temperatures (°C)	Avg. Influent NH_3-N, (mg/L)	Avg. Effluent NH_3-N, (mg/L)	Removal (%)
BREW Project, septic	15–20	73.0	1.0	98.6
BREW Project, leachate	15–20	713	76	89.3[a]
Maidstone leachate	16	51.4	1.0	98.0
Alexandria high temp	25	9.3	0.2	97.5
Alexandria low temp	6	14.7	0.4	97.0
RGM Reclaim Water	26	14.7	0.1	99.4

[a] For the Keele Valley leachate, part of the effluent from a three-cell EEB pilot test unit was recycled to feed, and the results of Table 6.1 for its first BBR cell reflect a 1.5:1 recycle rate.

As may be seen from Table 6.1, for NH_3-N influent concentrations of less than 100 mg/L, removals in HSSF BBR test cells were above 97%. Even with the very high ammonia nitrogen Keele Valley MSW landfill leachate (>700 mg NH_3-N per L), over 89% removal was obtained.* The nitrification results of Table 6.1 may be compared with the 30%–70% ammonia nitrogen removals typical of ordinary, unaerated SSF CWs (Kadlec & Wallace 2008).

The volumetric rate constants at 20°C for the oxidation of ammonia nitrogen (nitrification) in pilot-scale BBRs obtained during the various treatability tests of Table 6.1 are listed in Table 6.2. As may be seen from Table 6.2, not unexpectedly it is easier (higher k's) to treat domestic wastewaters and Reclaim Water than more highly contaminated leachates.†

As has been discussed, testing using pilot- and even demonstration-scale test facilities is not to be limited to EEB cells, but may (and often does) include upstream and downstream treatment cells and other components. This allows the test units to better simulate the situation to be expected for

* The diluted wastewater feedstock for the BREW Project leachate test contained an average ammonia nitrogen concentration of 713 mg/L, so its slightly lower removal percentage was not surprising. During this test, a second BBR cell in series reduced the effluent concentration to the same levels as achieved in the other tests.

† In comparison, for an ordinary HSSF CW, a widely accepted literature value of the PFR model ammonia nitrogen areal rate constant for treating municipal sewage is 34 m/year or about 0.3 d^{-1} (Kadlec & Knight 1996), so it can be seen that ammonia conversion is approximately an order of magnitude higher in BBRs.

TABLE 6.2

Ammonia Nitrogen EEB Volumetric Rate Constants @20°C (per day)

Project	k_{PFR}	k_{CSR}	k_{2TIS}
BREW Project septic	3.2	53.1	7.0
BREW Project leachate	1.5	5.6	2.2
Maidstone leachate	1.5	18.4	4.5
RGM Reclaim Water	1.4	34.2	5.7
Alexandria sewage	3.1	35.7	5.6

a subsequent full-scale EEB System. For example, the on-site treatability test unit used at RGM (see Section 6.6) included both a simulated Aerated Lagoon cell* before its two BBR cells, and a simulated Swamp-Forest Natural Wetland cell† after them.

Another example is the two demonstration-scale EEB Systems located outside beside the greenhouse at CAWT's facilities at Fleming College (see Figure 6.5), where the VSSF BBR cells have both primary and tertiary treatment components before and after them.

6.3 EEB Test Facilities

There were two central locations in Ontario where EEB treatability testing took place. One closed in 2015, while the other is still in operation.

The first was located at Campus D'Alfred of the UoG (*Alfred College*) in Alfred, ON, Canada (the *Alfred Test Unit*). Bioreactor test facilities there were located in a heated building. From 1998 until 2014, EEB treatability testing was carried out at Alfred College by ETDC, Nature Works, the *Ontario Rural Wastewater Centre* (ORWC) and Jacques Whitford (later Stantec). However, the UoG decided to withdraw from further involvement with Alfred College, and that college is now closed.

The second central EEB test site is located at the facilities of CAWT (the *CAWT Test Unit*) and still operates. Since 2006, pilot- and demonstration-scale EEB testing by CAWT has been carried out there by ETDC, Nature Works, Jacques Whitford (now Stantec) and others.

* Indeed, data from this pilot-scale test cell were used to design a full-scale Aerated Lagoon that was later constructed beside the mine's mill.

† It was intended that treated Reclaim Water from the EEB System be discharged into a local Swamp-Forest Wetland adjacent to the mine's mill for polishing before discharge into the receiving waters.

The *Centre for Alternative Wastewater Treatment* was founded in 2002 when the Canada Foundation for Innovation (CFI) awarded funds to Fleming College for its creation. The Ontario Innovation Trust matched the CFI funding and the college made additional contributions. CAWT's facilities were built as part of a new Environmental Technology wing at the Frost Campus of the College in Lindsay, ON (part of the City of Kawartha Lakes). Construction took place from 2003 to 2004 and the facility became fully operational during the 2004–2005 academic year. Under the CFI application, the goal of the proposal was to establish CAWT in a manner that would permit applied research in five key areas:

- The performance of constructed wetlands in cold climates with the objective of developing systems for small or isolated communities and specific industrial applications
- The development of effective constructed wetland systems for water treatment and reuse for the Canadian aquaculture industry
- The design of low-cost/minimal maintenance natural water treatment solutions for developing countries
- To investigate the ability of innovative wetland treatment system designs to remove problematic environmental contaminants such as cyanides, heavy metals, pathogenic bacteria, organic contaminants and excess nutrients
- To carry out R&D on on-site wetland treatment systems for individual homes

Since 2005, CAWT has secured over $15 million in resources, over 100 industry partnerships, and expanded applied research activities into many ecotechnologies and into technology development and commercialization services.

Pilot-scale EEB R&D and treatability testing at CAWT are usually carried out inside a large greenhouse there that has in it large environmental chambers and a walk-in refrigerator, allowing tests to be conducted over a range of temperatures.

At the CAWT Test Unit, there is also a much larger, two-train, demonstration-scale EEB System facility outdoors beside the greenhouse that has also been used for various tests.

Figure 6.5 illustrates this outdoor test unit.

As currently configured (early 2017), each train of the demo unit shown in Figure 6.5 involves one downflow VSSF BBR cell (middle) preceded by one HSSF CW cell (which can, if desired, be operated in an EEB mode and can be subdivided into three cells), and followed by a tertiary treatment engineered substrate cell. Effluent from both trains enters a small, polishing FWS CW (pond) cell before disposal.

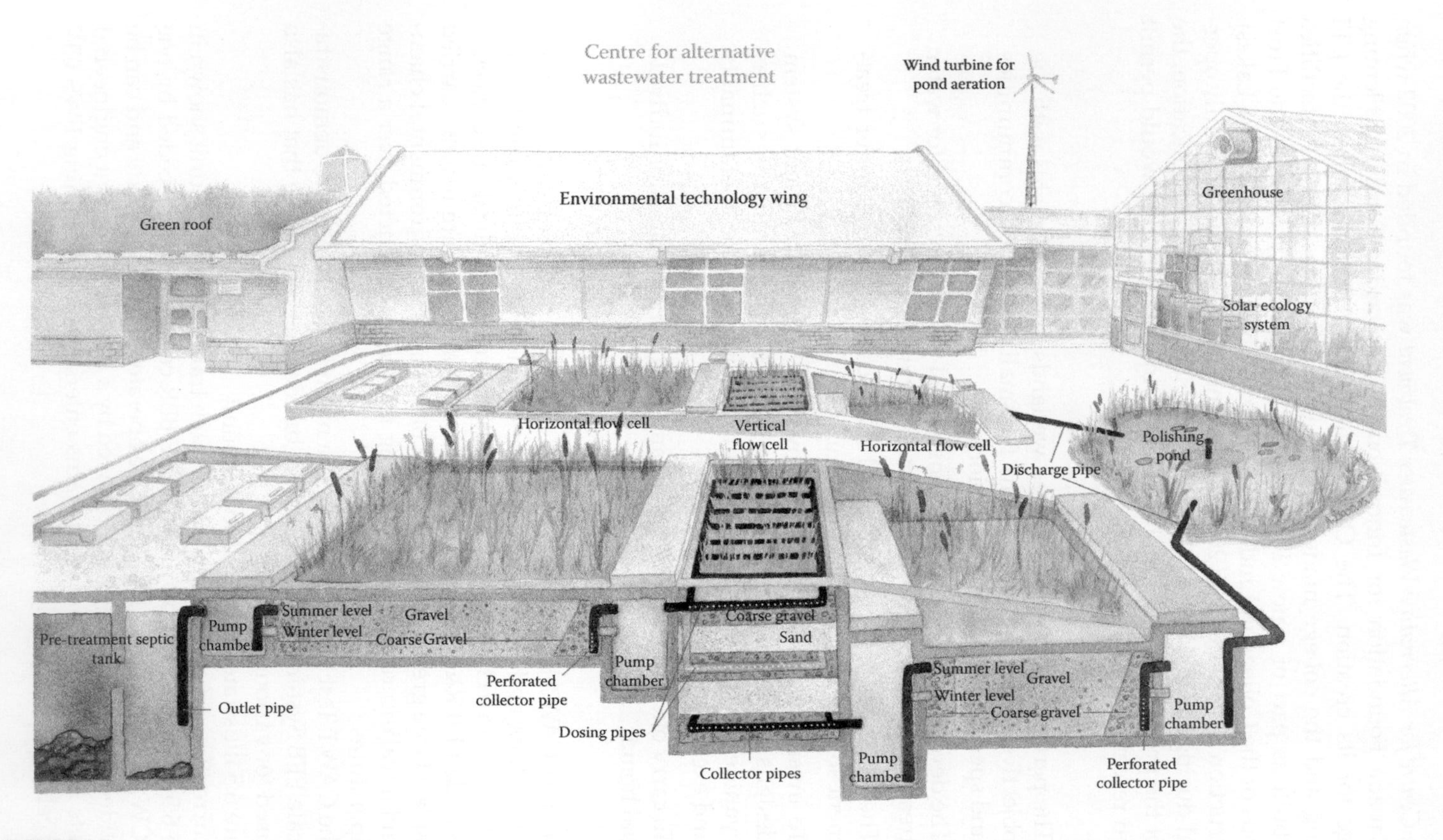

FIGURE 6.5
A two-train demonstration unit at Fleming College.

CAWT's facilities are complemented by an accredited on-site laboratory with advanced analytical instruments capable of carrying out a suite of organic, inorganic and biological tests.

At the CAWT Test Unit, pilot- and demonstration-scale EEB System test capabilities are augmented by a number of large outdoor ponds that can be converted to test cells for specific projects.

6.4 EWSWA Maidstone Landfill Leachate Treatability Testing

The *Essex-Windsor Solid Waste Authority* (EWSWA) is responsible for waste management activities for the City of Windsor and surrounding communities in southwest Ontario, Canada. Among the landfills it operates, the Essex Landfill complex #3 in the Township of Maidstone has at it an old municipal/commercial landfill, the *Maidstone Landfill*, that was closed in 1997. This landfill continues to generate approximately ~200 m^3/day of saline leachate that has to be trucked to a local municipal WWTP for disposal. The main CoCs (in 1998, 1999) were ammonia nitrogen (60–300 mg/L), COD (180–1000 mg/L), BOD (10–120 mg/L), TSS (10–300 mg/L), chloride (1000–4,000+ mg/L), iron (0.4–10 mg/L), boron (2–5 mg/L) and minor amounts of Al, Cd, Cr, Co, Cu, Ni and Zn. pHs ranged from 6 to 9 (Higgins et al. 1999; Higgins 2000a).

In 2000, Jacques Whitford was contracted by EWSWA to explore the potential of constructing an EEB System to treat the leachate on-site by the landfill so that it could be discharged to local stormwater ditches, eliminating the ongoing cost of the trucking. It was determined that an off-site treatability test would be carried out at the Alfred Test Unit to assess the treatability of the Maidstone leachate. For the treatability test, 10,000 L of the leachate was collected and transported to Alfred College where it was stored in a large plastic tank (see Figure 6.2).

A pilot-scale test unit was constructed there consisting of a preliminary HSSF CW cell piped to allow flow in series or parallel with a similar cell, but one which was aerated by a small blower providing air into the bottom of 1/4″–3/4″ gravel substrate 0.8 m thick via porous irrigation tubing in the bottom of the bed (i.e., a BBR cell, then called an EW cell). The third cell was filled with steel slag, 0.8 m thick. All three cells were vegetated with Reeds (*Phragmites* spp) transplanted from local ditches. The cells were acclimatized for 3 weeks with salty tap water and inoculated with raw sewage from a local WWTP prior to usage to treat the imported leachate. Treatment continued for 3 weeks at an influent flow rate of 100 L/min (Higgins et al. 2006b).

The results for ammonia removal during the Maidstone landfill leachate treatability test are shown in Tables 6.1 and 6.2 for treatment in the pilot unit's BBR cell where the nominal residence time was 2 days. Average influent TN and COD concentrations during the test were 115 and 320 mg/L, respectively, and the latter were reduced by only 20% on passing through the BBR cell, indicating that most of the organics in the leachate were relatively recalcitrant and not amenable to biodegradation. Chloride concentration in the feedstock was 3900 mg/L – brackish rather than fully saline – and did not change during the testing. Neither did the boron concentration. As is noted in Table 6.1, the NH_3-N concentration fell from 51 to 1 mg/L, and at the same time the NO_3-N concentration rose from 39 mg/L in the influent to 123 mg/L in the BBR cell's effluent.

The Maidstone landfill treatability test demonstrated that 98% of the ammonia could be removed from the leachate in an aerated (BBR) cell, and that moderate leachate salinity did not adversely impact CoC biodegradation.

6.5 Alexandria Pilot-Scale Treatability Testing

The Township of North Glengarry in Eastern Ontario, Canada owns and operates lagoon-based municipal WWT facilities at the Town of Alexandria in Eastern Ontario, Canada (Alexandria). A systems analysis (see Section 8.1 in Chapter 8) (Higgins et al. 2006c) was carried out to evaluate the potential of upgrading of the Alexandria WWT facilities to an EEB System, the Conceptual Design of which was envisaged to involve an Aerated Lagoon (see Section C.4 in Appendix C) followed by three Cattail-vegetated, VSSF BBR cells (then called EWs) in parallel.

The upgrade was also envisaged to involve sludge collection from the Aerated Lagoon and its treatment in four Reed Bed Biosolids De-Watering Wetland cells (see Section B.5 in Appendix B). Other facilities for the EEB System were to include inlet screening, a Sedimentation Pond (see Section C.2 in Appendix C), a balancing pond and effluent UV disinfection equipment. As part of a systems analysis, a treatability test was carried out at the Alfred Test Unit.

This controlled, indoor, off-site treatability test was carried out at both a high design basis temperature (25°C) and a low design basis temperature (6°C), values that were typical of sewage temperatures in summer and winter at Alexandria.

The purpose of the treatability test was to complement Conceptual Design and to provide Reaction Kinetic data and other scale-up parameters needed to carry out the Detailed Design and Engineering of the envisioned full-scale EEB System.

For the treatability test, the Alfred Test unit was configured as a 5000 L storage tank for holding raw sewage (imported by tanker from the sewage lagoons at Alexandria); an inlet Mixing Tank; one aerated gravel-bed EW (BBR) cell; a small air blower to provide aeration air to the bioreactor; an automatic sampler, refrigeration equipment that could be assembled around the BBR cell in order to carry out low temperature testing; and a grow light system (see Figures 6.2 and 6.3). The BBR cell was vegetated with transplanted common Reeds collected from roadside ditches nearby.

Both the Mixing Tank and the BBR cell had internal heaters so that treatment temperatures could be set for higher temperature runs.

Some of the objectives for the Alexandria treatability test were (1) to define the treatability of raw sewage using a saturated, downflow VSSF BBR (as earlier BBR testing had been carried out using HSSF test cells); (2) to evaluate the hydraulics of this VSSF BBR cell; (3) to determine whether minor contaminants in the sewage feedstock affected the microbial degradation of ammonia; (4) to define initial kinetics for the removal of contaminants (especially BOD and ammonia) at low and high design basis temperatures; and (5) to carry out a tracer test to determine the Reaction Kinetic model and expected residence time distribution of wastewater for subsequent full-scale facilities.

During the treatability test, hydraulic loading rates (HLRs, see Section 8.2 in Chapter 8 for parameter definitions) for the pilot unit's BBR cell ranged from 32.8 to 33.5 cm/day and average hydraulic retention time (HRT) was about 1.2 days.

After an initial acclimatization and calibration period, the pilot unit was operated for roughly 5 months over the summer of 2005, 24 h a day, 7 days a week. The grow lights were turned on between 6 a.m. and 6 p.m. During the treatability test, treated effluent samples were also taken for testing UV transmissibility (needed to design UV disinfection equipment for full-scale facilities); to test toxicity to trout and *Daphnia*; (there was none for the effluent) and to determine the amounts of nitrifying bacteria present.

Table 6.3 summarizes some of the results from this pilot-scale treatability test (Higgins 2000a).

TABLE 6.3

Alexandria Treatability Test Summary of Results (mg per L)

	High Temperature Runs		Low Temperature Runs	
Concentration	**Influent**	**Effluent**	**Influent**	**Effluent**
Ammonia-N	9.27	0.24	14.67	0.44
Nitrate-N	2.7	17.0	3.4	18.6
TKN	13.5	2.8	20.0	4.6
Org-N	4.2	2.6	5.4	4.2
DO	3.4	5.7	7.1	7.0
cBOD	16.0	4.7	13.8	3.0
TSS	26.7	11.4	29.7	8.9
Alkalinity	209	147	183	135

Volumetric removal rates for the high and low temperature results were calculated based on PFR, CSR and two tanks-in-series (2TIS) models.

Tables with the calculated average volumetric removal rate constants for treating imported Alexandria sewage are provided in Tables 6.4 and 6.5.

TABLE 6.4

Calculated Alexandria Volumetric Rate Constants at High Temperature (25°C)

Rate Constants	$k_{25°C}$, PFR (per day)	$k_{25°C}$, CSR (per day)	$k_{25°C}$, 2TIS (per day)
Ammonia (mg/L)	3.3	38.5	10.0
cBOD (mg/L)	1.1	2.1	1.5

TABLE 6.5

Calculated Alexandria Volumetric Rate Constants at Low Temperature (6°C)

Rate Constants	$k_{6°C}$, PFR (per day)	$k_{6°C}$, CSR (per day)	$k_{6°C}$, 2TIS (per day)
Ammonia (mg/L)	3.0	29.4	8.4
cBOD (mg/L)	1.3	3.1	1.9

From the above data, the Arrhenius constants (thetas) for the ammonia nitrogen and cBOD rate constants can be calculated to be 1.02 and 0.99, respectively.*

The Alexandria treatability test demonstrated that aerated VSSF BBRs can successfully treat raw sewage to very low pollutant concentrations at very high hydraulic loading rates, and work as well as or better than HSSF ones. Ammonia nitrogen in the wastewater was removed to almost non-detect levels at both the high and low design basis temperatures, and the rate constants found were very much higher than those achievable in ordinary CWs. It was also found that while BBR cells themselves provide some degree of disinfection for municipal wastewaters being treated in them, they also produce clear effluents that respond very well to downstream UV disinfection if such is mandated.

From the Alexandria treatability test, the basic concepts were developed that evolved into today's successful designs for BBRs and SAGR Bioreactors.

6.6 Rosebel Pilot-Scale Treatability Testing

6.6.1 Background

Rosebel Gold Mines NV (RGM)† operates an open pit gold mine (the Mine) with an associated mill (the Mill) in the Amazon jungle in Suriname, South America.

* Kadlec and Knight (1996) suggest Arrhenius constants for ordinary CWs of 1.04 and 1.0 for ammonia nitrogen and BOD, respectively.

† Originally, the majority interest in RGM was held by Montreal-based mining company Cambior Inc., but this firm was later purchased by Toronto-based Iamgold Corporation. A minority interest in RGM is held by the government of Suriname.

Gold is extracted from crushed and milled ore using sodium cyanide (see Sections 2.3.4 and 2.6.3 in Chapter 2), and tailings are disposed of in a large tailings pond (the Tailings Pond). An average of 650 m^3/h (15,600 m^3/day) of recycled water from the Tailings Pond (Reclaim Water) is pumped into a large Treated Water Storage Pond (TWSP) from which it can be reused in the Mill to slurry more crushed and milled gold ore from the Mine's several pits, or discharged periodically (during high flow periods) to the receiving waters, a small local stream, the Mindrineti River (Higgins 2006; Higgins et al. 2006c). However, the Reclaim Water contains contaminants that may inhibit its reuse or disposal.

The most important CoC in the Reclaim Water at the Mine is ammonia and this species results from the photodissociation of some of the residual cyanide in the tailings (leftover milled ore after gold removal – 99.9%+ of the original ore) in the Tailings Pond and from the influx of residual ammonium nitrate (from rock blasting in the Mine's several pits). The Reclaim Water also contains suspended solids and traces of metals and cyanide complexes (see Section 2.3.4). At the Mill, Reclaim Water is first treated in an existing active WWTP, the effluent treatment plant (ETP). In the ETP, suspended solids are precipitated out using ferric sulfate and lime using a large clarifier. Sludge from the ETP clarifier is pumped back to the Mill, while the ETP's output stream (ETP Effluent) is gravity discharged to the TWSP.

RGM contracted Jacques Whitford (later, after that firm's acquisition, its contracts were with Stantec) to proceed with a project to design and engineer an EEB System to further treat the ETP Effluent to remove the ammonia and other contaminants to very low levels so it could then be reused in the Mill to slurry fresh gold ore without adverse consequences (i.e., negative impacts on the Mill's gold cyanidation process), disposed of to the environment (a small river) without exceeding the mandated discharge criteria and/or be safely put to other uses (e.g., firewater, vehicle washing, sample bag washing). The undesirable contaminants in the ETP Effluent are small amounts of suspended solids (<5 mg TSS per L), ammonia (up to 15 mg NH_3-N per L, but usually somewhat lower), dissolved cyanides (up to 2 mg/L, mostly strong acid dissociated, or SAD cyanides) and small amounts of dissolved metals (see Section 2.3.5 in Chapter 2 for information on cyanide chemistry).

To define the optimal treatment, a systems analysis involving both on- and off-site treatability testing (including tracer testing) was first carried out to define scale-up and other design parameters needed to design the full-scale Aerated Lagoon and BBR cells (Higgins et al. 2006c).

The EEB System was designed to precede the TWSP and involve an Aerated Lagoon in series with BBR cells as well as the usual associated infrastructure (e.g., control structures, splitter boxes, pumps, air blowers, piping, valves and fittings, instrumentation, power supplies, housing for pumps, and blowers).

For this project, two separate pilot-scale treatability tests were carried out: a preliminary, off-site test at the Alfred Test Unit, and a subsequent on-site treatability test at the Mine in South America.

6.6.2 Off-Site Phase 1 Treatability Testing at the Alfred Pilot Unit

The first treatability test was carried out indoors at the Alfred Test Unit using: (1) a synthetic feedstock simulating a typical Reclaim Water and containing worse-case levels of the most important CoCs (the bulk of the test) and (2) imported Reclaim Water (one tote full of actual Reclaim Water was shipped from the Mine to the college).

Figure 6.6 shows the layout of the indoor, off-site pilot unit.

FIGURE 6.6
RGM Reclaim Water off-site treatment test unit.

As may be seen, the RGM pilot unit at the Alfred Test facilities involved a single, Reed-vegetated BBR cell plus the usual ancillary equipment as described in Section 6.2. The purposes of this first test were (1) to define proof-of-concept (i.e., that RGM Reclaim Water can be successfully treated in an EEB System), (2) to assess the basic treatability of the Reclaim Water (i.e., would contaminants in Reclaim Water inhibit treatment in a BBR?) and (3) to determine Reaction Kinetic rate constants for the most important CoC, ammonia, in such water treated in a BBR.*

As part of this first treatability test, the equipment for the subsequent second (on-site) treatability test also was accessed, assembled, tested and placed in a shipping container, and shipped to the Mine where it was reassembled for use.

* During this off-site treatability test at the Alfred Test Unit, Prussian Blue (see Section 2.3.4) was added to its synthetic feedstock in an attempt to provide cyanides in it so that cyanide degradation rate constants could also be measured, but this was not successful and these rate constants were not determined until the subsequent on-site testing where actual cyanide-contaminated streams from the Mine were the feedstocks.

The off-site test system at Alfred College was operated at an average temperature of 25°C, with hydraulic loading rates of 12.0–14.7 cm/day and nominal hydraulic retention times ranging from 2.8 to 3.5 days, both with and without aeration. For most phases of this off-site testing at the Alfred Test Unit, the synthetic feedstock was used, but for one phase of operation (Run C) the pilot unit was operated with the actual Reclaim Water from a tote of it imported to Canada from the Mine. The off-site pilot unit at Alfred College was operated with and without aeration to simulate operation with a BBR cell and an unaerated SSF CW cell. The operational phases of the off-site treatability test are summarized in Table 6.6.

TABLE 6.6

Test Runs during the Off-Site RGM Treatability Test at Alfred College

Run	A	B	C	D
Dates (2005)	15 Feb–11 Mar	11 Mar–18 Mar	23 Mar–29 Mar	29 Mar–6 Apr
Hydraulic loading, cm/day	14.7	14.7	12.0	12.0
Feedstock	Synthetic	Synthetic	Imported	Synthetic
Aeration	On	Off	On	On
Temperature, °C	24.0	24.9	24.3	25.3

Flow rates during the off-site treatability test at the Alfred Test Unit ranged from 139 to 160 L/day.

Rate constants for observed ammonia nitrogen removal were calculated for each of the four runs. Rate constants for ammonia were temperature corrected to 20°C using a theta (θ) factor of 1.02, as calculated from the Alexandria data set. Results are summarized in Table 6.7.

TABLE 6.7

Ammonia Nitrogen Removal Rate Constants for the RGM Off-Site Test

Run	Feedstock	Aeration	k_{2TIS} (per day)
A	Synthetic	On	4.56
B	Synthetic	Off	0.52
C	Actual	On	5.54
D	Synthetic	On	7.02

On runs where aeration was employed (Runs A, C, D), the 2TIS ammonia nitrogen removal rate constant averaged 5.7 per day (Kinsley et al. 2002). In contrast, without aeration (Run B), the rate constant dropped to 0.5 per day. For comparison, a standard area-based rate constant of 34 m/year from the literature (Kadlec & Knight 1996) yields an equivalent 2TIS volumetric constant of approximately 0.3 per day, which is comparable to the non-aerated rate observed in Run B. (These results for ammonia nitrogen treatment in a BBR can be compared with those during the treatment of other wastewaters in Table 6.2.)

The initial, off-site treatability testing at the Alfred Test Unit was highly successful, showing that: (1) the concept of using an aerated VSSF BBR to treat ammonia- and cyanide-contaminated waters from the tailings pond at RGM's Mine was indeed feasible, (2) aerated VSSF BBRs allowed just as much removals of oxidizable contaminants such as ammonia as did the aerated HSSF ones that had been tested in earlier treatability tests, (3) very high (>99%) levels of ammonia removal were possible with aerated VSSF EWs and (4) soluble cyanide species in simulated and actual Reclaim Waters did not seem to inhibit ammonia nitrification. On the basis of these indoor test results in Canada, later, outdoor on-site pilot-scale testing was carried out at the Mine in South America (Higgins et al. 2006c).

6.6.3 On-Site Phase 2 Treatability Testing at RGM

The on-site test unit consisted (see Figure 6.7) of a Mixing Tank (right bottom), an open, water-filled aerated tote (simulating an Aerated Lagoon), two BBR cells in series and a final FWS CW cell. (Feed tanks and empty totes are not shown in this picture.)

FIGURE 6.7
RGM Reclaim Water on-site treatment test unit.

The BBR cells were vegetated with local Cattail plants (*Typha domingensis*) harvested from the peripheries of the nearby TWSP. The final FWS CW cell was vegetated with biodiverse local wetland plants harvested from a local Swamp-Forest Natural Wetland nearby (and discharging to) the Mindrineti River, as it was intended that treated water from a full-scale EEB System that was not needed in the Mill would be discharged to that wetland where it would be further polished before entering the river.

Figure 6.7 shows this second on-site test unit reassembled and installed outdoors near the TWSP at a location beside the ETP at the Mill. (Part of the ETP's clarifier is shown on the right.)

During this subsequent *on-site outdoor treatability test* at the Mine, the first BBR test cell (EW Cell #2) was tracer tested using sodium bromide and was observed to operate as one completely stirred reactor (CSR) (Higgins et al. 2006c) as is shown in Figure 6.8.

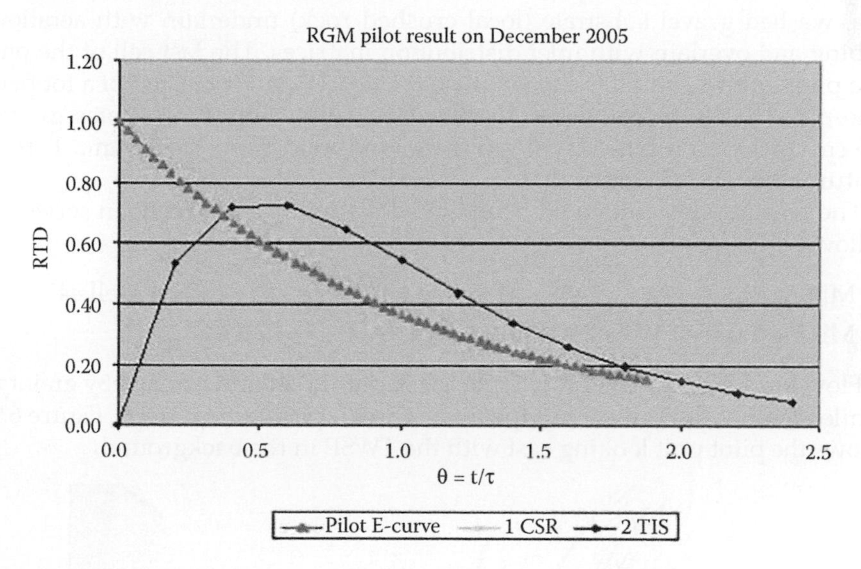

FIGURE 6.8
Results of RGM off-site treatability test.

As may be seen from the above figure (which shows normalized results, see Section 1.8), the BBR cell can be modelled by a CSR. However, since this degree of internal mixing might not occur in full-scale BBRs with thicker gravel substrates, conservative design is to use a 2TIS model for scale-up, and a 2TIS curve is shown in the above figure for comparison.*

In addition to the tracer testing, the purposes of the subsequent on-site treatability test at the Mine in Suriname were (1) to confirm that the results of the off-site, proof-of-concept pilot-scale testing at the Alfred Test Unit would also occur on-site with actual Reclaim Water being treated, (2) to pilot-test the efficacy of a proposed Aerated Lagoon which would precede BBRs in a full-scale system, (3) to test outdoors aerated VSSF cells vegetated with the locally growing Cattails and to assess the growth of the plants in them, (4) to test the

* The indoor tracer test in the Alexandria treatability test also indicated a CSR but such is not always the case for modelling BBRs, as may be seen from Figure 1.4 in Chapter 1.

efficacy of the Swamp Forest Natural Wetland (SFNW) to polish effluent from the EEB System and (5) to obtain rate constants for other CoCs that were not determined during the earlier treatability test at the Alfred Test Unit in Canada.

The on-site pilot unit for RGM was designed so that feedstock (either the ETP Effluent or the Reclaim Water) could flow to the two BBR cells (EW* cells #2 and #3) in series either with or without passing first through the preliminary Aerated Lagoon cell (AL cell #1) which consisted of a topped, water-filled tote (water 80 cm deep) with a weighed, diffused air grid placed in its bottom. Aeration air for the two BBR cells was taken from ETP's compressed air system.

The cells of the on-site pilot unit were filled with about 0.90 m^3 of 2.5 cm dia. washed gravel substrate (local crushed rock) underlain with aeration tubing and overlain with inlet distribution matrices. The last cell of the on-site pilot unit was intended to simulate the SFNW and consisted of a topped tote, cut in half and provided with three layers of saturated substrates: gravel (15 cm thick), saprolytic clay (5 cm thick) and mud from the Swamp Forest Natural Wetland (5 cm thick).

The on-site pilot unit could be operated as three or four cells in series as follows:

Mixing Tank→AL Cell #1→EW Cell #2→EW Cell #3→SFNW Cell #4

Mixing Tank→EW Cell #2→EW Cell #3→SFNW Cell #4

Flow from the second BBR cell into the Natural Wetland cell was by gravity, while all other flows were pumped using small peristaltic pumps. Figure 6.9 shows the pilot unit looking east with the TWSP in the background.

FIGURE 6.9
On-site pilot unit at RGM looking east.

* At this point, all BBRs were still referred to as Engineered Wetlands.

On-site pilot unit operations were categorized into four periods (runs) as follows, and the averaged data from these periods were selected for analyses:

Run	Feedstock	Cells in Operation	Dates
E	Reclaim Water	All four	May 18–July 12, 2005
F	ETP Effluent	All four	August 2–November 15, 2005
G	Reclaim Water	All four	November 16–December 13, 2005
H	Reclaim Water	3 (Cell #1 Bypassed)	December 27, 2005–January 17, 2006

The pilot unit did operate during the intermediate periods between the Runs E & F and G & H, and data were collected, but for various reasons (e.g., switching feeds and changes in the location of the Reclaim Water pump at the Tailings Pond) results were not used in the correlation of results presented herein. The average feedstock flow rate was 143 L/day and average nominal residence time in the EW (BBR) cells was 2.7 days at a loading rate averaging 13.2 cm/day. pHs ranged from 8 to 9. Reclaim Water influent TSS concentrations were 20–30 mg/L.

Water samples taken at the pilot unit were analysed for a variety of parameters at either the RGM laboratory at the Mill and/or sent offsite for analyses at the SGS Lakefield laboratory in ON, Canada. Parameters analysed for included ammonia (NH_3-N), total cyanide (CN_T), WAD cyanide (CN_{WAD}), free cyanide (CN_{free}), nitrite (NO_2-N), nitrate (NO_3-N), suspended solids (in mg TSS per L), alkalinity (in mg $CaCO_3$ per L), key ions such as Na, K, Ca, Cl, SO_4 and TDS (in mg per L), and Ni, Pb, Cu, Zn and Fe metal concentrations (in mg per L).

Samples of feedstock and effluents from each of the cells were analysed/sent offsite to be analysed twice a week for alkalinity, ions and metals, pH, NH_3-N, CN_T, NO_3-N, NO_2-N and conductivity. Water samples were taken from the same points where flow rates are measured. Sampling at the on-site pilot unit occurred at roughly the same time every day. SAD cyanide (CN_{SAD}) results were obtained using Equation 2.20 in Chapter 2.

Despite the outdoor conditions and different vegetation (Cattails instead of Reeds) from those of the off-site pilot testing, ammonia nitrogen results for the on-site testing were similar to those from the off-site pilot unit and ranged from 94% to 99% removals overall.*

Cyanides were mostly low in both the Reclaim Water (ETP influent) and the ETP Effluent, and were about half as much in the latter, although whether this reflects some additional removal in the ETP or some other variation is not known.

* Concerns about the measurement of ammonia nitrogen concentrations near the RGM laboratory's MDL for that species indicated that the more accurate ammonia nitrogen rate constants obtained earlier at the off-site treatability test at Alfred College would be used for design.

Owing to a change in ore type being sent to the Mill in mid-November of 2005, cyanide levels in the Reclaim Water rose dramatically. The new higher cyanide levels present in the Reclaim Water provided an opportunity to evaluate what an EEB System would do to such a wastewater and Runs G and H of the on-site TT were carried out with this new Reclaim Water.

The average total cyanide (CN_T) results for the on-site testing are presented in Table 6.8.

TABLE 6.8

Average Total Cyanide Concentrations during On-Site Tests (mg per L)

Run	Date	Feed Tank	AL Cell #1 Effluent	EW Cell #2 Effluent	EW Cell #3 Effluent	SFNW Cell #4 Effluent	Feedstock
E	18 May–12 July	0.082	0.038	0.019	0.012	ND	Reclaim
F	2 Aug–15 Nov	0.045	0.022	0.014	0.009	0.007	ETP
G	29 Nov–13 Dec	2.024	0.799	0.755	0.507	0.334	Reclaim
H	27 Dec–17 Jan	1.217		0.733	0.636	0.350	Reclaim

Removal % in		AL Cell #1	EW Cell #2	AL and EW Cells	Overall System
E	18 May–12 Jul	54%	50%	85%	88%
F	2 Aug–15 Nov	53%	51%	80%	84%
G	29 Nov–13 Dec	61%	6%	75%	83%
H	27 Dec–17 Jan		40%	48%	71%

Note: Run H was for EW cells # 2 and #3 + SFNW cell #4 only, and did not include the AL cell #1.

The first thing to note from Table 6.8 is that the aerated cell (AL cell #1) removed over half of the influent cyanides, and this occurred at both lower total cyanide concentrations in influent Reclaim Water (<0.1 mg/L, Runs E and F) and higher CN_T concentrations (1–2+ mg/L, Runs G and H). The BBR (EW) cells were less effective in removing the remaining cyanides, at best taking out up to half of what was left, either after the aerated cell or directly when it was not included in the process scheme (Run H). Small amounts of cyanides were removed in the simulated Swamp Forest Natural Wetland cell (SFNW cell #4) as well, suggesting that effective cyanide removal may require both aerobic and anaerobic conditions.

The BBR cells were very effective at removing suspended solids, getting very high removal percentages (>90% in most cases). Influent Reclaim Water TDS concentrations were 320 mg/L range during Run E and stayed about the same in the effluent from the aerated cell (AL cell #1). TDS levels dropped by 7% or so in the first BBR cell (EW cell #2), and by a similar amount in the second BBR cell (EW cell #3). The TDS levels then were about the same at the outlet of the last cell (SFNW cell #4). Run F results were approximately the same, although the influent values (~390 mg/L) were somewhat higher, reflecting the addition of some dissolved solids in the ETP.

The average drop in alkalinity between the inlet and outlet of the pilot unit for Run E was 30 mg/L (as $CaCO_3$).

The nitrification of ammonia consumes alkalinity: 8.64 mg of HCO_3- per mg of ammonia by stoichiometry (see Section 2.4.3). The average ammonia nitrogen removal in the on-site pilot unit was 4.85 mg/L (5.89 mg/L as ammonia), a value quite consistent with the drop in alkalinity (5.89×8.64 $\sim$51 mg/L as HCO_3-).

For Runs E, G and H, the iron levels of influent Reclaim Water (2–3 mg/L) drop significantly in the Aerated Lagoon cell (AL cell #1), the fall again by the outlet of the first BBR cell (EW cell #2), and then by smaller amounts in the second BBR cell (EW cell #3) and the Natural Wetland cell (SFNW cell #4).

During Run H when the aerated cell was not in the process train, total iron levels fell by 80% in the first BBR cell (EW cell #2), and by lesser amounts in the two remaining cells of the on-site pilot unit. In the case of Run F when the ETP Effluent was the off-site pilot unit's feedstock, the ETP already had removed much of the ETP Influent iron ($\sim$1 mg/L) iron, and that which was left was removed down to quite low levels (<0.2 mg Fe_T per L) in the pilot unit. Other potential SAD Cyanide complex ions (e.g., Ag, As, Co, Cr, Pb, Ni, Se) that had concentrations very much lower than iron to start with (only trace amounts were present) did not vary much at all across the cells of the pilot unit.

These data suggest that Fe-SAD CN complexes are more amenable to removal in an aerated environment than are other kinds of SAD CN complexes. This is fortunate, given that iron is a major CoC and its concentrations were orders of magnitude higher (up to a few mg per L) in the Mine's Reclaim Water than those of other metals that form SAD CN.

At the completion of the on-site testing at RGM, the second BBR cell was broken open and the roots of its cattail plants were compared with those of a cattail plant growing nearby in the shallow water periphery of the TWSP. Figure 6.10 compares the roots of that plant with one of those from the BBR.

As may be seen, the root zone of the cattail from the on-site pilot unit cell (right) at the Mine was much bigger and whiter-looking than the other (left), and its roots were longer ($\sim$0.8 m + vs. $\sim$0.4 m). The roots of the cattail from EW cell #3 extended downwards (towards the aeration header) while those of the control cattail from the TWSP were mostly horizontal and located near the wetland soil surface. Also, the rhizome of the cattail from the VSSF BBR cell was tilted vertically, while that of the control plant was oriented horizontally just under the soil surface. (These results are similar to ones obtained in column tests for NAWE by Matthys et al. 2000.)

In addition to the above, two entire cattail plants were collected and washed on March 10, 2006 for vegetation analyses. One was collected from EW cell #3 and was typical of those growing in the cell before it was cut open. Another was harvested from the local Swamp Forest Natural Wetland into which it was desired to discharge effluent from a full-scale EEB System.

FIGURE 6.10
Comparison of control and BBR cattail roots.

Both plants were collected with care to obtain all of their leaves and roots. Both were weighed and then shipped to SGS in Canada for analyses. At the laboratory, the collected vegetation was separated into roots and shoots, and then dried to constant weight in an oven at about 80°C for 48 h. Shoots and roots were analysed for TKN, nitrate-nitrogen, total phosphorus (TP), and copper (Cu) using standard methods.

Table 6.9 presents results of Cattail analyses.

TABLE 6.9
Cattail Analyses

		From BBR Cell #3			From SFNW		
Analysis		**Roots**	**Shoots**	**R/S**	**Roots**	**Shoots**	**R/S**
Wet Wt.	g	1612	542	3.0	566	202	2.8
Dry Wt.	g	213	257	0.8	59	31	1.9
TKN	%	0.54	0.45	1.2	0.69	0.78	0.9
NO_3	%	<0.05	<0.05	–	<0.05	<0.05	–
Cu	μg/g	25	5.3	4.7	19	7.2	2.6
TP	μg/g	480	120	4.0	320	440	0.7

The wet weight of the plant from the pilot unit was almost three times as much as the one from the SFNW, while the dry weights were over five times as much. These data probably reflect the more favorable environment in the pilot cell where the influent wastewater provided nutrients to the Cattails. The dry weight root-to-shoot ratio is an indicator of the health and nutrition status of a plant, and is unusually about 1:1. This was nearly the case (0.8) for the plant which grew in the pilot unit cell. However, the plant that grew in the Natural Wetland had a ratio of almost 2:1. The root–shoot ratio is known to increase if water is withheld from a rooting medium and this, or the fact that light intensity was probably less in the SFNW where trees and bushes provided some shading, was responsible for the difference. TKN's were slightly less in the pilot unit plant than those in the SFNW but the differences for these for the copper and nitrate nitrogen were roughly similar.

Total phosphorus was significantly higher in the roots of the plant from the pilot plant cell (root/shoot = 4) while roughly equivalent amounts were found in the Cattails that grew in the SFNW (root/shoot = 0.7). This suggests that, while the low levels of phosphorus that were present in the two feedstocks that flowed through the pilot unit (Reclaim Water and ETP effluent), it was not being taken up into the pilot plant cattail's aboveground tissues probably due to (1) competition with other ions present in the wastewaters and/or (2) competition for same resource by other organisms (e.g., the microbes of the much more pervasive biofilm in the pilot unit cell). Reduced plant growth (such as that of the Cattails from the Natural Wetland) and high phosphorus levels can induce deficiencies of Fe, Zn, Cu and Mn and this might have been the case for plants growing in the SFNW. (High levels of phosphorus also cause Cattails to crowd out other native plants as is often observed with their growth in stressed urban environments such as roadside ditches.)

As was noted above, from the results of the on-site testing at the RGM Mine, it was observed that the Aerated Lagoon cell of the pilot unit was behaving as a complete mix reactor and its removals could be described by the CSR model (see Section 8.5 in Chapter 8). Calculated volumetric rate constants for it ($k_{26°C}$) for ammonia nitrification were about 0.7 per day and using this and other scale-up data obtained (e.g., local precipitation and pan evaporation data), and prudently assuming that ammonia nitrogen conversion in it would be about 50% in a full-scale unit (compared to the 80% found during the pilot testing), a 4.5 m deep, 70 m × 150 m (surface area) full-scale Aerated Lagoon was constructed alongside the TWSP. This Aerated Lagoon now treats the ETP effluent (and camp sewage) prior to its discharge into the TWSP. Aeration air for the Aerated Lagoon is supplied by eight floating aerator–aspirators. Operations have been much as predicted.

As was mentioned, it was not possible to get accurate cyanide kinetic data during the Alfred College off-site test, but it was possible during the on-site testing and these can be used (Table 6.10).

TABLE 6.10

Calculated Volumetric Rate Constants for Total Cyanide (per day) in the First EW Cell during On-Site Treatability Testing

Run	k_{PFR}	k_{CSR}	k_{2TIS}	Feedstock
E	0.24	3.5	0.29	Reclaim Water
F	0.16	0.2	0.18	ETP Effluent
G	0.02	0.02	0.02	Reclaim Water
H	0.18	0.2	0.20	Reclaim Water

As may be seen by comparing the previous tables, Run G data are probably not indicative of total cyanides removal in a BBR as most of those cyanides amenable to aerobic transformation had already been removed in the Aerated Lagoon cell before its effluent entered the first BBR cell (EW cell #2). Accordingly, for the design of BBR to treat Reclaim Water contaminated with iron cyanide complexes, a k_{2TIS} value of about 0.2 per day could be considered but this is moot as design and sizing will be governed by ammonia not cyanides.

On the basis of the combined results of the off- and on-site treatability tests for treating RGM Reclaim Water, the required minimum size (floor area of all cells) of the full-scale BBR-based EEB System can be calculated. If is assumed that the design flow rate of the ETP effluent delivered to the BBR cells via the AL will be 15,600 m^3/day; that the pH of the water will be 8.5; that its temperature will be 26°C (the average water temperature at the Mine); that substrate in the BBR cells will be a crushed rock with a porosity equivalent to that of the gravel used in the on-site pilot unit (0.45); that the substrate in the EW (BBR) cells will be 1.5 m thick; that ammonia will enter the BBR cells from the Aerated Lagoon at a worse-case concentration of 7.5 mg NH_3-N per L (i.e., assuming a conservative 50% removal in the "upstream" Aerated Lagoon), and its effluent will contain 0.5 mg NH_3-N per L (100 ppb unionized ammonia); that the 2TIS volumetric rate constant for ammonia nitrogen removal (nitrification) at will be 5.7 per day and that the Arrhennius constant (theta) will be 1.02, the minimum area of the BBR cells will be about 2 ha as is shown in Table 6.11.

As was mentioned, PFR models represent the maximum size that an EEB might be, and in any case are not recommended for modelling aerated gravel bed ones such as BBRs (EWs). Alternatively, CSR models represent the minimum size for them but prudent design is to use a model that gives the next size up, the 2TIS model in this case. Here, given a water depth of 1.5 m, the minimum size for the BBR cells under the 2TIS model will be 20,376 m^2, suggesting two trains of two cells, each about 51 m × 100 m in floor area. RGM has sufficient heavy earth moving equipment to excavate the EEB System's Aerated Lagoon and BBR cells quickly and economically, and the saphrolytic (clayey) soil near the TWSP where the EEB System will be built precludes the need for liners on any of the cells.

TABLE 6.11

Ammonia Calculation Spreadsheet

Reclaim Water BBR								
Ammonia Calculation								
Ammonia nitrogen in influent $[NH_3\text{-}N]_i$	7.5	mg/L	Influent $[NH_{3\text{-}N}]_i$	9.1	mg/L	Unionized influent ammonia	9107	ppb
Target effluent ammonia $[NH_3\text{-}N]_o$	0.5	mg/L	Ammonia in effluent $[NH_3]_o$	0.62	mg/L	Unionized effluent ammonia	100	ppb
Nitrate nitrogen in influent $[NO_3\text{-}N]_i$	2.0	mg/L	Nitrate nitrogen in effluent $[NO_3\text{-}N]_o$	9.0	mg/L	Nitrate in effluent $[NO_3]_o$	39.8	mg/L
Influent feed rate	15,600	m³/day	Substrate porosity	0.45		Cell water depth, h	1.5	m
pH	8.5					Temperature, T	26	°C
Arrhenius constant	1.02	theta	Number of TIS cells	2				
Volumetric PFR rate constant at 20°C	1.41	per day	Volumetric TIS rate constant at 20°C	5.71	per day	Volumetric CSR rate constant at 20°C	34.2	per day
Volumetric PFR rate constant at T °C	1.59	per day	Volumetric TIS rate constant at T °C	6.43	per day	Volumetric CSR rate constant at T °C	38.5	per day
BBR area, PFR	39,125	m²	3.91		ha			
					Key		Input	
BBR area, CSR	8223	m²	0.82		ha		**calculated**	
BBR area, TIS	20,376	m²	2.04		ha			
			RGM design basis					

It is noted that the size calculations in Table 6.11 are very sensitive to influent pH. If the pH were 9.0 rather than the 8.5 assumed above, the BBR area using the 2TIS model would rise to 3.5 ha (2 ha using the CSR model) and an effluent of 0.2 mg NH_3 would be needed to achieve an unionized ammonia level of 100 ppb). In actual practice, the influent ammonia nitrogen will hardly ever be as high as 7.5 mg/L (15 mg/L into the Aerated Lagoon); usually, BBR effluent ammonia nitrogen concentrations will be at or near non-detect levels; the EEB System's effluent will be diluted when it is discharged into the Treated Water Storage Pond; and RGM will always have the option of diverting discharged water from the TWSP via the Swamp Forest Natural Wetland (which would reduce pH and hence residual ammonia toxicity).

6.7 BNIA Pilot-Scale Treatability Testing

As a preliminary to the design and engineering of the BBR-based EEB System at Buffalo Niagara International Airport (BNIA or BUF), an off-site pilot-scale treatability test was carried out at the Alfred Test Unit using actual stormwater runoff from BNIA exported by tanker into Canada. At the test facilities, the imported feedstock was spiked with ADAFs to assumed worse-case conditions.

During the treatability test, testing in one cubic meter BBR cells was evaluated against Aerated Lagoon cells and non-aerated SSF cells. The aeration in the BBR cells was found to profoundly affect treatment performance. When BBR cells were aerated at 0.85 m^3 air per h per m^3 of the EEB bed, the carbonaceous BOD_5 removal rate constant determined averaged, as was mentioned earlier, 5.4 per day with an Arrhenius temperature coefficient (θ) of 1.03, based on treatability testing carried out at 22°C and 4°C (Higgins et al. 2006c; Wallace et al. 2007; Kadlec & Wallace 2008). (The BOD removal rate constant for the unaerated SSF CW cells was $\sim<0.5$ per day.)

The kinetic and other scale-up parameters determined during the treatability test were used to design the full-scale EEB System for BNIA. Calculations at the end of the feasibility study indicated that about 3.4 ha of aerated BBR cells (with gravel 1.5 m thick) would be needed at BNIA to treat all of the spent ADFs and GCSW all of the time under worse case conditions.

Owing to funding constraints, the NFTA decided to proceed with a first phase involving four BBR cells of 1.9 ha of surface area, while retaining the option of impounding stormwater and still diverting some Spent Glycols to the local WWTP if extreme conditions merited doing so. Consideration will be given later to adding two to four new cells. (An RFP for this was issued in the Spring of 2017.)

Figure 6.11 shows one of the BBR test cells during the pilot-scale treatability testing for the BNIA project.

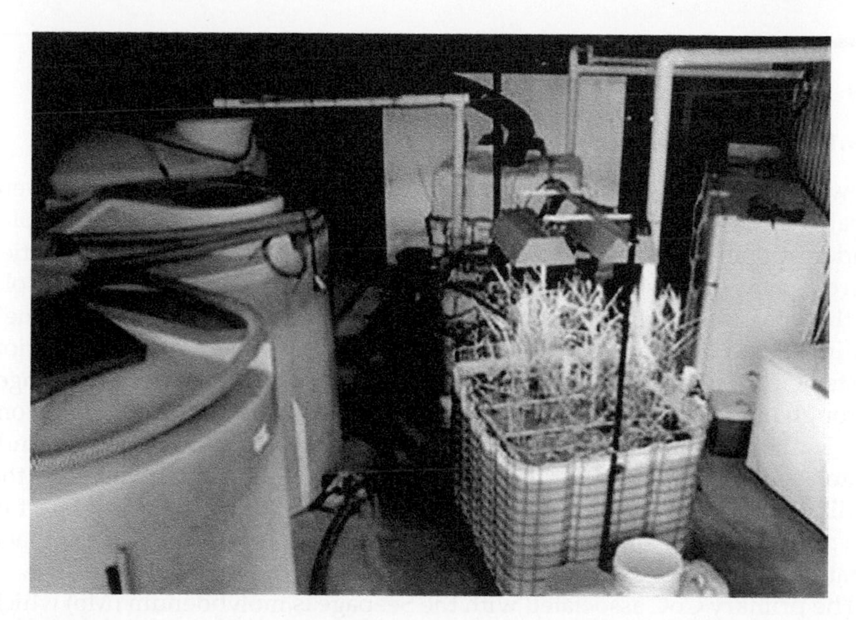

FIGURE 6.11
Pilot-scale EEB test unit for BNIA Project.

6.8 Casper Pilot-Scale Treatability Testing

Prior to the construction of the full-scale VSSF BBR-based EEB System at the former BP Refinery in Casper, Wyoming (see Section 3.4 in Chapter 3), a pilot-scale treatability test was carried out to evaluate the treatment of the contaminated groundwater from under the site and, in particular, to determine BTEX degradation rates for the proposed cold-climate application. The pilot-scale EEB test system consisted of four cells, each dosed at a nominal flow rate of 5.4 m^3/day, and it was operated between August and December 2002.

Areal rate constants (k_A values) were calculated based on an assumed Reaction Kinetics model for three tanks in series (3TIS). It was found that the presence of a wetland "sod" insulation layer on top of the BBR cells and aeration both improved treatment performance. Mean k_A values for BTEX removal were 244 m/year for cells without sod or aeration, and this improved to 356 m/year for cells with sod and aeration. On the basis of the results of this pilot-scale treatability test, scale-up parameters for design were determined system, and a full-scale The EEB System was designed and built. It is capable of operating at 6000 m^3/day and was started up in May 2003.

6.9 Golden Giant Pilot-Scale Treatability Testing

6.9.1 Newmont Golden Giant Mine Systems Analysis

Newmont Canada's *Golden Giant Mine* and associated mill facilities were located near Marathon in North-Western Ontario, Canada. This open pit/underground gold mine closed on December 31, 2005 and mining facilities were removed. Most of the property and its associated facilities were sold to the operator of a nearby mine, leaving only the old Golden Giant mine's tailings pond (the Tailings Pond) and an associated groundwater collection system (collectively, the Site) with Newmont. Groundwater (the Seepage) through one or more of the Tailings Pond's berms was percolating out from the Tailings Pond and had to be collected downslope in the Site's groundwater collection pond (GWCP). From there, it was pumped back into the Tailings Pond to prevent it contaminating local receiving waters. As part of the provincially mandated closure plan for the Site, Newmont desired that a treatment facility be built to treat the Seepage.

The primary CoC associated with the Seepage is molybdenum (Mo) which occurs in low to moderate concentrations (~1 mg/L), and was presumed present as molybdate (MoO_4). Other CoCs include antimony (Sb), cyanides (measured as CN_T and CN_{WAD}, see Section 2.3.5 in Chapter 2), ammonia and sulfates. Newmont wanted any contaminants in any water discharged from its closed Golden Giant facilities to meet or approach Ontario Provincial Water Quality Guidelines (PWQO criteria) before disposal into the receiving waters.

Newmont contracted Jacques Whitford to carry out a systems analysis to see if a "passive" natural treatment system could be defined to address the problem. The systems analysis consisted of a Phase 1 proof-of-concept feasibility study that included bench-scale testing, and a Phase 2 Conceptual Design that included a standard off-site pilot-scale treatability test (Higgins et al. 2007b).

6.9.2 Newmont Golden Giant Mine Phase 1 Bench-Scale Results

As part of the Phase 1 feasibility study, Jacques Whitford sub-contracted CANMET-MMSL to carry out some preliminary Phase 1 bench-scale testing to evaluate the suitability of both the anaerobic treatment of GWCP water passed through primary pulp and paper mill biosolids substrate (see Section 7.4 in Chapter 7) in a bench-scale column being operated to simulate an EEB ABR (a BCR in this case), and its treatment in another containing Steel Slag sorbent. Vertical flow column testing was carried out at ambient temperatures. The purpose of the BCR-simulating bench-scale column was to see if molybdenum could be removed under anaerobic conditions. The most desirable outcome would be Mo precipitation as a stable molybdenum sulfide (e.g., MoS_2) (Cotton et al. 1999), rather than its sorption to the organic substrate or iron.

The steel slag in the second column was intended to act as a sorption medium for various metals. This column was provided with aeration to enhance oxidation, based on previous work conducted by Jacques Whitford and its associates. Initially, it was also thought that nitrification of ammonia could be enhanced in the reactor (Wallace 2002). However, as it turned out, the pH of the steel slag column effluent was very basic, much too alkaline for microbially catalysed nitrification, and ammonia was most likely volatilized.*

Two cylindrical plexiglas columns at CANMET-MMSL were used for the Phase 1 bench-scale testing. A peristaltic pump was used to pump the imported Seepage from the Tailings Pond at the Golden Giant site. The physical dimensions of the slag column and the BCR column are listed in Table 6.12.

TABLE 6.12

Physical Parameters for the Golden Giant Bench-Scale Testing

Parameter	Steel Slag Column	BCR Column
Permeability	n.d.	0.052 cm/min
CS area	184 cm^2	184 cm^2
Height	60 cm	73 cm
Porosity	0.40	0.21
Voidage	~11 L	~13.4 L
Bulk density	1.9 kg/L	1.5 kg/L

Select chemical analyses for the solid substrates are listed in Table 6.13 (Higgins et al. 2007b).

As may be seen, neither the steel slag nor the BCR substrate (which was primary biosolids from a Kraft pulp and paper mill mixed with sand in a 2:1 ratio) contained appreciable amounts of Mo or other CoCs. A diagram of the test facilities is presented in Figure 6.12.

Although the columns were constructed so that they could be be operated in series or in parallel, in the end, only the latter was chosen. A summary of select data from the testing in the bench-scale BCR column is presented in Table 6.14.

The results indicated that molybdenum and other CoCs could readily be removed from the Seepage in an EEB BCR, and that Phase 2 pilot-scale treatability testing at the Alfred Test Unit with this EEB was justified.

* More details on the use of steel slag as an engineered substrate for EEB Systems are found in Section 7.6 in Chapter 7.

TABLE 6.13
Chemical Parameters for the Golden Giant Bench-Scale Testing

Parameter	Slag Concentration (%)	BCR Organic Substrate (%)
Al	1.2	6.3
Total C	0.4	33
CO_3-C	0.34	0.70
TOC	0.06	32.3
Ca	22	3.1
Fe	21	0.31
K	0.04	0.26
Mg	5.4	0.19
Mn	12.8	0.12
Mo	<0.002	<0.0007
Na	0.10	0.18
P	0.23	1.2
S	0.15	0.25
SO_4	<0.1	<0.1
Sb	<0.002	<0.002
Zn	0.005	0.03

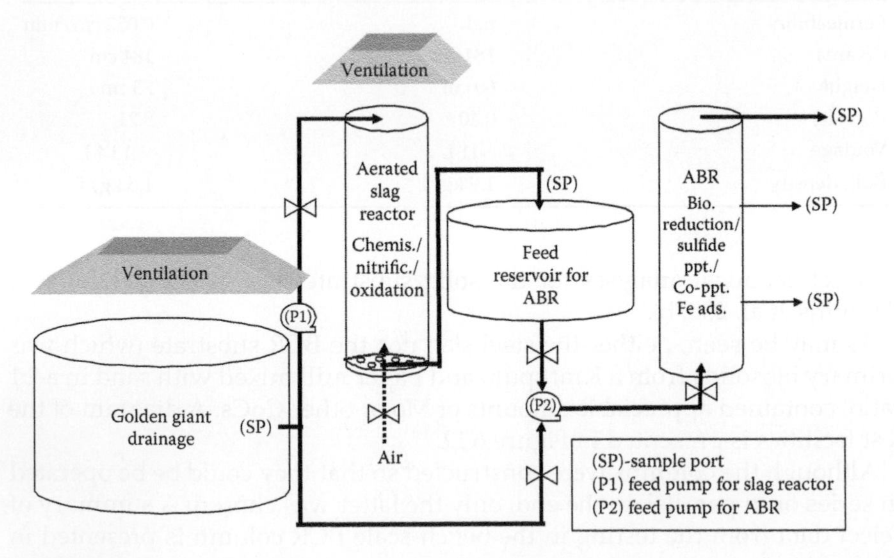

FIGURE 6.12
Sketch of Phase 1 bench-scale testing equipment at CANMET-MMSL.

The results for the bench-scale testing in the aerated steel slag column were less satisfactory. pH increased dramatically in the GWCP water passing through the column to a point where the effluent would not have met the PWQO criteria. This and other problems with the handling of the slag, the operation of the column, and the results from treatment in it led to a

TABLE 6.14
Golden Giant Bench-Scale BCR Column Results

Line Item	Influent	Effluent
Flow rate (L/day)	1.47	n/a
Retention time	1.9 days	n/a
Temperature (°C)	20	20
pH	8.0	7.2
Conductivity (mS/cm)	2.6	6.7
DO (mg/L)	6.7	0.8
ORP (mV)	+2	−260
NH_3-N (mg/L)	18	25
Total sulfur (mg/L)	466	553
Sulfate (mg/L)	1403	1010
Mo (mg/L)	1.05	0.0005
Sb (μg/L)	1.04	0.21
Alkalinity (mg/L $CaCO_3$)	139	550
TSS (mg/L)	9	144
Al_T (mg/L)	0.16	0.6
Fe_T (mg/L)	0.77	1.04

conclusion that steel slag sorption was not a viable option for the treatment of Golden Giant GWCP water, and further consideration of this option was dropped.

6.9.3 Newmont Golden Giant Mine Phase 2 Results

The second stage of the Newmont project involved a pilot-scale testing study carried out in 2006–2007 at the Alfred Test Unit (see Section 6.3) using Seepage collected after the Tailings Pond and shipped by tank truck to Alfred College. This treatability test included a 20-week two-part run at ambient temperatures as well as a further run at low temperatures. The first part (Phase 2A) of the treatability test was separated into two stages for reporting purposes as flow through the system was halted for a 2-week period during the Christmas break in 2006 and this led to changes in the system that reflected negatively on Mo removal efficiency.

At the conclusion of the high temperature run, a detailed examination of the bacterial population was carried out at CANMET-MMSL. The results showed that appropriate numbers of the three main types of bacteria that were involved in metal removal, APB, IRB and SRB, were present, and the most probable number of each species at three levels in the cell were reported. Samples for the enumeration study were taken as the BCR cell for the high temperature runs (Phase 2A) was deconstructed in the Spring of 2007. At the same time, sulfide ion traps that had been installed in the BCR cell during construction were removed and samples sent to the UoG where sulfide speciation studies were started. Facilities at the university were not capable of determining the presence of MoS_2 and a sample was then sent to a specialty lab at the Laurentian University in Sudbury, Ontario for final determination. The presence of MoS_2 was shown.

The results of the low temperature runs (Phase 2B) were disappointing and illustrate the difficulties of establishing a viable appropriate bacterial population in a short time frame.* However, despite the lack of Mo removal, a better understanding of the system's characteristics was gained. In addition, it was clearly demonstrated that low concentrations of many metal contaminants can be reduced to very low concentrations when an EEB System is used.

6.10 Pilot-Scale EBR Testing

The Landusky–Zortman mines in Phillips County, Montana, USA produced gold and silver. Although mining began over a century ago, most open pit mining followed by heap leach precious metal extraction occurred from 1977 to 1998. Cyanide spills and leaching from tailings and old heap leach pads lead to ARD generation and extensive surface water and groundwater contamination with CoCs including cyanides, nitrates, sulfates, arsenic and selenium.

In 1993, the US EPA required the reclamation of the 486 ha area and this led to litigation, the bankruptcy of the mining company operating the mines, the assumption of control by the US Bureau of Land Management and a reclamation/WWT program that will continue to 2018 (Williams et al. 2009). WWT at the sites of the two nearby mines is underway using a conventional (mechanical) biotreatment system (i.e., an active WWTP based on anaerobic bioreactors) to treat collected cyanide leach pad drainage wastewaters. (See Section 2.3.5 in Chapter 2 for a description of CN-contaminated MIWs.) The WWTP has three approximately 950 m^3 anaerobic bioreactor vessels operated in series with activated carbon as a microbial biofilm support medium in them. Design WWTP flow rates were about 1600 m^3/day, but following leach pad WW chemistry changes that dropped the pH to about 4 and raised the sulfate levels to over 5000 mg/L, the system had to be operated at about 550 m^3/day in attempts to meet the discharge criteria. Additional measures

* This treatability test was completed a few years before genomic analyses (see Section 1.11 in Chapter 1) became widely available and the kinds of microbial analyses available then, as is noted in the text, were not accurate enough to clearly define all of the microbes involved and their metabolic pathways. It is interesting to speculate what genomic testing might have shown but as Newmont has indicated that it is considering a Phase 3 for the project, this may be determined in future.

to try to meet the discharge criteria included mixing the WWTP effluent with approximately 50% clean site waters before discharge.

A pilot-scale test system was built and operated to assess the potential of conversion of the WWTP facilities to use EBR technology (see Section 4.4.2 in Chapter 4 for a background on EBRs). The pilot unit was built in a shipping container that was moved to the site for testing. It included three upflow, fixed-bed EBR-type ABRs in series and the final effluent being aerated. No filtration or other solids management post-treatment was required. The EBRs had a 3V potential across them at an average current of 0.46 Amperes. This provided the EBRs with a predetermined optimum electron supply of approximately 2.87×10^{17} electrons per second during the test period. Molasses was injected to provide the needed nutrients. Figure 6.13 shows the containerized, PLC-controlled, pilot-scale EBR test unit.

FIGURE 6.13
EBR pilot unit for selenium removal.

The EBR test unit was fed with a slipstream from the collected stream of leach pad wastewaters flowing to the WWTP and had a total HRT of approximately 20 h, approximately one-half that of the full-scale WWTP (see Equation 8.3 in Chapter 8). The pilot-scale EBR tests were carried out for approximately 3 months and during that period water temperatures ranged from 2°C to 13°C. A comparison of treatment results for the EBR system with those from the WWTP and (where applicable) regulatory limits is found in Table 6.15 and the data in the table represent averages for the entire test period and/or (for the WWTP) historical data available.

As may be seen from Table 6.15, all monitored parameters in the EBR pilot system effluent met the discharge criteria during the entire test period. Following the successful EBR pilot test, the anaerobic bioreactors of the full-scale WWTP were reconfigured to operate as EBRs.

TABLE 6.15

Comparative EBR Pilot Unit Results

	Influent	WWTP Effluent	EBR Pilot Unit Effluent	Discharge Limit
pH	6.84	7.14	6.98	6.5–8.5
TSS (mg/L)	<10	NA	18	20
NO_3-N (mg/L)	297	22	0.02	10
CN_T (mg/L)	0.0843	NA	<0.005	0.0052
CN_{WAD} (mg/L)	0.012	0.0723	<0.005	NA
Al (mg/L)	0.31	1.99	0.04	NA
As (μg/L)	14	NA	4	18
Cd (μg/L)	135	125	1	5
Cu (μg/L)	57	122	3	31
Fe (μg/L)	270	790	213	NA
Mn (mg/L)	56	55	28	NA
Ni (μg/L)	826	893	7	NA
Se (μg/L)	858	417	38	50
Zn (mg/L)	2.15	2.94	0.04	0.3

Additional EBR pilot system evaluations, in non-retrofit configurations, have been conducted on numerous other mining and coal-fired power plant wastewaters containing nitrates, selenium and sulfates as primary CoCs of interest. These systems have consistently demonstrated NO_3-N removals from 10 to 150 mg/L to below 0.02 mg/L, selenium removals from waters containing from 0.050 to 2.5 mg/L to produce effluents below 1 μg/L, and sulfate removals from an average 1000 mg/L to below 250 mg/L using retention times ranging from 2 to 8 h. Pilot and full-scale EBR systems have also demonstrated the removals of secondary CoCs such as Ag, As, Sb, Cd, Cu, Mo, Ni, Pb and Zn to below discharge criteria across different water chemistries and temperatures.

7

Eco-Engineered Bioreactor Substrates

7.1 Substrate Compositions

The most important part of an EEB is its substrate.* The substrate of an EEB is the bed of granular and/or other solid materials in it to which the microbial biofilms that cause most of the transformation of CoCs passing through it are attached.†

The first tasks when planning an EEB project is to define what materials will be used to make up their bioreactors' substrates; what sort of compositions are deemed suitable; how available substrate materials and their related bedding materials (e.g., sand and limestone) are locally to the proposed site of the facilities; how much they are expected to cost; and whether more economic alternatives are available. Whenever an EEB System project is being planned, detailed technical specifications should be prepared describing the sources, preparation and installation of both bioreactor substrate materials and the ancillary materials used.

As soon as possible during the early phases of an EEB project, the substrate(s) to be used in the EEB cell(s) should be decided on and samples of this (these) substrate material(s) should be obtained by those planning an

---

* The term "substrate" as used in this book does not include other layers of materials that may be in or on an EEB in addition to its active bed, and may also serve as locales for biofilm attachment. These include separate layers of reactive materials such as limestone found in some SAPS Bioreactors (below the substrate) and sometimes in smaller passive treatment BCRs (upstream of the substrate). The term substrate as used herein also does not include the relatively thin layers of coarse materials (usually sand) which may be used in many cases to separate the substrate from other beds in an ABR, or which are used as bedding material for inlet and outlet distributor piping, or aeration tubing in these or other kinds of EEBs. The term substrate as used herein also does not include the layers of insulating material that may overlay the substrates in many BBRs, or the overlying water layers used with some anaerobic bioreactors, or any material (e.g., earth and geomembranes) which alternatively may be used in some cases to bury some kinds of anaerobic EEBs (e.g., DNBRs and BCRs).

† The term "substrate" as used in this document should also not be confused with "active medium" which may be only one component of an anaerobic bioreactor's substrate, or with the word as used to describe a material acted on in a chemical reaction. Many literature articles on anaerobic bioreactors such as SAPS Bioreactors and BCRs are unclear on this aspect, and care should be taken when using them for reference.

EEB project and tested for a variety of parameters (e.g., purity, pH of water when they are put in it, robustness, porosity and tendency to compact).

Except for cases where similar WWs are to be treated using the same kind of substrate materials of the same qualities for a new project as are already being treated successfully for existing projects, a treatability test will always be required using the planned substrate in the test unit (see Chapter 6). It is emphasized that just because a particular Reaction Kinetic model describes a particular WW/substrate combination for one situation (e.g., the CSR model for the BBRs at BNIA as is reviewed in Section 6.7 of Chapter 6) does not mean that the same model will necessarily be applicable for a somewhat similar WW at another location where the substrate composition(s) and other factors (e.g., soils, climate, facilities, and operating methods) will not be quite the same.

7.2 Substrate Logistics

Obtaining, transporting, preparing, mixing and installing the substrates and other material layers of EEBs is the most time-consuming and onerous aspect of constructing an EEB. Even though EEB Systems are generally much more economical to build than alternative mechanical WWTPs, the costs for substrates and related bedding materials still are a major part of an EEB System's CAPEX, amounting to 50% (or more) of total costs in some cases.

Generally, the aggregate materials such as gravel used as substrates in BBRs and SAGR Bioreactors and in semi-passive treatment BCRs are available from suppliers ready to use, and can be put in place immediately. Aggregate materials used in such substrates should be specified as washed and screened as doing so will remove fines and ensure better permeability and compactability.

With the granular substrates of aerated bioreactors, inoculants (sources of the required microbial consortia) might be added as solids (e.g., manures) but mixing them in would be onerous and they would contribute to fines, so with these types of bioreactors, if at all possible, inoculants are almost always added in liquid form to initiate the growth of the microbial biofilms on the substrate particles. For example, for the very large CW system at Edmonton International Airport (now upgraded to an EEB System, see Section 3.7 in Chapter 3), mud from local stormwater ditches in which glycol-contaminated runoff from de-icing pads had flowed was collected in water-filled barrels; fertilizers were added to promote the growth of indigenous microbes in the mud; the water was stirred and the mud allowed to settle; and the decant water was added to the feedstock to the wetland cells. This successfully inoculated the wetland with a glycol-acclimatized microbial consortium.

Preparing substrates for passive treatment ABRs is much more complex and expensive than is the situation for EEBs with gravel substrates. Not only are the carbonaceous active media more difficult to manage (e.g., materials such as composts, wood chips and biosolids are harder to handle than aggregates), but the substrates of such bioreactors usually also involve more than one kind of material (see Chapter 4). For example, in addition to the various components of the active media, these substrates often also include support materials of one or more kinds such as sand and gravels as well as reactive media such as limestone. Each of these materials must be separately accessed, brought to a mixing site, stored, mixed homogeneously and only then moved to the bioreactors for installation. Accordingly, the preparation of most ABR substrates (i.e., those for SAPS Bioreactors, some DNBRs and all passive treatment BCRs) requires a large area on which to assemble the component materials, mechanical equipment to take up and mix them in the predefined amounts, and trucks or loaders to transport the mix to the bioreactors. The preparation area may be on-site near the ABRs, or at some off-site location.

Those involved with full-scale aerobic and anaerobic bioreactor projects often greatly underestimate the complexity and time required to prepare their substrates. This results in misleading estimates of costs for them. Wherever the substrate preparation site is, provision has to be made at it to protect the local environment around it from dust, odours, and contaminated runoff from the piles of component materials that may have to be stored and mixed there. There has been a tendency for researchers evaluating contaminant removals in bench- and pilot-scale Stand-Alone, passive treatment BCRs to carry out their testing using substrates which are complex mixtures involving several kinds of active media and support materials, giving little consideration to the ramifications of building full-scale bioreactors using these materials. The KISS principle should be applied: use as few kinds of materials as possible and seriously consider inoculating with a liquid inoculant added after construction rather than including a solid one such as manure as part of the active media.

In addition to the above, when selecting substrate materials for either aerobic or anaerobic EEBs, careful consideration should be given to using materials that

- Are low in cost
- Are easy to transport
- Are readily available at or near the location where the EEB System will be constructed
- Are easy to replace
- Have long-term effectiveness
- Are not prone to degradation

Also, the impact of delivering the substrate materials on the communities that they must traverse to get to the bioreactor site needs to be carefully considered when planning an EEB System project. For example, the total amount of gravel required for the substrate for the four very large BBRs for BNIA (see Section 3.6 in Chapter 3) was ~62,000 tonnes (68,000 short tons). The gravel trucks owned by the quarry firm that won the public bid to supply this gravel each had a carrying capacity of 33 tons per load. The location of the supplier's quarry to the airport allowed each of its 10 trucks to deliver up to 4 loads per day, a total of 1320 tons/day. This meant 80 passages of large trucks through the local communities every day to supply all of the gravel for the four BBR cells being built. The minimum time that could have been needed for the trucks making 40 trips per day to deliver all the substrate would have been 51 days (68,000/1320). In practice it took about 60 days, a long period of disruption for the communities the gravel trucks had to travel through.

Each EEB project will be unique but the point here is that in designing EEB Systems, substrate logistics have great impacts on costs, environmental effects and other aspects, and need careful attention by design engineers at the Feasibility Study stage.

Figure 7.1 shows the placement of gravel substrate into a BBR cell (in this case one of the two at Casper, WY, see Section 3.4 in Chapter 3).

FIGURE 7.1
Placement of gravel substrate in a BBR cell at Casper.

7.3 Substrates for Aerobic EEBs

Substrates for aerated EEBs such as BBRs and SAGR Bioreactors are made up of aggregate materials such as quartz gravel, crushed rock, shale or limestone and dolomite gravels.* In many aerated EEBs, a single kind of substrate material is used throughout (e.g., the BBRs at BNIA used nominal 10–15 mm dolomite gravel aggregate for their substrates, see Section 3.6 in Chapter 3). However, in some downflow VSSF bioreactors, the substrate can consist of a top layer of sand underlain by two or more progressively coarser layers of gravel or crushed rock.

Figure 7.2 shows the dolomite gravel used at BNIA.

FIGURE 7.2
Gravel used in the BNIA BBR cell substrates.

A semi-passive aerobic EEB (BBR) substrate needs to

- *Have the properties of construction aggregates:* A practical substrate will have a constitution similar to that of a roadbed aggregate (ideally, the substrate bed should have road-grade gravel compaction, but practically this is rarely the case).
- *Be robust:* The substrate should not break down or solidify in service.
- *Have sufficient permeability:* To allow free passage of the WW being treated, but enough compactness to ensure contact of the targeted contaminants with the microbial biofilms.
- *Have good hydraulic properties:* The types of gravels and other aggregates commonly used in BBRs should have hydraulic conductivities in the 10,000–50,000 m/day range.

* Where wastewater pHs are appropriate. However, be careful, as some wastewaters (e.g., airport glycols) can break down into acidic compounds such as VFAs (see Section D.1 in Appendix D) that can degrade limestone and dolomite.

- *Be available in large quantities:* Even though EEBs are much smaller than equivalent SSF CWs, they still require relatively large amounts of substrate materials, and to be feasible, the substrate materials have to be readily available in sufficiently large amounts convenient to prospective EEB System project sites.
- *Should not itself cause further pollution problems:* In addition to removing certain pollutants, an aerated EEB substrate should not add to the pollution problem in the WW being treated by breaking down and releasing and/or leaching into the WW being treated excessive quantities of other COCs that are difficult to polish.
- *Be available economically:* In most applications,* the gravels used for aerated EEB substrates are expensive, and constitute a large part of the CAPEX of EEB System projects.
- *Be readily inoculated with an established microbial consortium:* It mostly involves aerobic bacteria in biofilms on the surfaces of substrate particles and other materials (as was discussed above, this inoculation can be implemented quite simply pumping in a liquid inoculant after construction is complete).

Generally, the Technical Specifications for the supply of the aggregate substrates of aerated EEBs such as BBRs require that the materials: be graded to meet a pre-selected sieve analysis; be free from organic impurities (a typical criterion might be less than Plate 1 as per ASTM C40); not contain any sand, silt or clay and have porosities of greater than some fixed target (e.g., the porosity specification for the gravel substrate for the BBRs at BNIA was that it be greater than 38%). Generally, ordinary pit-run gravels will not be suitable for use as substrates in BBRs as their particle sizes are too variable and they contain fines.†

The bigger and more complex the EEB System project, the greater is the need for comprehensive and tight Technical Specifications for the substrates. Civil engineering tests such as sieve analyses; porosity measurements; determinations of specific gravity; measurements of hydraulic conductivity; mason jar tests for fine particles; aggregate breakage/wear tests (ASTM C131); California bearing ratio tests (ASTM D1883); soundness tests (resistance to breakdown under freeze–thaw conditions) and Micro-Deval tests (wear resistance in a wet environment) are just some of the substrate gravel tests which may be needed. Testing may need to be carried out not only on the gravel at the supply source (e.g., a quarry) but also on each truckload delivered to a project site.

* Except perhaps at many mine sites that already need to crush large amounts of rock for use as gravel for other purposes, and therefore already have the equipment and capabilities to produce large volumes quite economically.

† Pit-run gravels can, however, be used if they are first screened and washed.

The most desirable substrate material for BBRs is sieved and washed quartz gravel, but this material may not always be available economically or in sufficient quantities locally to an EEB System project site. Providing that the pH of the WW being treated is not low, agricultural limestone or dolomite aggregates can also be considered (if a low pH WW such as an ARD needs to be treated in a BBR, it must first be neutralized in a limestone drain or a SAPS Bioreactor before it reaches the aerated cell). Figure 7.3 shows the placement of gravel in a BBR cell.

FIGURE 7.3
Gravel substrate placement in a BBR cell.

Usually, the substrate surfaces of BBRs (and some other aerobic bioreactors) are covered over with a layer of insulating material from 0.1 to 0.5 m thick. A variety of materials can be used but they must not contain fine materials that could wash down into the underlying gravel, causing plugging. Peat and ordinary composts or mulches are not suitable. The woody reject by-products from the preparation of many composts are good choices.

In addition to the insulating effects of in-ground morphology and overlying beds of insulating materials in BBRs, it should be noted that the aerobic degradation reactions in most aerobic EEBs are exothermic. For example, the ~4500 kg of bacteria in the gravel particle biofilms of the four BBR cells at BNIA cause the waters passing through them in winter to rise in temperature by 15–20°F.

Substrates for EEBs, especially aerated ones such as BBRs, are usually relatively thick (1.5–2 m or more) and have to be laid down in a series of "lifts." Great care must be taken in the laying down of these lifts to ensure that areas of uneven compaction do not occur as these would lead to preferential flow areas, causing reduction of bioreactor effectiveness.

The need to ensure even, uncompacted substrate placement is particularly important in HSSF varieties. Where some BBRs are vegetated with terrestrial grasses, it is important to realize that they are not playing fields and no activities should be permitted on their surfaces other than grass-cutting with light equipment.

Figure 7.4 shows the placement of gravel in one of the cells at EIA.

FIGURE 7.4
Gravel placement in lifts at EIA.

7.4 Substrates in Passive Treatment Anaerobic EEBs

Substrates for passive treatment EEB ABRs generally consist of the active medium or media (the carbon source); inactive support material(s) to provide permeability to the substrate (e.g., sand, quartz, and gravel); in some cases (e.g., some BCRs), reactive support materials (e.g., limestone as well) and an inoculated or otherwise supplied microbial consortium, which may come from a solid active medium such as manure, but, as was mentioned above, can also be (preferentially) a liquid inoculant solution added after construction.

In some cases where the active medium is itself a material with structural properties (e.g., some kinds of wood chips), only it may make up the substrate.

As with aerobic EEBs, in addition to the substrates, passive treatment anaerobic bioreactors such as some BCRs may contain separator layers, bedding layers for piping, and, in some cases, separate layers of reactive materials such as limestone gravel or cobbles.

An anaerobic EEB (e.g., a DNBR and a SAPS Bioreactor or a passive treatment BCR) needs to have substrates containing

- An active medium or media which provide a labile carbon source for microbes.

- An established microbial consortium mostly involving anaerobic bacteria in biofilms on substrate particles and other surfaces (as mentioned, this may be a solid part of the active media or added later in a liquid).
- Nitrogen and phosphorus sources (which usually come from the active media).
- A support medium (e.g., gravel and/or sand) to allow permeability.

The carbon source for a passive treatment anaerobic EEB can be any type of carbonaceous material: the decaying roots/detritus of plants, or a layer/cell fully or partially filled with a microbially available carbonaceous mat such as pulp and paper mill biosolids or wood waste (wood chips and sawdust) (Mattes et al. 2010).

In general, mixtures of several kinds of active media in substrates, ones that provide immediate, intermediate-term and long-term sources of labile carbon, perform better in anaerobic bioreactors than do ones using a single waste (Zagury & Necultia 2007), but this should be balanced against the increase in complexity and logistical problems too many media cause, as was mentioned above (Béchard et al. 1994).

Table 7.1 shows (Gusek & Wildeman 2002) a few of the kinds of materials that have been used as solid active media in ABRs (mostly BCRs).

TABLE 7.1

Some Solid Active Media Used in Passive Treatment ABRs

• Wood chips (various tree species)	• Bagasse (sugarcane processing waste)
• Wood waste	• Cotton waste
• Sawdust	• Corn stover
• Rice hulls	• Alfalfa (various forms)
• Yard waste	• Beet pulp
• Spent mushroom compost	• Municipal biosolids
• Raw and composted manures	• Pulp and paper mill biosolids
• Nut shells	• Paper recycling mill biosolids
• Cardboard (shredded)	• Peat
• Chicken litter	• Organic soils
• Marsh sediment	• Soya bean hulls
• Hay and straw (chopped and various kinds)	

Figure 5.8 in Chapter 5 shows the placement of pre-mixed substrate in a BCR cell during the construction of the Teck Demo EEB System. It contained 65% pulp and paper mill biosolids, 20% limestone and 15% sand. Although the EEB System was only demonstration-scale and treated only 20–25 m^3/day of arsenic-, cadmium- and zinc-contaminated WW, each passive treatment BCR still required 900 tonnes of substrate!

Biosolids from pulp and paper plant and paper recycling plant WWTPs are particularly good active media for passive treatment anaerobic bioreactors. Such biosolids often are supplied as mixtures of two components: *primary biosolids* from the WWTP's 1° treatment stage (e.g., cellulosic materials which will degrade in the intermediate to long terms), and *secondary biosolids* from a WWTP's 2° treatment stage (largely dead bacteria which are short- and intermediate-term sources of labile carbon).

There are various kinds of biosolids depending on the kind of WWTP from which they are sourced; municipal WWTP biosolids and biosolids from pulp and paper plant WWTPs make good active media for passive treatment ABRs, although care must be taken as to what contaminants the biosolids might introduce due to the nature of their sources. Table 7.2 presents an analysis of a pulp and paper biosolids material.

TABLE 7.2
Chemical Parameters for a BCR Cell's Biosolids

Parameter	%
Al	6.3
Total C	33
CO_3-C	0.70
TOC	32.3
Ca	3.1
Fe	0.31
K	0.26
Mg	0.19
Mn	0.12
Na	0.18
P	1.2
S	0.25
SO_4	<0.1
Sb	<0.002
Zn	0.03

In addition to biological alkalinity generation, a BCR's substrate may also contain a reactive neutralizing agent (e.g., limestone) that supplements it with chemical alkalinity generation. However, as was mentioned in Section 4.3 of Chapter 4, a SAPS Bioreactor's substrate should not contain reactive agents, or involve active media that might lead to the premature addition of alkalinity (e.g., spent mushroom compost and many manures) that can promote ferric iron hydrolysis and the armoring of the underlying limestone. Candidate SAPS Bioreactor substrate materials (active media, support materials and other additives) should be tested for pH rises (undesirable) before being selected.

In addition to any reactive materials forming part of the substrate matrix in BCRs, as was mentioned, one or more separate layers of them may be present in these bioreactors upstream of the substrate. In some SAPS Bioreactors (which operate downflow VSSF), there may be an underlying layer of neutralizing limestone.*† As with some aerobic EEBs, separate layers of non-reactive materials (e.g., sand and/or gravel) can be used to separate the substrate from reactive media layers, to provide beds for influent and effluent distributors and to separate the substrate from an overlying water layer. Although judicious selection of materials and proper design should prevent any armoring of the limestone particles in the separate reactive material layers of anaerobic EEBs, one study (Sasowsky et al. 2000) found that adding crushed sandstone to the beds of limestone drains inhibited armoring, and such might be considered for bioreactors as well.

As was mentioned, if adequate amounts of sulfate are not available in the WW, a solid sulfate source such as gypsum may be added to the substrate.

For passive treatment BCRs containing solid active media, the characterizing kinds of bacteria (e.g., DNB, IRB, and SRB) present in the substrates cannot directly use (oxidize) the complex and recalcitrant organic matter involved (e.g., carbohydrates, polysaccharide, proteins, lipids, cellulose, hemi-cellulose, and lignins) and prefer to metabolize low-molecular-weight organics and H_2. Accordingly, the complex organics must be broken down and this begins with anaerobic degradation which involves two steps: (1) their conversion into new microbial cells and lower molecular weight organic acids and alcohols by acid-forming bacteria (acetogens) and (2) the continuing oxidation of breakdown products into methane, H_2S, ammonia and carbon dioxide by methane generating bacteria. Simultaneously, anaerobic and facultative fermenting microbes also break down the intermediate-molecular-weight material into waste products such as the lactates, ethanol, acetates and formates that can be used by the characterizing bacteria such as SRB. For example (Wildeman et al. 2006), cellulose, a major component of many kinds of active media used in anaerobic bioreactors, is first broken down by the hydrolytic acetogens (CDB) into cellobiose, then into glucose, which is in turn broken down by fermenting bacteria into hydrogen and the small organic acids such as lactates which can then be still further converted by other fermenters and bacteria such as SRB. The final result is carbon dioxide and methane.

As was mentioned, active media for many bioreactors may be added as a liquid solution with the bioreactor's feed WW. The list in Table 4.1 in Chapter 4 shows the kinds of liquid active medium in LABRs, mostly in BCRs.

---

* Alternatively, for larger, field-scale passive treatment SAPS Bioreactors, this limestone may be provided in a downstream, buried ALD.

† Dolomite can sometimes be used instead of limestone.

7.5 Clogging in EEB Substrates

WW flow in both HSSF and VSSF CWs may be impeded and their beds may clog if organic and/or inorganic solids accumulate in the pore spaces between their substrate particles. The tendency of these SSF ABRs to clog will be influenced by the characteristics of the substrate (e.g., grain size and distribution, porosity and the shape of the particles) as well as those of the system the EEB is in (e.g., level of loading, WW flow rate, any plant root zone effects, the potential for chemical precipitation in the cells, availability and effectiveness of pretreatment and the availability of rest periods when the EEB is off-line). The problem of clogging in these EEBs may be addressed by periodically removing and replacing and/or cleaning up the substrate beds, or washing them (and their related piping/distributors with hydrogen peroxide or acid wash solutions (Nivala et al. 2009; Knowles et al. 2011).

Substrate and other beds in EEBs may become reduced in hydraulic throughput, or even clog, if: (1) particle sizes are too small; (2) inorganic material accumulates in or on them; (3) chemical precipitation leads to the deposition of material between the particle pores; (4) sliming occurs and/or (5) the rate of organic sludge (microbial biomass) accumulation exceeds the autolysis rate (sludging).

Some early kinds of HSSF CWs used soil as their substrate but the small pore spaces that this medium involves often led to clogging and/or channelling (Kadlec & Knight 1996). BBRs use gravel or similar higher pore space aggregate as their substrates, and clogging due to too small pore spaces is no longer a problem in well-designed systems of this sort.

Inorganic material (i.e., suspended solids) will only clog EEB beds if it is not first removed to low levels upstream of the bioreactor cells and the prevention of such situations is a matter of judicious design. Generally, the use of Sedimentation Ponds and/or rock filters as early 1° treatment cells of EEB Systems will prevent this problem.

Similarly, reduced hydraulic conductivity/clogging of anaerobic EEB substrate beds due to chemical precipitation also can be prevented/mitigated by proper consideration of chemical and process conditions. An example of a situation where chemical precipitation can be a problem is in the treatment of ARD containing dissolved ferrous iron that can oxidize to ferric hydroxide precipitates if exposed to oxygen at certain pH levels. Even in such situations, advanced system design methods can deal with these kinds of problems (see Chapter 8).

The last potential reason for reduced hydraulic conductivity/clogging in EEBs can occur due to the formation of so much excess organic sludge (i.e., microbial biomass) in the microbial biofilms on the substrate particles that these begin to bridge the gaps between the particles, plugging the beds. As was mentioned, EEBs may be regarded as kinds of attached growth systems. However, the comparison with other kinds of attached growth systems should not be carried too far as semi-passive EEBs are generally much more massive than other attached growth systems such as active treatment biofilters, and usually operate at very much lower local loading conditions. Up to certain limits, any "excess" sludge formed in an EEB's substrate will be quickly degraded by microbial organics degradation due to enhanced populations of heterotrophic bacteria.

Accordingly, the accumulation of excess sludge in the beds of well-designed EEBs does not usually occur at levels that will clog the beds, or even seriously impede hydraulic conductivity for many decades. Even fears that such excess sludge will slough off continually into effluent from the bioreactor where it will report as TSS and BOD have proved to be unfounded. Beds in EEBs can be easily replaced or beds and piping purged with hydrogen peroxide or acid if organic material clogging occurs, but this should not be expected to be required under normal conditions in well-designed systems.

7.6 Engineered Substrates

As was mentioned in Section 1.3 of Chapter 2, one way to "create" an EEB from an SSF CW is to replace some or all of its substrate with a material with specific properties that allow it to sorb, precipitate and/or otherwise remove certain colloidal or dissolved CoCs.

WW CoCs in suspended solid form can be removed by sorption, filtration or flocculation/settling in the primary cells of an EEB System. For relatively small concentrations in a WW of colloidal or dissolved CoCs, removals can occur in the substrates of secondary treatment EEB cells, but for larger concentrations of such CoCs, especially where bulk removal is needed, the route is to use contact with *engineered substrates* in tertiary cells of EEB Systems. Examples of the CoCs that can be removed in this manner include compounds of phosphorus, aluminum, fluorine and arsenic, with phosphorus removal being the most interesting.

The following kinds of engineered substrates have been suggested/tested for removing phosphorus compounds from WWs: pumice, certain sands that have iron in/on them, fly ash, blast furnace slag, activated alumina, artificial materials such as LECA (a reactive porous medium), zeolites and natural materials such as wollastonite (Brooks et al. 2000).

Not all substrates are equal as phosphorus sorbents. Some are not able to sorb practical amounts of phosphorus; some are too expensive for large-scale use; others may come in forms that are not permeable enough; some may be of limited local supply; others may be prone to breakdown or degradation; and/or still others may only be suitable for treating relatively small volumes of WWs.

"Steel" slag is an example of the kind of material that can be used as an engineered substrate for treating elevated concentrations of phosphorus in a

WW. Several kinds of slag are available in the by-products of iron- and steel-making and recycling (e.g., blast furnace slag, basic oxygen furnace [BOF] slag and electric arc furnace slag) and these slags may be available as fine powdery materials, as materials with the consistency of sand, as mixtures containing large chunks of hard slag and as materials that can be screened to produce aggregates similar to gravel.

The University of Waterloo patented the *Phosphex*™ process for using alkaline slags as a phosphorus sorbent. Jacques Whitford agreed to test its sorbency with a treatability test at the CAWT facilities (see Section 6.3 in Chapter 6) where pilot- and demonstration-scale EEB Systems treating septic tank overflow consisted of a 2° treatment BBR cell followed by 3° treatment cells which were filled with blast furnace slag and other substrate materials in various proportions (Wootton et al. 2010). While the results of the treatability testing showed that blast furnace slag was able to almost stoichiometrically remove phosphorus from the WW, problems with the medium (e.g., breakdown, sludge formation, channelling, the formation of areas of "hard pan," a tendency to form "reaction fronts" rather than allowing reactions throughout the media; heterogeneous sources of supply and the production of very high pH effluent requiring still further treatment) indicated that this kind of slag was not an appropriate sorbent material for use in EEB System.

However, there are WW streams containing phosphorus that are particularly difficult to manage and different kinds of slags have been considered to treat them. An example of the latter is the high phosphorus concentration leachate from the gypsum stacks left over from the manufacture of phosphate fertilizers. Phosphate fertilizer operations react sulfuric acid with phosphate ore to produce fertilizers and this results in a waste by-product, gypsum (calcium sulfate) sludge. This sludge is low in pH (2–5) and is contaminated with phosphates (100–1000 mg TP/L) and other species from the ore and the acid (e.g., fluorides and sulfates). It is usually disposed of on-site in large ponds. Over time the level of the accumulated gypsum sludge in these ponds rises to the point that they become large landfills called gypsum stacks.

Although the operators of phosphate fertilizer facilities usually try to limit the influx of water into their gypsum stacks, practically percolating precipitation and groundwater will sooner or later find their way into them and result in the movement of pore water out of them. This exfiltration is usually collected in drains around the stacks and has to be treated before the leachate can be discharged to the environment. One common method for doing so is the use of active WWTPs (lime plants) in which the WW is contacted with lime to precipitate out the major inorganic CoCs. Generally, leachate from a gypsum stack is first directed to large gypsum ponds where it mixes with rainwater and the gypsum sludge from production operations. If there is a lime treatment plant on the site, and there usually is, calcium hydroxylapatite sludge from that plant is also disposed of in the ponds. Excess water from the ponds is sent to the lime plant for treatment where it is mixed with lime to form the hydroxylapatite.*

However, since the generation of leachate from gypsum stacks will continue for long periods, this means that phosphate fertilizer manufacturers have to continue operating lime plants long after the phosphate fertilizer plants that resulted in the creation of the gypsum stacks are closed. Such can be highly expensive.

An example is the former phosphate fertilizer plant of the International Minerals and Chemical Corporation (Canada) Limited (IMC) in Port Maitland, ON which closed in 1989 leaving a large gypsum stack to be managed as part of a site decommissioning process. The typical leachate that was generated from its gypsum stack had pHs in the 3–3.5 range and contained 400–500 mg TP/L.

In 1997, Jacques Whitford and SESI were contracted by IMC to seek a more economic method of dealing with the leachate from the Port Maitland gypsum stack. Table 7.3 summarizes the results of the pilot-scale tests with engineered substrates (in this case gravel-like BOF slags from the Dofasco steel mill in Hamilton, ON) for two WWs from IMC: gypsum stack leachate, and a lime plant feedwater containing diluted leachate. Both were treated at a rate of 200 mL/min in a single pilot-scale HSSF engineered substrate cell (slag 0.8 m thick) with residence times of about 1 day. Table 7.3 shows the results on the treatability test.

TABLE 7.3

Results for Treating P-Contaminated Streams in an Engineered Substrate Cell

	Leachate	Lime Plant Feedwater
Average influent o-PO_4 concentration (mg/L)	460	162
Average effluento-PO_4 concentration (mg/L)	0.7	0.6
Removal (%)	99.9	99.6
Indicated PFR rate constant k_v ($year^{-1}$)	1687	1246

As may be seen, almost all of the phosphate was removed in a single pass. Some hydroxylapatite precipitate was observed to accumulate on slag particles near the test cell inlet, but this did not lead to clogging or interfere with operations.

The first-order volumetric plug flow phosphate rate constants of Table 7.3 may be compared with average literature values of 53 $year^{-1}$ for municipal WWs treated in ordinary SSF wetlands. Fluorides in the two streams (38 and ~5 mg/L, respectively) were also reduced dramatically to 0.2 and 0.1 mg/L, respectively, yielding the indicated fluoride rate constants of 1361

* See Equation 2.18 in Section 2.5.5 of Chapter 2.

and 887 year^{-1}, respectively. Test cell effluents were quite alkaline (pH > 11), so in an actual operating situation, any EEB System would have to include downstream pH adjustment and probably polishing in a BBR or an Aerobic Wetland cell.

This treatability test resulted in a concept to replace the IMC's lime plant with on-site trains of EEB cells based on "sacrificial" tertiary treatment cells filled with BOF slag 2–3 m thick that would be sized to remove phosphorus from the leachate for 5–10 years, after which the engineered substrate cells would be closed and capped, with new engineered substrate cells being built and operated nearby connected to the rest of an EEB System. Such a treatment option would be much more economical to operate, would not require operators 24/7 (as does the lime plant) and would probably result in discharged effluent of better quality than the lime plant.

8

The Design of EEB Systems

8.1 General Design Considerations

EEB Systems are designed in manners similar to those used for designing passive treatment WWT systems and mechanical WWTPs. While the sizing of CWs and Stand-Alone ABRs (determining required surface area needed for a particular WWT situation) is often carried out using empirical correlations (e.g., some predetermined m^2 of area or m^3/day of WW), EEBs are sized using Reaction Kinetics and designed using modern civil and chemical engineering methods (Kadlec & Knight 1996; Kadlec & Wallace 2008; Higgins 2014).

To design an EEB System, information is required on the flow rates of the WWs involved (base and peak rates); the contaminant loadings in the WW (again, average and peak rates); information on the (geo)chemistry of the major CoCs to be removed; the location, nature and condition of the receiving waters (or aquifer for infiltration disposal); discounted* expected target effluent quality; site aspects (space, soils, climate and geology); expected regulatory effluent discharge criteria and any relevant economic and logistical considerations.

For most EEB Systems, influent WW flows to EEB cells are equalized in some manner (e.g., in upstream 1° Sedimentation Ponds or balancing ponds, tanks or vaults), and introduced into the downstream EEB and other cells of treatment trains at controlled rates. The design approach has to take peak flows into account.

It is, however, noted that EEB Systems have high turndown/turn-up ratios and, unlike many active WWTPs, can be operated successfully over ranges from a few percent of design flow rate to several times that rate for short periods with no adverse effects. Some EEB Systems can even be "turned off," that is, left with no flow, for months at a time.†

* For example, if the regulatory discharge criterion for a CoC was 30 mg/L, conservative design might be for an effluent concentration (C_o) of 15 or 20 mg/L.

† Except during heavy rainfall, usually three of the BBR cells at BNIA are left idle full of water outside of the de-icing season and this has no effect on them as they are easily started up again when the next season begins.

The kinetic and other scale-up parameters needed to design EEBs should be determined during pilot-scale treatability testing (see Chapter 6), either at an on-site location or at a central EEB test facility (or both). During such treatability testing, the same kind of substrate that will be used in the later full-scale EEB System should be used to treat an imported or artificial WW spiked with the major CoCs to have a quality indicative of worse case conditions. A tracer test carried out in conjunction with the treatability testing will identify the Reaction Kinetic model that will best simulate the operation of the EEB (see below).

The cells of full-scale EEB cells are usually lined with geomembrane sheeting (HDPE, PVC or bituminous lining) but in some cases, where soils or local sources are appropriate, they may have compacted clay liners.*

Water levels in EEB cells can be controlled by swivel standpipes or adjustable weirs located inside concrete control structures located at the far ends of the cells. Generally, piping is designed to allow cleaning, flow reversal if needed and bypassing during periods of extra high flow. The surfaces of BBR cells are usually built flush with ground level, but in some cases they can be designed to allow the impoundment of water above them.†

EEB Systems can be designed to meet even the most stringent effluent performance criteria for contaminant concentrations and toxicity, including, if required, even producing effluents meeting drinking water quality guidelines. The ability to meet these high standards can be demonstrated by on- or off-site treatability testing, and this has been proven for many successfully operating facilities.

As is the case with many other kinds of WWT projects, EEB Systems projects are civil/chemical/ecological engineering projects that can consist of eight phases in series from the perspective of the ecological engineering firm designing them:

Phase 1: Project Preparation
Phase 2: Conceptual Design
Phase 3: Detailed Design and Engineering
Phase 4: Tendering and Contractor(s) Selection
Phase 5: Construction Supervision
Phase 6: Start-Up and Commissioning
Phase 7: Operations and Maintenance
Phase 8: Decommissioning

* The walls of the BBR cells at Heathrow Airport are made of concrete.

† For example, one of the BNIA BBR cells shown under construction in Figure 3.3 in Chapter 3, the one in the foreground, was constructed with low berms around it so that it also functions as an emergency stormwater pond. Doing this in no way adversely affects the performance of the BBR.

Treatability testing is almost always recommended as an early stage in EEB System project development. Sometimes this treatability testing is carried out before Conceptual Design (i.e., as part of Project Preparation) and at other times as the first part of Phase 2. Often, in addition to the activities described below, Project Preparation may involve carrying out a Feasibility Study to compare various technologies (including the use of EEBs) that might be used to treat a WW.

Project Preparation (Phase 1) should always involve a site visit or visits; data collection and analyses; GAP analyses and evaluations of local situations and conditions. In some cases where new and different WWs are being evaluated as to whether they can be treated in EEB Systems, Phase 1 activities may also include a variety of other project scope, cost, alternatives, economic, locational, operational impact and scheduling evaluations as well as a Feasibility Study.

Conceptual Design (Phase 2) generally involves more detailed site evaluations; the analyses of alternatives for parameters such as EEB and associated equipment locations; actual conceptual-level design and engineering and initial estimations of project capital (CAPEX) and O&M (OPEX) costs. Generally, cost estimates at this phase are based on data from earlier projects, although some polling of certain costs (e.g., those for substrates and major pieces of equipment such as air blowers) may also be included (Conceptual Design should not occur for new potential EEB System projects until site visits have occurred).

Often, the initial stage of an EEB System project is referred to as Systems Analysis and such involves both Phase 1 Project Preparation and Phase 2 Conceptual Design.

Phase 3 Detailed Design and Engineering often consists of two stages: *Phase 3A Preliminary Design and Engineering* and *Phase 3B Final Design and Engineering*. These can be shortened and combined for smaller EEB System projects.

For larger EEB System projects, the bulk of the design and engineering is carried out during *Phase 3A Preliminary Design and Engineering*. In such a case, it usually involves two major submissions by the design engineering firm to a client: that for the completion of 30% of the design and engineering work (*30% Design*) and that for its *50% Design* completion. Preliminary Design and Engineering also often involves preparing applications to regulatory and other authorities for needed permits and licences (based on either an earlier Conceptual Design, or an early stage of Preliminary Design and Engineering [e.g., that at 30% Design]) and assisting the client in obtaining these.

Phase 3B Final Design and Engineering involves finishing of the design and engineering for an EEB System. It may also involve carrying out *Value Engineering* analyses to ensure that the design so far is the best, most economical possible. For larger projects, Phase 3B generally consists of three stages: those at *70% Design*, at *95% Design* and at *100% Design*. Generally, *100% Design* Contract Documents (e.g., contract conditions, restrictions, details on required submissions, *Technical Specifications and Construction Drawings*) are also used for *Tendering*.

Combined CAPEX and OPEX estimates (referred to as *Engineer's Cost Estimates*, ECEs) should be prepared at each stage of an EEB Systems project. ECEs should progress from about ±50% accuracy for Conceptual Design to about ±30% accuracy at the Phase 3A 30% Design stage to ±95% accuracy at the Phase 3B, 100% Design level. ECEs involve obtaining progressively more detailed quotations and better cost information on project components.

Phase 4 Tendering and Construction Contractor(s) Selection for EEB System projects usually has the EEB design engineering firm assisting a client in preparing Tender documents using the 95% or 100% Design documents as bases; in some cases, managing the distribution of these documents to prospective construction contractors if the client so wishes; assessing construction proposals received; and recommending to the client which construction contractor(s) it may wish to select.

Phase 5 of a project, *Construction Supervision*, usually consists of two simultaneous activities, *Construction Monitoring* and *Design Support*. Both involve representing a client while the construction of an EEB System is underway by a selected construction contractor (or contractors in some cases where there are separable aspects), either as the Owner's Representative (the *Engineer*) or as support for a client Engineer.

Phase 5A Construction Monitoring can involve activities such as project administration; inspection; materials testing; contract documents interpretation; inspection of work-in-place; progress schedule review; maintaining records; reviewing progress payments to the construction contractor and its sub-contractors; serving as a licensed professional for certain aspects; reviewing the activities and progress of the contractor, suppliers and subcontractors; progress reporting; overseeing change orders; ensuring that As-Built documents are being prepared; assisting with subsystems start-ups and testing and ensuring project completion.

Phase 5B Design Support may consist of activities such as submittals reviews; schedule review; testing and inspection of project components; review of change orders; review and approval of As-Built drawings and, most importantly, ensuring that the EEB System is built as specified.

Phase 6 of an EEB System project, *Start-up and Commissioning*, usually consists of preparing O&M plans; assisting client personnel with starting up and operating the EEB cells and other facilities; and training client personnel on the operations and maintenance that will be required. Phase 6 activities may also include setting up monitoring activities; assisting in taking samples for process control or regulatory requirements during start-up and early operations and interpreting and advising on analytical results.

Phase 7 involves *Operations and Maintenance*. One of the great advantages of EEB Systems is their ability to operate virtually unattended without problems for extended periods of time. This is a particular advantage at locations (e.g., remote mine sites) where local conditions (e.g., winter snow) may make access difficult at certain times. However, it should be noted that EEB Systems are LOW maintenance systems, not NO maintenance ones. Routine monitoring is required with them, and periodic inspections are needed to ensure that associated mechanical components (e.g., pumps, blowers and instruments) continue to operate normally.

While active WWTPs rarely have operating lifetimes of more than 20 years before requiring complete replacement, EEB cells should have operating lifetimes of 25 years or more, although any associated mechanical components (e.g., pumps and blowers) may require more frequent upgrading/replacement. Operators of EEB Systems should plan to cleanse and/or replace the substrate beds of their EEB cells every decade or so.* Although cleaning or replacing the substrates in EEB cells might prove time-consuming, the costs for doing so will be small on a net present value basis in comparison to the savings in normal ongoing OPEX costs associated with alternative active WWTPs that would be realized if these were operated instead over the same periods.

Where EEB cells have gravel substrates (e.g., BBRs, semi-passive treatment BCRs), the removed gravel may be washed and reinstalled, although some associated piping may need to be replaced at the same time. In the case of passive treatment ABRs which contain substrates involving active media that may be used up during operations, consideration should be given every decade as to whether to replace them completely.

It should be noted that, like most other kinds of WWT systems, EEB Systems should be designed with upstream diversion chambers containing orifices or diversion ponds with gates which will allow the diversion of excess flow around them should storm conditions lead to the exceedances of their design flows.†

It is also noted that some components of EEB Systems may have high O&M requirements. These include the routine sludge and sediment removal tasks that will be associated with components such as Sedimentation Ponds, dosing ponds and Aerated Lagoons. As is mentioned in Appendix D (EEB Systems and Airports), the reformulation of de-icing fluids may now result in turbid influents to airport EEB Systems contaminated with particles of

* While a number of EEB cells (especially some test cells as discussed herein) have been shut down and their operations discontinued for various reasons, none has yet operated long enough to require major maintenance involving substrate replacement, However, it is noted that neither has substrate replacement yet been found to be necessary with well-designed SSF CWs that have been operating for many years, and the same may be expected for SSF EEB cells.

† Generally, an EEB System is designed (sized, see Equation 8.1) to treat water resulting from wastewater flows plus the influx of runoff from a design storm of some preset magnitude (e.g., a 25-year storm). However, care is needed in calculating the flows from such a storm using historical data as, owing to global warming, precipitation that might have been designated from, say, a 50- or 100-year storm in the past may be now much more common.

sludge, and primary treatment equipment will now have to include a sediment filter that will also require routine maintenance and regular clean out.*

As with any WWT project, planning activities for EEB Systems should consider their eventual *Phase 8 Decommissioning*. As has been mentioned, EEB Systems can be operated for extended periods after the closure of a WW-generating facility (e.g., a mine and a landfill), and can be designed such that, as CoC concentrations in the WW decline over time, to evolve first into CWs,† and over the longest time periods, so long as WW flows continue, into Created Wetlands. This can offer full walk-away closure.

8.2 The Modelling of EEBs

8.2.1 Bioreactor Hydraulics

To design an EEB, a hydraulic mass balance must be performed in order to define the water flow rate used in determining how big it needs to be. In the absence of storage or in/exfiltration, the amount of WW leaving an EEB (Kadlec & Knight 1996) is given by

$$Q_o = Q_i + Q_b + (P - ET).A \tag{8.1}$$

where the subscripts i and o define EEB influent (inlet) and effluent (outlet) conditions, respectively; Q is the flow rate of the WW passing through it (m^3/day); Q_b is the amount of precipitation that falls on the top and inside surfaces of any berms (banks) around the EEB cell above its surface (and hence flows into the cell, m^3/day); A is the surface area of the cell (m^2); P is the precipitation rate (m/day) and ET is the evapotranspiration rate (m/day).

Generally, for Conceptual Design, it can be assumed that the design WW flow rate can be given by:

$$Q = (Q_i + Q_o)/2 \tag{8.2}$$

The residence time of water in an EEB cell is given by:

$$\tau = \varepsilon.h.A/Q \tag{8.3}$$

* In any case, the high sludge-generating propensities of Aerated Lagoons treatment of glycol-contaminated streams at airports (to say nothing of their potential for causing bird strike problems) make them poor choices at these facilities, and designs involving them should be avoided.

† For example, turning off the air to a vegetated BBR will leave it to operate thereafter as an ordinary SSF CW, and this could occur once CoC levels in the WW had become low enough.

where τ is the residence time, ε is the porosity of the substrate medium (dimensionless) and h is the substrate thickness (m). τ is also sometimes referred to as the hydraulic* residence time (HRT) or the nominal residence time of an aliquot of water.

Another important parameter used in EEB design is the hydraulic loading rate (q or HLR) that is often quoted in units of cm/day.

$$q = Q_i/A = \varepsilon.h/\tau \tag{8.4}$$

The loading of any specific CoC being treated in an EEB is given by:

$$L(kg/day) = Q(m^3/day).C(mg/L).10^{-3} \tag{8.5}$$

The critical loading parameter for aerated EEBs is usually that for organics, measured as BOD, COD or TOC. The trend with the design of new airport BBRs (see Chapter 4) has been to use loading rates as high as 1500 kg BOD/day, but in doing so great care has to be taken to avoid excess biomat formation near the inlets of the bioreactors. If the concentration of BOD in an EEB cell's feed increases, the flow rate has to be decreased to maintain the loading in the desired range. This can be accomplished by monitoring feedstock flow and the COC concentration into an EEB cell using a SCADA system that will automatically adjust the flow rate and path as required.

Often, existing CW cells can be upgraded to EEBs, but in doing so physical constraints in the construction of the cells may limit loading. For example, the design loading rate (L_{BOD}) for two new BBR cells at EIA is 700 kg BOD/day, but this was for converting two existing HSSF CW cells (of 12 in six trains of two cells each) whose earthen berms limited the depth of gravel that could be added to them when upgrading them to the VSSF BBR service (the previous CW gravel depth was 0.7 m and the maximum that could be added given the existing cell berms only raised the thickness to 1 m, see Section 3.7 in Chapter 3).

Flow through a permeable medium such as the substrate of an EEB can be given by a form of Darcy's Law as

$$Q = -K.dh/dx.A_c \tag{8.6}$$

where K is the hydraulic conductivity of the substrate (m/day), dh/dx is the hydraulic gradient in the direction of flow and A_c is the cross-sectional area perpendicular to flow. As the flow rate increases in an HSSF EEB cell,

* Refer to the definitions in the Glossary to distinguish between hydraulics and hydrology.

it becomes progressively more difficult to "push" water through the substrate, and the risk of water "surfacing" and flowing over the top of the bed increases. However in a downflow VSSF EEB such as a BBR, the hydraulic head, dh/dx, is 1, A_c becomes the surface area of the EEB cell (A) and the maximum flow rate can be given by Q = A.K. Accordingly, VSSF BBRs can allow much higher flows than the HSSF ones and are generally preferred unless some specific project aspect indicates otherwise.

When designing pilot-scale EEB cells for treatability tests, and even smaller commercial-scale ones, provision has to be made internally to prevent wall effects (i.e., the bypassing of some of the WW being treated due to it running down the bioreactor walls and not being exposed to the microbial biofilms).

8.2.2 The Sizing of EEBs

There are various methods for determining the size of a natural WWT basin (cell) such as an EEB. All begin by determining the area needed to achieve some predetermined removal of a particular CoC that is the most germane (e.g., BOD, NH_3-N, a particular dissolved metal [Me]). These include the following:

- Areal Loading Methods
- Pond Models
- Multiple Regression Models
- Reaction Kinetic Models

8.2.3 Areal Loading Methods

Areal loading methods to determine size (cell surface area) are empirical in nature and relate some parameter to the area of a cell. The parameter may relate to flow rate (e.g., A/Q, the area [A] compared to a design WW flow rate [Q] which can be expressed in m^2/m^3/day or ha/1000 m^3/day); to the loading rate of a particular CoC (e.g., kg/day of the contaminant per hectare) or, for municipal WWs, to how much treatment area is required per person equivalent of sewage [Pe] generated daily (e.g., m^2/Pe, conservatively 3–5 for an SSF CW).

An example of the areal loading method used in the past in the coal mining industry was to size anaerobic (compost) wetlands (see Section B.7 in Appendix B) for treating NMD using correlations for metal loading, pH and alkalinity. Typical values for their sizings in the past were 10 g Fe/m^2/day or 2 g Mn/m^2/day and the toleration of up to 3.5 g of acidity/m^2/day (Brown et al. 2002).

For treating domestic sewage, Table 8.1 provides some typical sizing estimates based on hydraulic loading.

TABLE 8.1
Areal Loading Method Correlations for Sewage Treatment

Treatment Method	Minimum Area (ha/1000 m³/day)	Maximum q (cm/day)
Natural treatment wetland	5–10	1–2
FWS CW	3–4	2.5–5
SSF CW	1.2–1.7	6–8
BBR	0.4–0.8	20–40

8.2.4 Pond Models

Pond models assume that the area of a natural treatment system is only related to the flow rate of WW being treated in it.

$$A = kQ \tag{8.7}$$

Pond models may be used to size simple Pond Wetlands that might form primary or tertiary treatment cells in EEB Systems. For example, for treating dilute WWs, a k value of 4.3×10^{-3} can be used with a depth (h) of 0.3 m and a residence time of 7 days (Reed et al. 1995).

For the use of a BBR to treat municipal sewage, a very rough first guess is to assume a sizing of 0.12 $m^3/day/m^2$ of the bioreactor surface area (~3 USG/ft^2).

8.2.5 Multiple Regression Models

Multiple Regression models are the results of the assembly of large amounts of empirical data on WWT in a natural treatment system. Although not often used for detailed design, multiple regression methods can prove useful in some situations. Kadlec & Knight (1996) provide that many such empirical correlations relating various wetlands engineering parameters (e.g., the concentration of BOD out of a SSF CW [C_o] is given as: $C_o = 0.33.C_i - 1.4$) are the results of a multiple regression analysis and were used by some to size Stand-Alone (i.e., non-EEB type) SAPS Bioreactors for treating coal mine ARD.

8.3 Reaction Kinetic Models

Modern EEB design is carried out using chemical engineering models based on Reaction Kinetics.* With them, contaminant transformations are assumed

* Reaction Kinetic models are also called mass balance models, and their use is also referred to as the Rational Method (Kadlec & Knight 1996; Kadlec & Wallace 2008).

to follow first-order kinetics and it is assumed that hydraulic performance in them can be modelled by one of the following Reaction Kinetic models.

- Plug Flow Reactor (PFR)
- Continuously Stirred Reactor (CSR)
- Tanks-in-Series Reactor (TIS)
- Some Combination of the Above

CSR models are sometimes also referred to as complete mix (CMR) or completely stirred treatment (CSTR) reactor models.

In the particular case of BBRs, modelling using Reaction Kinetics may be supplemented by aeration mass transfer methods (WEF 2010; ACRP 2013; Liner 2013). In any case, WEF aeration mass models require the input of data from a treatability test (i.e., Reaction Kinetic information). Accordingly, since every aerated EEB use situation will be different (e.g., WW, climate, soils and local design constraints), such models should only be used at the ongoing Phase 3 Detailed Design and Engineering phase of an EEB project to cross-check and supplement Reaction Kinetic data after Conceptual Design involving a treatability test (see Chapter 6) has been completed.

Often, it is known which CoC transformation will govern sizing and the equations for the selected model (as determined by a tracer test) need only be solved for that particular contaminant. However, this is not always the case and where one is unsure which CoC will result in the largest sizing, the equations have to be solved for each of the CoCs of relevance and the dominant one selected (e.g., this is sometimes the case where elevated concentrations of ammonia nitrogen are found along with BOD.)

In addition, some CoCs may be transformed into others during treatment in an EEB (e.g., organic nitrogen may be mineralized to ammonia, and ammonia nitrified to nitrate in a BBR) and such will cause concentrations of others to change continuously in such a manner that the direct application of a model sizing equation will not be possible and an iterative approach will be required. Kadlec & Knight (1996) have suggested a Microsoft EXCEL™ spreadsheet method using its solver routine to carry out such iterative approaches.

The spreadsheet in Table 8.2 follows their method and is for the hypothetical treatment of leachate resulting from stormwater infiltration into and through a pile of chicken manure.* As may be seen from it, even though BOD is moderately high (200 mg/L), and ammonia is relatively low (15 mg NH_3-N/L), it is the ammonia that dominates sizing.†

* This example is based on an actual confidential project.

† In this case, the solved minimum sizing of 16,623 m^2 suggested using several BBR cells in parallel to achieve the treatment.

TABLE 8.2

Use of a Spreadsheet for BBR Sizing

BREW Bioreactor Sizing

Parameter	Symbol	Units	BOD	TSS	TP	NH_3-N	Org-N	NO_x-N	TN	Fecal Coliforms
Input Concentration	C_i	mg/L	200.0	60.0	5.00	15.00	4.00	1.00	20.00	5.0E+04
Background Concentration	C^*	mg/L	14.1	14.7	0.05	0.00	1.50	0.00	1.50	100
Target Output Concentration	Desired C_o	mg/L	15.0	15.0	2.00	3.00	3.00	10.00	16.00	1000
Estimated Outlet Concentration	Est. C_o	mg/L	39.3	14.7	4.65	10.47		1.02	11.23	1.4E+03
Calculated Output Concentration	C_o	mg/L	14.2	14.7	0.4	0.27	1.6	0.19	2.01	100.0
Removal	R	%	92.9%	75.5%	92.9%	98.2%	61.2%	n/a	89.9%	99.8%
Areal Rate Constant @ 20 °C	K_{20}	m/y	171	1498	61	88	86	171		375
Volumetric Rate Constant @ 20 °C	K_v	y^{-1}	120	1051	43	62	60	120		263
Arhennius Coefficient	Theta		1.06	1.00	1.00	1.048	1.05	1.09		1.00
Areal Rate Constant @ T °C	K_T	m/y	171	1498	61	88	86	171		375
Indicated BREW Bioreactor Area	A	m^2	11378	1223	5549	16623	2181			3911
Pollutant Loading Rate at A_{max}	L	kg/ha.d	120.3	36.1	3.0	9.0	2.4	0.6	12.0	n/a
					Un-Ionized	0.39%				

Customer	Confidential
Location	Newfoundland
Wastewater	Manure Leachate
Date	

Feed Rate				
1000	m^3/d	183.3	USgpm
365,000	m^3/yr	1.32	MMGPD

KEY
Input Data
Calculated
Estimate

Water Temperature, T	20	°C
	68	°F
Water Depth, h	1.5	m
Water pH	7	

EB Area	16,623	m^2
	1.7	hectares
	4.1	acres
Driving Contaminant		
NH3-N		

Hydraulic Loading Rate, q	
25.00	cm/day
39.97	ha/1000 m^3/d

Porosity	95%	
Hydraulic Loading Rate c	6.02	cm/day
Residence Time	23.7	days

The rate constants (k's) for the transformation of a particular CoC (e.g., ammonia in the example) may be presented either in their areal versions (m/year) or their volumetric versions (day^{-1} or $year^{-1}$) and the relationship between them is given by the following equation:

$$k_A(m/year) = k_V(1/day) \times 365\,day/year \times \varepsilon h(m) \tag{8.8}$$

For spreadsheets such as the one illustrated in Table 8.2, expected outlet concentrations (C_o) can be determined during a treatability test or estimated from available multiple regression data.

The model equations that follow assume that first-order kinetics apply (i.e., removal rates are functions of inlet concentrations) and this is usually the case. However, for the EEB treatment of some CoCs in some WWs, other kinetic orders (e.g., half order, second order) may apply, and if it is believed that such situations might lead to undersizing an EEB, then the following equations may need to be modified accordingly (the kinetic order of the transformation of any CoC during EEB treatment can be determined during a treatability test by measuring its concentration at various levels in the test reactor).

8.4 The PFR Model

The simplest Reaction Kinetics model is the Plug Flow Reactor (PFR) model that assumes that aliquots of water move through the EEB as intact "slugs" or aliquots.

The removal of any WW CoC (e.g., ammonia nitrogen) using a PFR model is given by:

$$C_o/C_i = \exp(-\varepsilon.h.k_{PFR}.A/Q) \quad (8.9)$$

Here, k_{PFR} is the volumetric PFR rate constant (day^{-1}) for the removal of the CoC. Once a rate constant is known, it can be used to solve for the minimum area of a full-scale EEB cell treating the same feedstock and substrate.

Using a PFR model, this can be calculated as:

$$A = -Q.\ln([C_o]/[C_i])/\varepsilon.h.k_{PFR} \quad (8.10)$$

where ln indicates the natural logarithm and other parameters are as described earlier.

Each CoC in a WW will have its own distinct rate constant. As was discussed, the removal of the particular contaminant or chemical that gives the largest area controls and dictates the size (area) of the EEB cell required.

Up until recently, the recommended method for sizing CWs was to assume that they could be modelled by PFRs, but this approach has now been shown to lead to erroneous design and, although older texts and sources still present it, it now has been recommended that the PFR model not be used for wetland sizing (Kadlec 2003). Similarly, it is recommended that the PFR model not be used for sizing BBRs, although it may be used for sizing some kinds of ABRs.

The problem with the PFR model is that most bioreactors do not behave as ideal plug flow reactors in which influent WW passes through the cell as an unmixed "plug."

Figure 8.1 sketches the kind of response curve a tracer test on a real EEB cell (therein referred to as an EW Bioreactor) would generate, and compares it to that of a PFR.

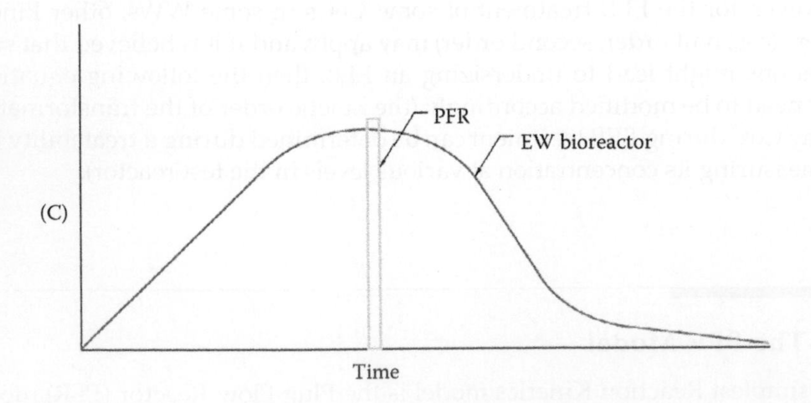

FIGURE 8.1
Tracer test response curve for an EEB and a PFR.

As may be seen from Figure 8.1, there is little correlation. In a real system, the concentration of the tracer measured in the cell's effluent will trace out a bell-shaped response curve, not the sharp spike expected of a PFR. This is because lateral and transverse mixing, backflow and a variety of other inefficiencies will cause water flowing through an EEB cell to deviate from plug flow conditions. PFR models can be amended to allow for such inefficiencies but the mathematics of such become quite complex.

However, simple (ideal) PFR models are still useful as they represent the most conservative conditions (i.e., if used in design, they result in the maximum size for a cell given the operating conditions). Often, sizings based on PFR models are used in fast initial calculations of an EEB cell's surface area. It is fully recognized that using PFR model-based sizings leads to over-sized bioreactors but they define upper boundary conditions and more sophisticated methods (see below) can be used during later stages of design to allow one to zero in on a more appropriate (smaller) size.

8.5 The CSR Model

Another type of Reaction Kinetic model that can be used in the design of EEBs is by assuming that the bioreactors behave as continuous stirred reactors. CSR models are often used in the design of conventional activated sludge WWT plants.

The removal of any CoC using a CSR model is given by:

$$C_o/C_i = 1/(1+\varepsilon.h.k_{CSR}.A/Q) \tag{8.11}$$

Here, k_{CSR} is the volumetric CSR rate constant (day^{-1}) for the removal of any CoC as determined during a treatability test. As with PFR models, once the CSR rate constant is known (again from a pilot-scale treatability test), the area of an EEB cell using the CSR model can be calculated as:

$$A = Q.([C_i]/[C_o] - 1)/(\varepsilon.h.k_{CSR}) \tag{8.12}$$

Intuitively, a CSR model might be expected to simulate an EEB better than a PFR model as WW being treated in them in some cases may be very well mixed (e.g., such as in a BBR cell where aeration provides good mixing). Just as PFR models define the upper boundary of sizing for bioreactors, CSR models allow the definition of the lower (minimum) sizes that the cells might have (the optimum size of real EEB cells will lie somewhere in between those predicted by PFR and CSR models.)

Figure 8.2 sketches the kind of response curve a tracer test on a real EEB cell would generate, and compares it to that of a CSR model.

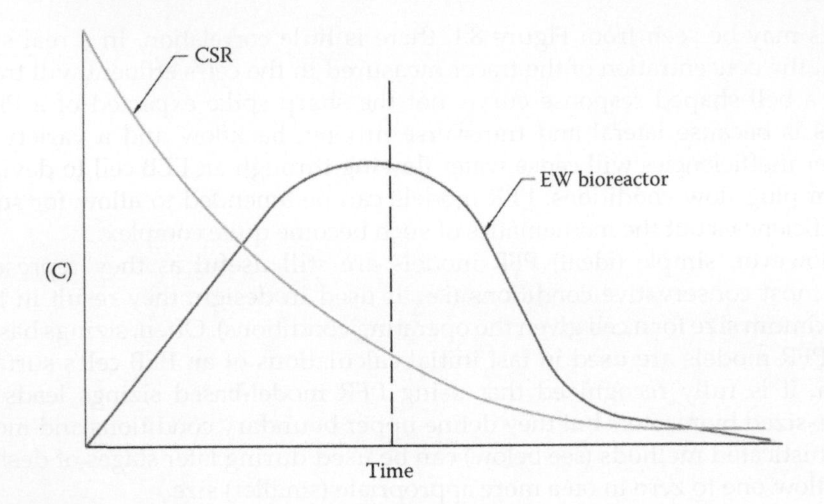

FIGURE 8.2
Tracer test response curve for an EEB and a CSR.

Again, for most EEBs, a CSR model does not accurately simulate reality. However, in some cases, it may do a better job than a PFR model. As was mentioned, CSR models are also useful as they represent the least conservative conditions (i.e., if used in design, they often would result in undersized cells) and again such can be useful in defining lower design limits.

As a first approximation, one might calculate an EEB surface area required based on sizings determined using both PFR and CSR models, and average the two (again, refer to the spreadsheet in Table 8.2.)

8.6 The TIS Model

There is a more complex version of a reactor-based model that has been shown, based on actual operating results from well-designed modern EEBs, to provide much better simulations for some of them. This is called the Tank-In-Series (TIS) model. It assumes that these bioreactors can be simulated by a series of CSR cells.

For this kind of model, which is also widely used in designing some kinds of lagoon-based WWT systems, the concentration of any contaminant is given by:

$$C_o/C_i = 1/(1+\varepsilon.h.k_{TIS}.A/N.Q)^N \quad (8.13)$$

The value of k_{TIS} will vary with N and will range from being identical with that k_{CSR} when N = 1 and to approach k_{PFR} as N becomes very large.

Figure 8.3 illustrates the concept.

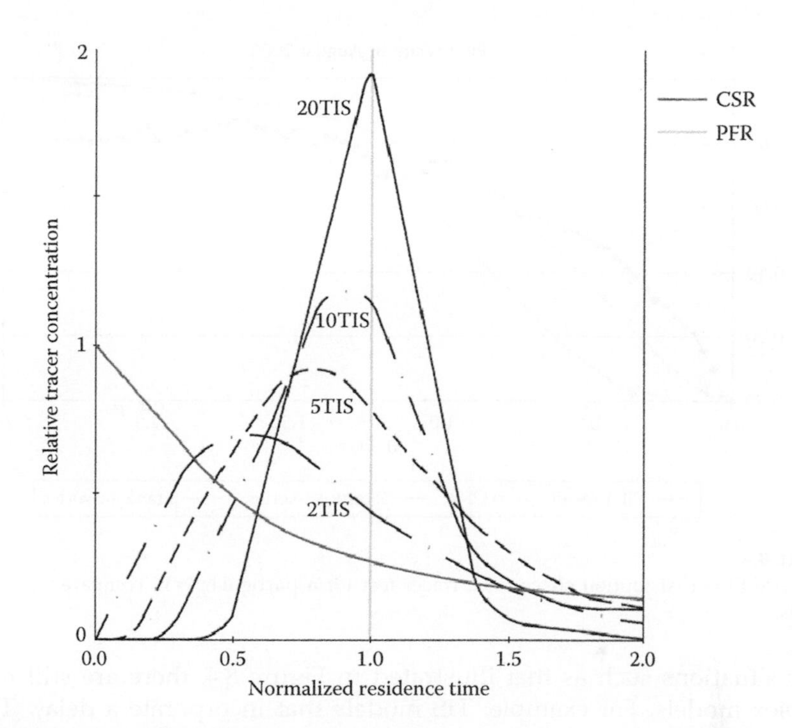

FIGURE 8.3
Some EEBs may be modelled by a series of CSRs (TIS).

For example, larger FWS CWs can be sized using the TIS model and assuming an N of 3 (i.e., the wetland is able to be accurately simulated by a 3TIS model, Kadlec & Knight 1996). For an EEB cell, flow patterns are expected to be complex, and although, intuitively, a CSR should be a closer match than a PFR, even this model is not expected to accurately model many EEBs.

For TIS models, the required area of an EEB to remove any contaminant to outlet concentrations is given by:

$$A = N.Q.\left(([C_i]/[C_o])^{1/N} - 1\right)/(\varepsilon.h.k_{TIS}) \quad (8.14)$$

8.7 The TIS with a Delay Model

Often, tracer test results indicate that simple Reaction Kinetic models will not apply for a particular WW/substrate/location/microbiome situation as is illustrated in Figure 8.4 graph (whose axes are normalized).

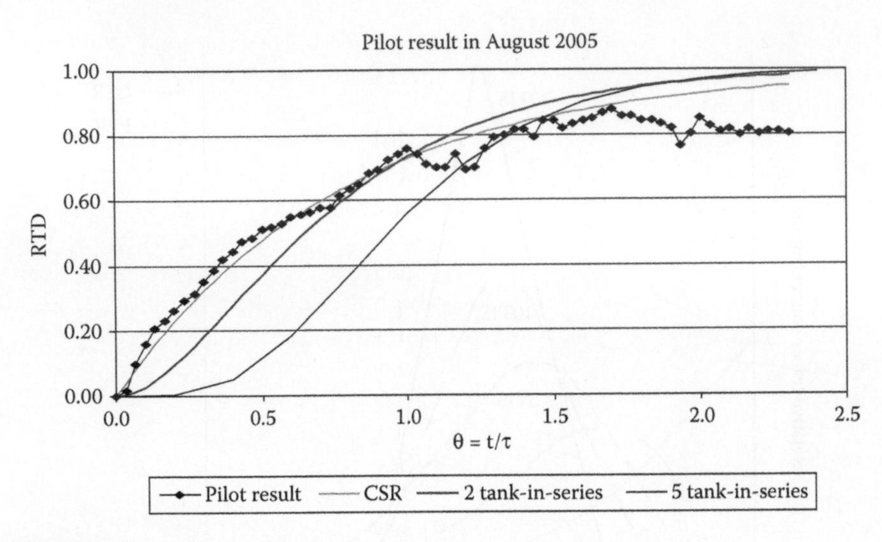

FIGURE 8.4
Residence time distribution curve for a tracer test for a particular WW compared with TIS Models.

For situations such as that illustrated in Figure 8.4, there are still more complex models. For example, TIS models that incorporate a delay (PFR) element are more appropriate. For N = 4, such a model can be given by:

$$C_o/C_i = (1 + \varepsilon.h.k_{CSR}.A/4.Q)^{-4}.\exp(-\varepsilon.h.k_{PFR}.A/Q)^{-1} \quad (8.15)$$

In this type of model, the value for the area (A) can be solved for by iteration.

8.8 The Impact of the Mass Flux Rate

There is a limit as to how much organic sludge can be tolerated. The substrates of aerated sub-surface flow EEBs (and CWs) must not clog, yet are vulnerable to doing so in both VSSF and HSSF morphologies. Excess biofilm growth occurs when production of new bacteria cells exceeds factors causing the death, consumption or decay of bacteria cells. If biofilm growth is so large that it extends beyond individual substrate particles, this microbial capsid material (the EPS or simply sludge) may bridge the gaps between them, forming biomats that can impede flow into and through bioreactors. Therefore, the design must be keyed to avoid this. Empirical design criteria have been developed to avoid clogging based on areal mass loading rates or long-term hydraulic conductivity (EPA 2000b; Lanfergraber et al. 2003) and support for this has been determined theoretically using *Monod growth kinetics*, a method that has been common since the mid-20th Century for the modelling of activated sludge-based WWTPs (Austin et al. 2006).

Since most clogging will occur in the inlet zone of a gravel-bed bioreactor, Wallace & Knight (2006) recommend limiting inlet BOD concentration loading perpendicular to flow to *250 g BOD/m^2.day* for HSSF CWs using gravel substrates in the 15–25 mm size range. Prudent design suggests paying attention to this criterion, the mass flux rate (F) even for aerated HSSF and VSSF EEBs, although much higher values (up to 500 g BOD/m^2.day) are allowable with them. If it determined that loading for a specific EEB exceeds a set target mass flux rate, EEB Systems operations should be keyed to reducing influent flow rate until it is met.

Practically, aerated gravel-bed (e.g., BBR) design can avoid the impact of mass flux rate periodically exceeding this limit by using substrates with larger diameter particles at the inlet to an EEB, using multiple inlet distributors, and/or using a VSSF morphology.

8.9 Rate Constants and Temperature

Generally, rate constants are recorded as their value at 20°C, and their value at any other temperature, T, can be calculated using the Arrhenius equation:

$$k_T = k_{20}.\theta^{T-20} \tag{8.16}$$

where θ (theta) is the Arrhenius coefficient. T is any other temperature for which rate constant data are available. In EEBs and CWs, the values of theta are usually determined during treatability tests by calculating the values for the k's at two or more temperatures. In CWs, θ is 1.048 for ammonia nitrogen according to Reed et al. 1995 and 1.04 according to Kadlec & Knight 1996. Table 8.3 outlines some values of theta for BOD from the literature and an earlier Stantec treatability test (Wallace et al. 2006).

TABLE 8.3
Arrhenius Theta Values

WWT Method	Arrhenius Theta
Activated sludge units	1.04
Aerated Lagoons	1.08
Trickling filters	1.035
Treatment CWs	0.98
BBRs, BNIA treatability test	1.03

8.10 The Influence of Background Concentrations

The concentration values used in the various reactor models mentioned above give reasonable results when the values of the influent and effluent concentrations of a contaminant used in the C_o/C_i ratio are relatively large. However, where concentration values are lower, they need to be modified (Kadlec & Knight 1996) by deducting C*, the irreducible background concentration of each CoC. The ratio then becomes $(C_o - C^*)/(C_i - C^*)$. Table 8.4 presents the range of and typical values for C*.

TABLE 8.4

Values for C* (mg/L)

CoC	Range	Typical C*
BOD	1–15	5
cBOD	1–5	2
NH_3-N	0–1.5	0
Org-N	1–3	2
TSS	1–6	3

Using C* values and the TIS model,* the surface area (A) of an EEB cell can be calculated using the following version of Equation 8.14:

$$A = N.Q.\left(([C_i - C^*]/[C_o - C^*])^{1/N} - 1\right)/(\varepsilon.h.k_{TIS}) \tag{8.17}$$

where N here is a value calculated during a treatability test; C* is a value from Table 8.4 or as calculated during a treatability test; Q is the expected design flow rate at the target loading condition (m^3/day); C_i is the anticipated "worse case" influent concentration for the parameter (e.g., BOD) for which design is being carried out (mg/L); C_o is the discounted effluent concentration for design; ε is the measured permeability of the substrate to be used and h is the operating depth of water in the EEB cell.

8.11 Design Suggestions

The following are a number of comments, observations and recommendations (*Do's and Don't's*) that should be considered when designing EEBs and

* Reaction Kinetic models that include C* values are sometimes also referred to as kC*P models.

the EEB Systems of which they form parts. These are suggestions only and are not absolute; there may be (and probably will be) exceptions for any of them.* For these Do's and Don't's, apply the KISS principle and use common sense in all cases.

- Do plan carefully before proceeding with an EEB System project. Information will be required on hydrology, WW quality, soils, site conditions, local flora and fauna, climate, weather, local and state/provincial regulations (e.g., discharge criteria), the permit/licence situation, who are the decision-makers and influencers for the client, who the major stakeholders are, and a host of other matters.
- Do include in your project team, or have available for advice as required, experts and support staff in areas such as civil and chemical engineering, (geo-)chemistry, hydrology, hydrogeology, microbiology, botany, geotechnical testing, ecology, permitting, aboriginal affairs, human relations, public relations (yes: PR!) and meteorology.
- Do not assume that environmental conditions (weather, climate and terrain) will not have a major impact on your project.
- Do not get involved with Stand-Alone bioreactor projects. All projects involving EEBs should be for their use as parts of more comprehensive EEB Systems.
- Do not assume that you know all of the technical terms associated with high technology projects such as those for EEB Systems. Check them out.
- Do always carry out a site visit as one of the first tasks of any EEB System project.
- With EEB projects, do not expect too much too soon. Biofilms take time to get going.
- Do give project management and site supervision priority for EEB System projects.
- Do demand and get As-Builts from construction contractors, and carry out thorough inspections before, during and after construction of an EEB System.
- Do request and get in writing appropriate performance guarantees from the suppliers of major pieces of equipment and instruments used in projects.
- Do realize that, although EEB cells are relatively robust and have high "inertia" (resistance to flow variations, CoC concentration spikes

* EEBs can range in size from the very small (e.g., ones occupying a single tote) to huge ones such as the existing BBR cells at BNIA. What may be an eminently suitable design or some other aspect for a small EEB may not be sensible for a large one, and vice versa.

and other upsets), EEB Systems are nevertheless complex, artificial, highly stressed natural systems and have to be regarded as such.

- Do realize that while EEB Systems are low maintenance systems, they are not "no maintenance" ones, and at least periodic attention to them will still be required.
- Where several or complex feedstock streams are involved, do always prepare a detailed Hydraulic Model for them before proceeding further with the design of an EEB System.
- Do always determine the relative values of TOC, BOD and COD in the feedstock stream of a proposed new EEB System, and use these data during design to help program its SCADA system.
- Do realize that many WWs (e.g., stormwaters and airport WWs) will not contain enough nutrients to maintain EEB cell substrate biofilms and that these nutrients will have to be continuously added to influents if EEB Systems are to be used to treat them (these nutrients are added as urea and phosphate fertilizers at BUF).
- Do not assume that feedstock pH will not affect the operations of an EEB. This is because some desired treatments to remove certain CoCs (e.g., ammonia using BBRs) are affected by WW pH and pH-adjusting chemicals (e.g., caustic soda) may also need to be added continuously to influents to allow optimal treatment.
- Do very carefully consider the costs and logistics of accessing, preparing, blending, transporting and installing the substrate and related materials (e.g., sand) for full-scale EEBs when projects involving EEB Systems are considered.
- Do carry out an upfront treatability test including tracer testing for almost all projects involving EEBs. Kinetic and genomic testing under "worse case" conditions is required to determine needed scale-up parameters. While indoor or outdoor testing can be carried out, it has been demonstrated that indoor pilot-scale testing under controlled conditions can provide the design parameters needed for successful full-scale design.
- Do design all EEB Systems to allow high-flow bypass.
- Unless it cannot be avoided, do not build one-train EEB Systems. If a one-train EEB System has to be considered initially, design with the aim of twinning it later and have it include at least two EEB cells that are piped up so that they can be operated (under the control of a SCADA system) in series or in parallel in such a manner that the CoC loading levels may be maintained by varying the flow path and feedstock rate.
- Unless project-specific matters dictate otherwise, do not use the HSSF BBR cells if "strong" (i.e., high CoC concentration) WWs are to be treated in EEB Systems.

- Do not undersize your EEB cells. Sizing using PFR models should only be used for ranging.
- Do try to design symmetrical EEB cells (e.g., rectilinear and circular) so far as site constraints allow. Do not use irregularly shaped cells if it can be helped.
- Do not use high aspect ratio (length:width) rectilinear EEB cells (>3–4:1). 1:1 or 2:1 is usually adequate unless site conditions dictate otherwise.
- Do design the piping around EEB cells to allow occasional flow reversal for clean out and other maintenance access.
- In areas with severe winter conditions, do heat trace piping into and out of EEB cells with heating tape, and provide stock heaters for inlet and outlet control vessels.
- Where operation in northern locations is involved, do cover the surfaces of BBR cells with insulating material. This material must be selected with great care so that it, any fines in it or its degradation products do not move down into the underlying granular substrate. If possible, underlying the insulating layer with a layer of geotextile should be considered to prevent this.
- Do not use aeration mass transfer models (WEF 2010; ACRP 2013) for the Conceptual Design of aerated EEB cells (e.g., BBRs). These may, however, once Detailed Design and Engineering commences, be useful for cross-checking sizing determined using Reaction Kinetics.
- Do realize that unless such is required for aesthetic or social reasons, the surfaces of BBRs and SAGR Bioreactor cells need not be vegetated at all, and if vegetating is wanted, it can be with terrestrial plants (e.g., road mix grasses) instead of wetland plants.
- Do not use the term "ABR" to define only BCRs; instead, use it as a general anaerobic bioreactor descriptor that includes DNBRs, SAPS Bioreactors, BCRs, Specialty BCRs and EBRs.
- If at all possible, do design the BCR cells of EEB Systems to be semi-passive in operation (i.e., with aggregates substrates and liquid active media addition) rather than using passive treatment (organic substrate) cells.
- Do not confuse the terms "active media" and "substrate." The active media may (or may not) be part of the substrate.
- Except in rare occasions for smaller EEBs, do not combine in a single cell the functions of two or more kinds of ABRs (e.g., DNBRs and BCRs), but instead do use separate EEB System cells for each function.
- So far as is feasible and practical, where solid active media are used in passive treatment EEB ABRs, do ensure that an absolute

minimum number of components is involved, and these should also be selected with due consideration to their cost in full-scale anaerobic EEBs, and the requirements for managing them at that scale.

- Do ensure that the proposed solid active media materials are carefully tested (e.g., mixed with water and assessed for pH, robustness, odours and breakdown products) before being considered as possible components of passive treatment EEB ABR substrates. Alkalinity-generating materials (e.g., limestone, un-composted manure and mushroom compost) should not be used in the substrates of SAPS Bioreactors. Wood chips and paper recycling plant biosolids have proven to be desirable kinds of active media to include in the substrates of all kinds of ABRs.
- Unless layouts to conform with terrain are required, do try to design BCR cells to be flat-bottomed and rectilinear.
- Do bury BCR cells or locate them at the bottom of impoundments to maintain anaerobic conditions in them. If the latter, design should include provisions for the bypass and draining of the impoundments for those times (rare) when access to the cell is needed for maintenance or to replace the substrate.
- Do design BCR-based EEB Systems to incorporate provisions to allow the periodic sampling of the substrate from at least one of the BCRs.
- Do ensure that mixtures of solid active materials used in ABR substrates (active media) are selected to include the readily bioavailable materials (e.g., "quick sugars") needed when the bioreactor first comes into operation, plus materials that biodegrade in the medium to longer terms.
- Do understand that biofilms in EEBs such as BCRs (and even BBRs) take some time to become established and permeate all parts of their substrates. Builders of EEBs must allow reasonable periods of time after inoculation (from 2 weeks to 2 months) for such *acclimatization* to occur before full operations can commence.
- Where the WW influent to a SAPS Bioreactor (or another kind of ABR) contains excessively high concentrations of suspended solids and/or dissolved iron, manganese or aluminum, do ensure that the concentrations of these CoCs are first greatly reduced in upstream primary treatment cells such as sedimentation or dosing ponds.
- Do design so that SAPS Bioreactor operations are such as to maintain the pH of a WW passing through it at a value below 3 until it reaches limestone treatment.

- While such may be acceptable for pilot- and demonstration scale EEBs such as SAPS Bioreactors and BCRs, for full-scale EEBs do try to carry out limestone neutralization in separate, buried ALD cells rather than in layers of limestone located inside the EEBs (i.e., KISS). Such anoxic limestone drain cells should be covered with plastic sheeting and earth, and buried near the EEB cells in such a manner that they will be accessible for opening and maintenance if required.
- Where a WW being treated in a SAPS Bioreactor contains more than 1 mg/L of dissolved aluminum, do design it to be operated in a pulse flow mode using a Dosing Siphon on its outlet.
- While effluent from a SAPS Bioreactors might be directed into a downstream Sedimentation Pond or CW cell (where ferric oxyhydroxide sludge will settle out), do try to design them so that their effluents are directed into BCR cells instead, as doing so will minimize sludge management requirements and potential CoC remobilization problems.
- Do design BCRs for downflow VSSF operation, and either buried or subaqueous. HSSF BCRs are not recommended. Reactive media such as limestone can form part of a BCR's substrate, allowing the bioreactor to supplement biological alkalinity addition with chemical alkalinity addition.
- Where a WW does not contain sufficient sulfate to allow sulfate reduction in a BCR, do supply the required sulfate in a liquid solution added with the influent and/or by adding a sulfate source (e.g., gypsum) to the BCR's substrate.
- Some CoCs (e.g., Cr) do not form sulfides in BCRs but do form oxides, hydroxides and/or carbonates in them under reducing conditions. Specialty BCRs can be designed to promote such reduction reactions. Since the resulting reductants may not be as insoluble as the sulfides formed when other metals (e.g., Cu and Zn) pass through a BCR, do design the compositions of their substrates such as to allow the (hydr)oxides/carbonates that do form in them to be strongly bound and rendered immobile.
- BCRs are not limited to sulfate reduction. Where other kinds of microbes are present such a selenium-reducing bacteria, do design a Specialty BCR to remove this element.

9

Summary and State-of-the-Art

9.1 Summary

9.1.1 The EEB Ecotechnology

EEBs are kinds of attached growth, in-ground, natural WWT methods in which morphology, substrates, operating methods, microbiomes, flows and/or other process conditions are manipulated and controlled to allow superior performance. They can successfully treat WWs of all sorts, even those containing particularly recalcitrant CoCs. In EEBs, microbes in bacterially generated biofilms on substrate particles are largely responsible for biodegradation. EEBs were once referred to as EWs or engineered bioreactors, but their designation as EEBs is now preferred. There are many kinds of aerobic and anaerobic EEBs.

The basic kind of aerobic EEB is the *BBR*, a kind of aerated SSF gravel-bed EEB (there are other kinds of aerobic EEBs as well). While many of the early BBRs had HSSF flows and sometimes this kind of flow path is still used, in most of the larger recent BBRs, WW flow is downflow VSSF. Flows of up to 10,000 m^3 or more are possible with this morphology. The surfaces of earlier BBRs were vegetated with wetland plants, but more recent ones have been planted with terrestrial plants such as grasses or left un-vegetated. The trend with BBRs is to use them with higher CoC loadings and to have thicker gravel substrates in them (2 m or more). In addition, instead of having one set of perforated inlet distribution pipes in each BBR near the substrate surface, a second set is now recommended as well, and this is located further down in the gravel substrate. Adding this second set of inlet distribution piping allows a better use of the substrate volume for CoC biodegradation, as well as the backflushing of the area around the upper set, and, if desired/needed, the "priming" of the bioreactor with an inoculant to generate increased microbial biomass before cold weather occurs. With BBRs that treat WWs deficient in nutrients (e.g., airport stormwaters), these compounds need to be added to the feedstocks to the bioreactors. BBRs represent proven and demonstrated ecotechnology.

The basic kinds of anaerobic bioreactors (ABRs) are *BCRs* and these are used to remove dissolved metals and metalloids from WWs such as MIWs and leachates. There are also other kinds of ABRs including DNBRs for removing nitrates and *SAPS bioreactors* for managing ARDs, as well as various types of *Speciality BCRs* for managing WWs with specific targeted CoCs (e.g., As, Se, and Mo).

There are two kinds of BCRs: (1) *semi-passive treatment BCRs* and (2) *passive treatment BCRs*. The former are gravel-bed bioreactors for which the organic active media (e.g., methanol) are liquids that are added separately to influents. Semi-passive treatment BCRs can treat large volumes of WWs (up to 10,000 m^3/day or more) and their flow regime is only downflow VSSF. Passive treatment BCRs are usually much smaller than semi-passive ones and cannot treat WW flows as high as those of semi-passive BCRs. Passive treatment BCRs can operate either up- or downflow VSSF and contain amended solid organic active media, the degradation of which leads to metabolites that are used by characterizing bacteria and other microbes to carry out the removals of the targeted CoCs. BCRs and other kinds of ABRs are usually either buried or located at the bottoms of impoundments to maintain anaerobic conditions in them. BCRs are proven and demonstrated ecotechnology.

EEBs form the *secondary WW treatment* components of EEB Systems and these are complemented and supplemented by *upstream primary WW treatment* cells and other components (e.g., ponds, filters and limestone treatment systems), and by downstream *tertiary WW treatment* cells and other components (e.g., filters, dosing siphons, CWs, and disinfection). In addition, EEB Systems can include a variety of associated equipment and infrastructure including diversion chambers/ponds, pump- and blower-houses, utility buildings and sophisticated monitoring and control instrumentation. EEB Systems can be designed to operate at high levels of effectiveness and efficiency at locations from the tropics to the arctic, and to operate with minimum attention for extended periods.

The following sections in this chapter review the kinds of WWs that can be treated in EEB Systems, and overview the perceived advantages of EEB Systems in general, and of the two main kinds of them, BBRs and BCRs. Also presented are suggestions for future R&D and for designing EEB Systems, BBRs and BCRs utilizing many of the results of testing and experience with actual ones presented in this book (state-of-the-art).*

9.1.2 WWs Treatable Using EEBs

The EEB System ecotechnology for natural WWT allows high levels of transformations of even recalcitrant CoCs using largely in-ground facilities that can operate at very high efficiencies whatever the ambient water and air temperatures (Higgins et al. 1999; Higgins 2003). Most aspects of the EEB

* Readers are cautioned that designing, building and operating EEB Systems requires not only "know how" (which can be gained from this book and other sources) but also experience and "know what" (which is not presented herein). The Co-Authors take no responsibility for the design or performance of EEB Systems that others might construct based on the information in this book.

ecotechnology are fully developed and proven, with over a hundred operating and pilot facilities using the this kind of natural WWT system to treat a wide variety of contaminated WWs, stormwaters and groundwaters including municipal sewage (Wallace 2004; Higgins et al. 2009c); municipal stormwaters (Higgins et al. 2009d); northern community WWs (Higgins & Liner 2007); aquaculture water (Higgins 2009); landfill leachates (Higgins 2000b; Nivala 2005); runoffs from landfarms and industrial waste tips (Higgins 2000a,b); mining influenced waters (Higgins & Mattes 2003; Higgins et al. 2003a, 2004, 2013); hydrocarbon-contaminated groundwaters (Wallace 2004; Wallace & Kadlec 2005); process waters from oil and gas production facilities (Higgins et al. 2006a,b); glycol-contaminated stormwater runoffs from cold weather aircraft de-icing operations (Higgins et al. 2007a,c, 2010a; Liner 2013) and many others (Wallace et al. 2006; Mattes et al. 2007; Higgins et al. 2007b).

9.1.3 Treatment in EEB Systems

EEB Systems allow the treatment throughout summer and winter in economical systems that can consistently meet stringent WW discharge criteria over extended periods. Their cells and other components can be "mixed and matched" to allow the treatment of WWs containing any combination of CoCs. EEB Systems are licensed in the same manner as are active WWTPs in a number of jurisdictions in Canada, the United States and elsewhere. EEB Systems are highly modular, allowing easy expansion (just add a train or cells), and should more stringent criteria ever be required in future, these systems can be relatively easily upgraded to meet them.

The following points summarize some overall EEB System design recommendations:

1. EEB Systems should have at least two trains, with each being designed to be capable of handling 80% of the design feedstock flow rate.* Where three trains are involved, each should be capable of handling 60% of the design flow rate.
2. EEB Systems should be designed so that major pumps and flow paths can be computer controlled using a SCADA system.
3. EEB Systems should always have primary treatment cells and components upstream of the secondary treatment EEB cells, and tertiary treatment cells and components downstream of them.
4. All EEB Systems should have diversion chambers/ponds near the front of the primary treatment cells/components. Usually emergency feed bypass control (e.g., flow control orifices) are located in

* The design flow rate will be the one used with "worse-case" WW quality settings and other information (e.g., scale-up data from treatability tests) to size an EEB cell. It does not represent the maximum hydraulic flow that can be sent through such a cell as an EEB System's SCADA computer control can set much higher throughputs, albeit under parallel cell operations.

them to allow the diversion around the rest of the EEB System to receiving waters of excess water during large precipitation events (although manual bypassing may be used in some cases). In addition, the WWs flowing into Diversion chambers/ponds normally should be monitored* to make sure that CoC concentrations always exceed some predetermined level, also allowing their diversion to receiving waters if they are found to be too dilute to treat.

5. In addition to diversion chambers/ponds, for most EEB Systems, settling/Sedimentation Ponds, cascades, dosing ponds, sediment ponds, rock filters, limestone drains, oil–grit separators and/or oil–water separators should form the primary treatment part of the process configuration, the exact components being dependent on the WW to be treated, its quality and specific project aspects.
6. While there are kinds of EEB Systems† that can handle varying feedstock flow rates, most EEB Systems should also have as the last step in their primary treatment configuration (just before secondary treatment EEBs) some sort of balancing method (e.g., impoundments, vaults, buried, or surface balancing tanks) to allow feedstocks to be fed to downstream, secondary treatment EEB cells at controlled rates. In EEB Systems with multiple treatment trains, the primary treatment components can be left as a single train that only divides into the multiple trains immediately before the EEB cells and their ancillaries.
7. For most EEB Systems, sand and/or carbon filters, polishing and/or dosing ponds, CW cells, Stormwater Wetlands also handling site runoff and (for WWs such as sewage containing potentially pathogenic microbes) disinfection equipment/vessels should form the tertiary treatment part of the process configuration.
8. If the EEB Systems are to be located in areas that experience harsh winter conditions, the surfaces and walls of EEB System cells should be insulated. In addition, piping between EEB cells as well as between them and primary and tertiary treatment cells should be buried at least 1 m, insulated and heat traced (in areas of extreme winter conditions, utilidors may be used). Inlet and outlet control structures should be similarly insulated and contain stock heaters.

* Often, this monitoring is done on-line using a TOC meter or an RI/TOC meter combination calibrated against desired parameters such as percent glycol, BOD or COD. For example, for an airport WW in Canada where the target maximum effluent concentration in WWs from major airports is 100 mg EG/L (see Appendix D), the meter could be calibrated to have the SCADA system open bypass valves if the EG concentration falls below this level. This would then allow most of the influent flow to enter pipelines leading directly to the discharge point at the receiving waters, although that fraction of the flow allowed through the diversion chamber's orifice would still proceed to the rest of the EEB System for treatment.

† See ESWs in Section 3.10 of Chapter 3.

9. At least one EEB cell of each type in an EEB System should be designed so that sampling can take place at various levels inside of it (in the case of the passive treatment alternative involving BCR cells with solid substrates, design also should be such that samples of the substrate can also be taken periodically).

   Unless specific project aspects dictate otherwise (and prospective sites have lots of space), EEB cells should be constructed with vertical sidewalls.Unless specific project aspects dictate otherwise (terrain, sub-surface conditions), EEB cells should be flat-bottomed and operate in VSSF modes.
10. Unless specific project aspects dictate otherwise (e.g., a need to have emergency stormwater storage above an EEB cell), the surfaces of BBR cells should be designed to be flush with ground surfaces.

9.1.4 Advantages of EEB Systems

In addition to specific advantages that EEBs such as BBRs and BCRs have (see Sections 9.1.4 and 9.1.5 below, respectively), EEB Systems *per se* have a number of overall advantages in general over other kinds of WWT, both natural WWT methods such as CWs and active mechanical WWT facilities. These include the following:

1. EEB Systems can be designed to operate at high efficiency whatever the ambient air and water temperatures.
2. If necessary, EEB Systems can be left to operate virtually unattended for relatively long periods, even during the harshest winter weather.
3. In many cases, the EEB Systems ecotechnology will be the *Best Available Technology Economically Available* (BATEA) for treating WWs, stormwaters and groundwaters contaminated with otherwise recalcitrant CoCs.
4. The cells and other process components of EEB Systems can be "mixed and matched" to allow the treatment of almost any WW, no matter how recalcitrant its contaminants might be.
5. When not needed for some reason, all or some of the EEB cells of an EEB System can be left idle, full of water, with no adverse impacts on performance after a future start-up, and no complicated shutdown or mothballing procedures are needed with them.
6. EEB Systems can be designed to produce effluents that meet even the most stringent discharge criteria.
7. Biocontrol is possible with EEB Systems.
8. Not only can EEB Systems successfully treat WWs containing high concentrations of CoCs, but they can also treat ones containing low, but not low enough, concentrations of them, something that is very difficult to do in many other types of WWT processes.

9. EEB Systems have a lot of "inertia." and process upsets (flow upsets or surges, contaminant concentration spikes, temperature swings, etc.) that might knock an active WWTP treatment offline for a period hardly affect their operations.
10. Generally, the CAPEXs of EEB Systems will be lower than those of active WWTPs treating the same WWs* (in some cases only half as much).
11. Since EEB Systems often do not require very much staff attention (as compared to that needed 24/7 to operate most mechanical active WWTPs), their OPEXs usually will be much lower than those of active WWTPs.
12. EEB System trains are highly modular, allowing their construction to be staged over time and/or for their facilities to be positioned at a number of locations instead of all at central locations.
13. Some EEB Systems with semi-passive treatment components (BBRs, semi-passive treatment BCRs) can be designed to evolve over time into more passive systems, and in some cases even "walk away" site closures may be possible with them.
14. EEB Systems can be designed to successfully treat WW flows from a few m^3/d to over 15,000 m^3/day.
15. The costs and time required to build and operate pilot-scale EEB test systems to demonstrate the treatability of any specific WW will usually be much lower than doing so for a comparable active WWT process.

EEB Systems are demonstrated and proven ecotechnologies for the treatment of glycol-contaminated WW streams at airports, and can manage both Spent Glycol and/or GCSWstreams from them, as well as any other CoCs (e.g., pavement de-icing chemicals) that may also be involved.

EEB Systems can be designed to treat acidic, neutral and alkaline MIWs and have already been demonstrated to be effective in treating many of them. EEB Systems for mines† will be more complex than those for treating municipal and airport WWs, and may involve both aerobic (e.g., BBR) and anaerobic (e.g., BCR) secondary treatment cells.

EEB Systems are proven and demonstrated ecotechnologies for the treatment of all sorts of municipal WW streams including sewage, stormwater, septage and CSO waters. Both HSSF and VSSF BBR morphologies can be specified as the secondary treatment component of them but the trend is for using un-vegetated downflow VSSF cells with thicker substrates.

Leachate treatment represents an area of wide potential for EEB Systems but each of the many kinds of leachates will be unique so extensive WW data

* Usually the CAPEX of an EEB System based on semi-passive BBRs is much lower than that of fully passive systems such as those involving SSF or FWS CWs.

† And associated facilities such as smelters and metal refineries.

collection and treatability testing will usually be required before projects are undertaken (see Table 2.5 in Section 2.3.5 of Chapter 1). It has already been demonstrated that MSW leachates can be readily treated in EEBs (see Chapter 6). EEB Systems treating leachates will often have to include, in addition to specific primary and/or tertiary treatment components, DNBR, BCR and BBR cells for secondary treatment.

Before any project involving EEB Systems proceeds very far, treatability testing should be carried out (see Chapter 6).

The EEB Ecotechnology is not complete but is continually evolving and improving. Section 9.3 lists the recommended (and in some cases already ongoing) R&D and treatability testing for doing so.

9.1.5 Advantages of BBRs

All BBRs are semi-passive (hybrid) in operation and require a source of aeration air (usually from nearby blowers). Downflow VSSF BBRs are the preferred morphology for the EEBs; gravel is usually used for their substrates; their cells are usually rectilinear; and loadings up to 1000 kg BOD/day.m^2 are now common. The advantages of BBRs over passive treatment alternatives such as CWs and active (mechanical) WWTPs treating the same WWs of the same volumes include

1. BBRs can be designed to treat most kinds of CoCs amenable to aerobic transformations and, in doing so, they often are as or more efficient than active WWT processes. Removals/degradations for many CoCs being treated in BBRs are usually in the 95%–99%+ range.
2. BBRs are very much more efficient than are ordinary CWs (rates up to an order of magnitude higher, often reducing CoCs in effluents to non-detect levels) and have very much smaller surface areas (down to 1/10 as much, as much thicker substrates [1–3 m thick] can be used with them).*
3. BBRs require much less aeration energy than that required by an equivalent mechanical, activated sludge process-based plant (again, about 1/10 as much).
4. BBRs can be built in a wide variety of sizes and successfully operating examples are already being used from those that treat small flows such as domestic septic tank overflows (where they replace leach beds) to ones treating many thousands of m^3/day of WWs.
5. BBRs can operate efficiently whatever the ambient air temperatures and with WW temperatures as low as 0.1°C. They can be used from the Arctic to tropical locations.

* The thicker substrates of BBR cells are deeper in the ground and thus retain heat better. Recall that the aerobic biodegradation processes in them are exothermic as well.

6. Nominal residence times in BBRs for aliquots of WW passing through them are usually measured in days (vs. only hours in most active WWTPs), and this allows better contact and better performance.
7. BBRs need not be vegetated with wetland plants (unless an operator so wishes), but instead may be vegetated with terrestrial plants (e.g., road mix grasses) or left un-vegetated.
8. BBRs can be configured to allow the luxury uptake of condensed phosphates from influent WWs.
9. When treating sewage and other WWs containing pathogenic organisms, BBRs will disinfect them and separate downstream disinfection may not be required.
10. Unlike many active WWTPs and Aerated Lagoons, BBRs do not produce by-product biomass sludges that require further management.*
11. Genomics can now be used to identify biofilm bacteria and other microbes that allow superior CoC transformations in a BBR, and populations of desirable ones may be grown using biostimulation or bioaugmentation, and inoculated into the BBR's substrate to enhance performance (biocontrol).
12. BBRs represent demonstrated and proven ecotechnology.

As with any EEB, the Detailed Design and Engineering of BBRs should be based on Reaction Kinetics, and scale-up information should be determined during a treatability test.

9.1.6 Advantages of BCRs

There are several types of passive and semi-passive treatment EEB ABRs but BCRs are the most common. ABRs can be designed to serve a variety of purposes, including: BCRs to remove metal(loid)s; DNBRs to remove nitrates, and SAPS Bioreactors to safely neutralize ARDs. All can be used as parts of 2° treatment step(s) in EEB Systems, and almost always some form of 3° treatment is provided downstream of them to remove degradation products when small amounts of the active media materials in them degrade.

The advantages of BCRs over passive treatment alternatives such as CWs and active (mechanical) WWTPs treating the same WWs at the same volumes include

1. BCRs are much more efficient at treating CoCs susceptible to anaerobic transformations than are passive WWT systems such as CWs, and often are as or more efficient than active WWT processes.

* BBR cells are designed such that the biomass produced in them exceeding that required for biofilm maintenance is removed by biodegradation and rhizodegradation as fast as it is produced.

2. Depending on particular project circumstances and the WW to be treated, BCRs can be configured either for passive WWT (organic substrates) or for semi-passive WWT (gravel or crushed rock substrates like BBRs).
3. For some WWs with lower pHs treating relatively small volumes of WW, some smaller (and usually passive treatment) BCR cells may have inside them separate, non-substrate layers of limestone before the substrate beds *per se*.
4. BCRs can be built in a wide variety of sizes ranging from small ones that treat small flows to ones treating thousands of m^3/day.
5. Semi-passive treatment BCRs with gravel substrates are capable of treating large volumes of WWs.
6. BCRs can be designed so that their substrate beds include limestone, thereby allowing chemical as well as biological alkalinity increases to occur in them.
7. Passive WWT BCRs can be designed such that the replacement of their organic substrates need not occur for extended periods (usually 5–20 years).
8. The substrates (gravel) in semi-passive treatment BCRs may never need replacement, although prudent project planning should allocate funds for replacing/cleaning the gravel after 10 years of operations.
9. Passive treatment BCRs can be located downstream of SAPS Bioreactors, allowing precipitable metal(loid) sulfides and other compounds (hydroxides and carbonates) to be deposited in their substrate beds as less voluminous and much more resistant to re-mobilization materials.*
10. Specialty BCRs can be used where the WW being treated does not contain sufficient sulfates to allow adequate sulfate reduction (in which case sulfate must be added to the influent or a source provided as part of the substrates), or where BCRs are required to remove specific, targeted metal(loid)s such as arsenic, chromium and selenium (Oremand et al. 1989), in which case other kinds of characterizing microbes than SRB may be important (Sonstegard et al. 2010; Walker 2010).
11. Genomics may also be used to identify biofilm bacteria and other microbes that allow superior CoC transformations in a BCR, and populations of them may be grown using biostimulation or bioaugmentation and inoculated into the BCR's substrate to enhance performance (biocontrol).
12. BCRs represent demonstrated and proven technology.

---

* This obviates the need for operators to have to manage sludge, which might be the case if effluent from a SAPS Bioreactor flowed into a downstream pond or CW instead of a BCR.

As with any EEB, the Detailed Design and Engineering of BCRs should be based on Reaction Kinetics, and scale-up information should be determined during a treatability test.

In summary, BCRs should preferably be designed for semi-passive treatment VSSF operation and can have substrate beds up to several meters thick, the maximum thickness of which will depend on specific project considerations and the nature of the substrate materials available/selected. In some cases, passive treatment BCRs may also be considered. Passive treatment BCRs whose substrates contain solid organic active media are suitable for lower flow rates and may be designed for either up- or down-flow operations. Semi-passive treatment BCRs should only be designed for downflow operations. The make-up of BCR substrate beds may include limestone to allow chemical as well as biological alkalinity increases in them. While layouts to conform with terrain may be required in some cases, in general BCR cells should be flat-bottomed, lined, rectilinear and have aspect ratios less than 3. BCR cells should either be buried or located at the bottom of impoundments to maintain anaerobic conditions in them. Designs should incorporate provisions allowing the periodic sampling of the substrate.

9.2 State-of-the-Art

9.2.1 State-of-the-Art and EEB Systems

On the basis of the material presented in this book, *EEB Ecotechnology*, and the experience of the Co-Authors and their associates, as well as the know-how presented in this book, the attributes of EEB Systems can be suggested that would allow the most up-to-date EEB designs possible. Two cases are discussed. The first, Case 1, is for a *BBR-based EEB System treating glycol-contaminated wastewaters at a hypothetical airport* located near an urban centre. The second, Case 2, is for a *project where large volumes of MIW at a remote northern lead-zinc mine are treated sustainably using a semi-passive treatment BCR-based EEB System.* Although the designs discussed below for these two hypothetical cases involve EEBs treating certain kinds of WWs, the information presented can be easily adapted for the same kinds of EEBs treating other kinds of WWs at different locations.

9.2.2 Case 1 State-of-the-Art for an Airport EEB System

Advances in BBR ecotechnology can be illustrated by considering the design of the following *Case 1* airport EEB System and its associated facilities.

Figure 9.1 shows a process flow diagram (PFD) for a single train, BBR-based EEB System treating Spent ADAFs at a northern airport where EG-based

ADAFs are used.* The EEB System at this hypothetical airport will treat Spent Glycols plus GCSW and other airport WW streams from collection systems around it.†

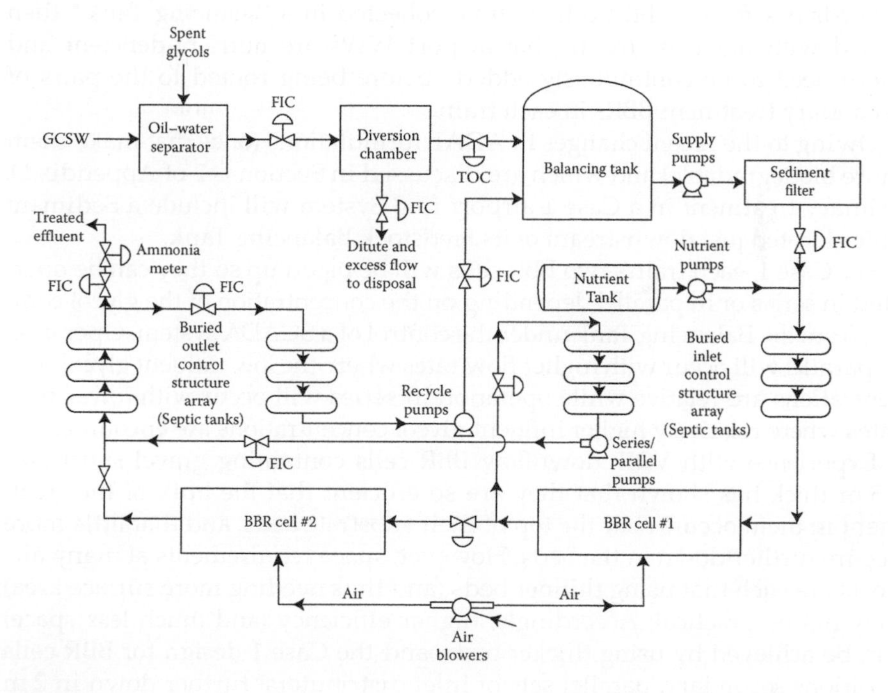

FIGURE 9.1
Case 1 PFD.

The process configuration for the EEB System at the Case 1 hypothetical airport will also allow, if feedstock EG concentrations become too high, the recycling of some treated effluent back to the Balancing Tank to keep ethylene glycol levels in it within a reasonable range.‡

* That is, the system will be in Canada but the logic presented herein will also apply to airports in the United States where PG-based ADAFs are used (see Section 2.3.2 in Chapter 2, Appendix D, and the Glossary to review technical terminology for airport de-icing WWT).

† See the possible slate of CoCs in the first paragraphs of Section 2.3.2 in Chapter 2.

‡ Typically, BBR cells reduce influent organic (e.g., glycol and BOD) concentrations by an order of magnitude. Very roughly, if the target effluent discharge criterion was 100 mg EG/L (see Table D.2 in Section D.3 of Appendix D), the maximum glycol concentration allowable in the feedstock tank would be 10,000 mg/L and the SCADA system would be programmed to operate the two BBR cells in series at a reduced feedstock flow rate for such conditions. If the EG concentration in the feedstock tank was above 10,000 mg EG/L, some treated effluent would be pumped back into it, diluting its contents until it reached that value. If the EG concentration in the feedstock tank were much lower than 10,000 mg EG/L, at some point the pre-programmed SCADA system would allow the two BBR cells to operate in parallel at higher feedstock flow rates. As was mentioned in the Design Suggestions of Chapter 8, a detailed feedstock Hydraulic Study should always be a part of the early phases of any new airport EEB project.

Depending on the size of the hypothetical airport, the Case 1 EBB System might have more than the single, two-BBR train indicated in the PFD. Indeed, for most larger airports, two or more trains will be desirable.

Feedstock for two BBR cells will be collected in a Balancing Tank,* then dosed with nutrients (recall that airport WWs are nutrient-deficient and these need to be continuously added), before being routed to the pairs of secondary treatment BBRs in each train.

Owing to the recent changes in ADAF formulations (ones that make them more biodegradable) and which are discussed in Section D.2 of Appendix D, primary treatment in a Case 1 airport EEB System will include a Sediment Filter located just downstream of its feedstock Balancing Tank.

For Case 1, each train's two BBR cells will be piped up so they can be operated in series or in parallel depending on the concentration of the glycol coming from the Balancing Tank under the control of a SCADA system. Operation in parallel will occur with higher flow rates where the low influent glycol concentrations are relative while operation in series will occur with lower flow rates where relatively higher influent glycol concentrations are encountered.†

Experience with VSSF downflow BBR cells containing gravel substrates 1.5 m thick has shown that they are so efficient that the bulk of the treatment in them occurs near the top of their substrate beds, and that little more occurs further down in the beds. However, space requirements at many airports are such that using thinner beds (and thus needing more surface area) may not be practical. Accordingly, higher efficiency (and much less space) can be achieved by using thicker beds, and the Case 1 design for BBR cells positions secondary, parallel sets of inlet distributors‡ further down in 2 m thick beds below the 1 m level. These will be piped up so that they can be operated individually or both at the same time. Having two sets of inlet distributors will allow better control and more effective use of bed space and, with appropriate piping, even the ability to back-flush areas nearer to the top of the bed should any bed problems ever occur there.

In the BBR cells of Case 1's EEB System, air lines (with appropriate check valves) will also be attached to the lower BBR cell influent distribution grids to facilitate backflushing and bed purging with air if needed.

For Case 1, the substrate of the BBR cells will be 15–20 mm screened and washed gravel. Cost and the availability of convenient local sources will be the deciding factors as to what kinds of gravel should be used.§

* It is noted that while the Case 1 mixed glycol-contaminated streams' feedstock could be assembled in a large aboveground tank, condition/situations at some airports might dictate underground feedstock tankage or partially buried vaults such as that at BUF (see Section 3.6 of Chapter 3).

† Already at BNIA and other airports with EEB Systems, BBR feedstock flow rates are varied depending on influent glycol loading, so this simply will extend an existing practice.

‡ SAGR Bioreactors often also have lower sets of inlet distributors as well but in a different morphology for different reasons.

§ Readers are reminded that the costs of buying and managing gravel substrates are some of the largest CAPEX items for EEB Systems (see Sections 7.2 and 7.3 of Chapter 7).

For Case 1, it is assumed that genomic testing of BBR cell biofilm microbial communities will be carried out regularly and that biocontrol will be used to optimize them (see Section 1.11 in Chapter 1).

Since the generation of VFAs (see Equation 2.7 in Section 2.3.2 of Chapter 2) can lower airport WW stream pHs and reduce the efficiency of some processes (e.g., nitrification), it is noted that, in addition to dosing needed nutrients into the feedstock of the Case 1 EEB System, some method of keeping feedstock pH circum-neutral will have to be employed as well. Injected caustic solution is assumed for Case 1.

For Case 1, it is envisaged that electrical power will be supplied from a local grid, and a transformer and motor controllers will be located in a heated utility building positioned near the BBR cells. The blowers and most of the pumps for Case 1 will also be located in this building which will also house the major instruments (TOC meters, an effluent ammonia meter and other instrumentation) as well as a small office and storage space.

For Case 1, it will be assumed that needed nutrients (urea and phosphate fertilizer) will be provided on contract as nutrient solution, and a tank for received volumes of this (plus dosing pumps and other hardware) will also be located in the utility building.

In summary, the major, state-of-the-art advancement represented by Case 1 will be the two-inlet BBR cell design and thicker substrates (2 m) in them. Another will be the use of biocontrol using genomic testing to identify and promote optimal biofilm microbial communities.

9.2.3 State-of-the-Art for a Semi-Passive BCR-Based EEB System

In northern Canada and Alaska, there are various base metal mineral deposits that have been found to be economic to mine using underground mining.* Typically, the ores involved are those containing cobalt, copper, lead, nickel and/or zinc, and minor constituents also present in the ores may include arsenic, iron, magnesium, manganese, selenium, silver and many more metal(loid)s.

Projects exploiting these kinds of ore bodies are major undertakings often involving thousands of people during construction and hundreds during operations. Mines for base metals involve extensive equipment and support systems (e.g., mining equipment, vehicles, headworks, hoists, crushers and grinders, conveyers and feeders, ventilation and air conditioning systems, compressed air systems, safety systems, bins and chutes, hydraulic systems, pumps, blowers and electrical systems to name just a few), as well as related infrastructure including haul roads, parking areas and garages,

* Some northern mines use open pits for mining, either as an alternative to underground mining or to supplement it, and there are many similar precious metal mines in the north that are somewhat similar but the Case 2 example will focus on an underground base metal mine, not one with open pits or involving precious metals.

maintenance areas and shops, storage buildings, slurry pipelines, water lines, raw mined-ore piles, waste rock accumulations, tailings ponds and sludge ponds, along with facilities to provide and store needed fresh make-up water, and to clean up and recycle reclaim water (e.g., pump stations, pipelines and ponds). All of these are generally located on level (or levelled) areas (process areas).

Advances in the design of BCRs can be illustrated by considering Case 2 for *a hypothetical underground base metal mine* (the Mine) at a semi-remote northern site. Ores from mines of this sort may or may not be acid-generating, but in most cases at least some acidic rock will have be managed. For Case 2, it will be assumed that the MIW from the Mine will be acidic. For Case 2, it will also be assumed that this acidic MIW will be treated in a semi-passive treatment BCR-based EEB System both during mining operations (Mining) and after mining ceases (after Closure).

Case 2 assumes that at the Mine there will be a cleared and levelled process area, the location of headworks, buildings, etc. (the Process Area) and that it will be located at a higher point above the terrain that falls steeply to a nearby reasonably wide river valley, and the valley's small river (the River, the receiving waters) will have wide ascending banks that are or can be surrounded by Natural Wetland features.

Most northern mines have large mills at their process areas that are used to upgrade the raw ore to the desired product: metal concentrates that are shipped out to smelters and metals refineries further south. The mills consist of large, heated buildings housing SAG mills, ball mills, flotation equipment and concentrators. Associated ore management equipment to get the ore into the mills involves haul trucks and other ore hauling equipment, crushers and grinders and conveyers. Mills usually also have associated fire water supplies and equipment, as well as locker rooms, storage areas, laboratories and related test facilities in them. Case 2 assumes that such a mill (the Mill) and associated buildings will be located on the Process Area.

Usually the same trucks that are used to transport out concentrates bring in on their trips north camp and mining supplies such as lime (in bags) as well as the liquids (many of them in totes) needed for operations such as methanol, flotation chemicals, glycols, propane and fuels, etc. This will be assumed to be the situation for Case 2, and most of these supplies will be stored in buildings near the Mill.

Mining and milling are energy intensive and while some northern mines generate their power on-site using diesel generators located in buildings at their process areas, many can access power from generators at nearby hydroelectric dams or from transmission lines carrying hydroelectric power south. Case 2 assumes that the Mine will have access to local electric power supplies.

Most mine sites in the north are accessible at least part of each year by existing roads or railroads and/or by ice roads in winter. Nevertheless, building roads from them to the sites of northern mines, and in and around the mine sites, as well as preparing operating areas, are major activities. The operators of most northern mines make their own gravel to facilitate doing so and they have dedicated level (or levelled) areas (gravel pads) nearby used for making the gravel from excavated rock transported there (e.g., equipment at the gravel pads includes crushers, dump trucks, loaders, mixers, screens, and conveyers) and associated equipment for accessing, storing and handling the gravel. Generally, other materials such as sand, limestone, stripped soil and rock are also stockpiled at or near the gravel pads. Case 2 assumes that there will be such a gravel pad (the Gravel Pad) in the area of the Mine.

Most northern mines in remote and semi-remote areas have large on-site camps that include sleeping areas, change and shower areas, recreational facilities, cafeterias and building infrastructure areas. These camps usually are associated with facilities for fresh water supply, sewage treatment, solid waste management, as well as roads, vehicles and airstrips used to rotate crews in and out. Case 2 assumes that such a camp (the Camp) will be located in the vicinity of the Process Area.

Environmental protection is a major activity at all northern mines and facilities and equipment are present for environmental monitoring, air pollution control and to handle process water, deal with Camp and Process Areas' sewage and treat other WWs.

Many existing mines in the north have dedicated or mixed use heated buildings onsite in which active WWT equipment such as those based on membrane bioreactors or SBRs are located. However, these require 24/7 operating personnel, have by-products of their own that need treatment, and their use begs the question of what will happen when the ore deposit is mined out and Closure occurs, since, after mines are decommissioned, leachates, contaminated runoff and other MIWs can continue to flow for decades and even longer. Case 2 assumes, that instead of considering an active WWTP at the Process Area, the hypothetical Mine's owners will opt *ab initio* for the use of semi-passive treatment BCRs forming part of a comprehensive and sustainable EEB System for MIW treatment both during Mining and after Closure.

For Case 2, it is also assumed that extensive treatability testing will have been carried out prior to designing the EEB System, with pilot-scale test units having been operated both at an off-site central test unit location like CAWT and later on-site at the Mine to determine the scale-up parameters needed to design the EEB System.

It will be assumed that the terrain at the Mine site, as is often the case in the north, will be bisected by several steep "canyons" leading down to the River in the valley bottom and one of these (the Canyon) will be wide enough and suitable for locating a sustainable, gravity flow, BCR-based EEB System in it.

Figure 9.2 illustrates the Case 2 process.

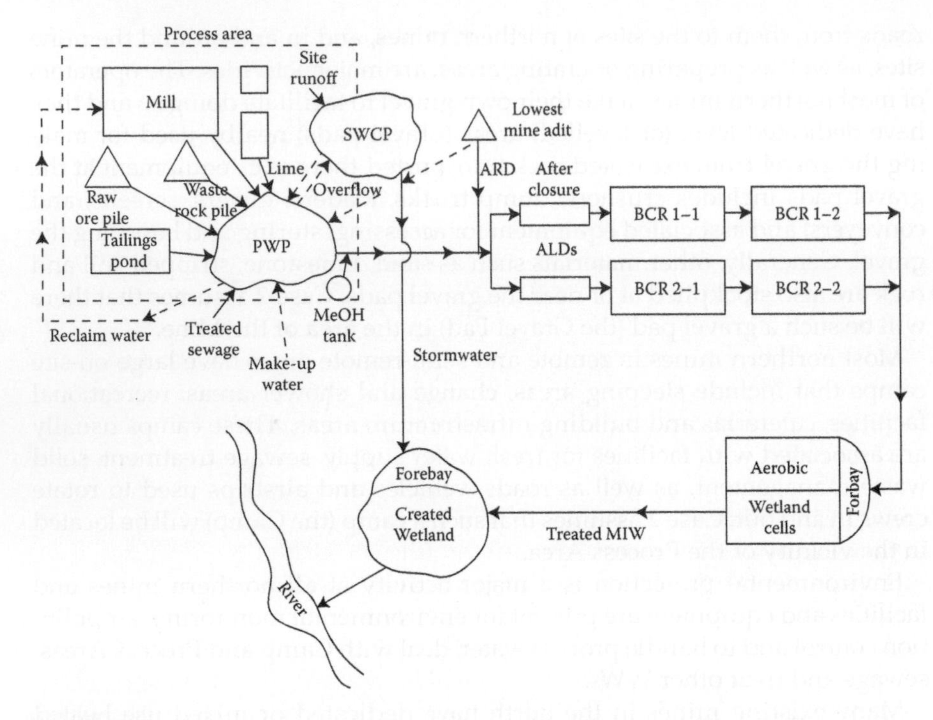

FIGURE 9.2
Case 2 PFD.

As illustrated in Figure 9.2, stormwater from in and around the Process Area will be collected in a stormwater control pond (SWCP) which will drain down via ditches and pipelines to a Created Wetland located beside the River. Overflow from this Stormwater Wetland will flow into the River.

For Case 2, it is assumed that the disposal of much of the tailings from the Mill will be by paste backfill into mined-out shafts, leaving only working volumes of tailings in a tailings pond located on or by the Process Area.

Case 2 also assumes that runoff and drainage from the tailings pond, from a raw ore pile and a waste rock pile by the Mill on the Process Area, will be assembled by buried toe drains and ditches around them and that this MIW will be directed into a nearby process water pond (PWP). During Mining, seepage and drainage into the underground workings will be pumped out into the PWP, from which water may be reused in the Mill or used to slurry ore being delivered into the Mill.

Make-up water from a pump station by the River in the valley bottom below the Mine site will, as required, also be pumped into the PWP. During mining, the PWP will have lime dosed into it to neutralize the MIW flowing into it and precipitate out metal(loid)s and other CoCs (e.g., suspended solids) as sludge. Treated water will overflow into a separate part of the PWP from which it may be recycled to the Mill and/or used to slurry more ore.

Sewage from Process Area buildings may also flow into the PWP* to also be treated in the EEB System, as will sewage from the Camp.†

Case 2 envisages building one or more trains of semi-passive BCR cells on existing or excavated and levelled shelves on the sides of one of the aforementioned Canyon, below the Process Area. These BCR cells will be roughly rectilinear, with approximately 2:1 surface aspect ratios and filled with 20–30 mm screened gravel made at the Gravel Pad. The outsides of BCR cells will be surrounded on three sides (the fourth will be the Canyon wall) by berms of non-acid generating waste rock covered with stripped soil, with 3:1 slopes on their insides, and slopes below the angle of repose on their outsides, as dictated by the particular Canyon. As available, stripped soil accumulated on the Gravel Pad during site preparation will be used to cover the BCR cells, and this soil will be vegetated with biodiverse local grasses.‡

Case 2 assumes that flow through the EEB System will be by gravity, and that the active medium for the BCR cells will be methanol which will be imported by truck and stored in a tank (the Methanol Tank) located on or just below the Process Area above the Canyon in which the EEB System will be located.

For Case 2, it is assumed that after the cessation of Mining and the shutdown of the Mine (Closure), the PWP will have much of its deposited sludge removed and either encapsulated in the Waste Rock pile and/or disposed of into a mined out underground shaft.

During mining operations, the acidic MIWs will be managed by liming the PWP and neutralized, partially treated MIW in excess of that required in the process will flow from it to the downstream trains of BCR cells. Since the liming of the PWP (with its attendant sludge formation) will not be feasible for long periods after Closure, just before Closure the EEB System will be expanded by having anoxic limestone drain cells (ALDs, ones built in much the same manner as the BCR cells discussed above) constructed on the cleared shelves just above (upslope from) the BCR cells. Thereafter, MIW from the PWP will be directed through them before it enters the trains of BCR cells. These ALD cells will be filled with locally mined limestone or dolomite gravel and will neutralize any acidity in the MIW after Closure (i.e., liming will be replaced after Closure with downstream limestone neutralization in ALDs.)

* In practice, such a PWP might be divided up into a sludge pond, a diversion pond and/or a balancing pond, but for simplicity in describing this case, it will be assumed that the PWP will be a single pond, albeit one which might have several chambers.

† If the Camp is at an inconvenient location to pump its sewage to the PWP, a small, separate BBR-based EEB System will be located near the Camp's main building and its treated effluent directed into the stormwater system.

‡ As required, the steep outer walls on some of the enclosed BCR cells may need to be vegetated with LIVING WEB mats (also see Section C.8 in Appendix C).

For Case 2, it is assumed that all adits to the underground working will be sealed after Closure, except for one at the lowest elevation that will open (after an appropriate internal dam) into the Canyon above downslope ALD and BCR cells.

MIW to be treated will flow by gravity to the BCRs from the PWP (directly during Mining and via the ALD cells after Closure), entering them near their tops and flow in them will be downflow VSSF. Effluent from the first BCR cell of each train will flow by gravity into the top of the second BCR cell that will be located downslope of the first. Methanol from the upslope methanol tank will be also flow by gravity, joining the feedstock MIW flowing into each of the BCR cells. Instrumentation and any required associated monitoring and active medium metering and injection equipment will be contained in small shelters alongside and just upslope of the BCR cells, with any required electricity requirements provided by solar cells on their roofs and batteries inside them.*

Depending on the volumes of MIW to be treated and its quality, there will be at least two trains of BCR cells (possibly three if the design MIW flow rate is above 10,000 m^3/day), with each train consisting of two BCR cells in series. Inside the BCR cells at their tops, MIW and methanol will enter inlet distributor piping grids having piping spaced at least 1 m apart. The main runs of each infiltration grid will be covered with half-pipe infiltration chambers.† All of the BCR cells will be lined on their floors and sides using bituminous liners‡ placed over (for their floors) layers of sand.

The exact thickness of the BCR cells' gravel substrates will depend on specific Mine and MIW types, as well as the anticipated amount of MIW to be treated, but it will be assumed for Case 2 that 3 m of substrate thickness will be used. As with the BBR cells of Case 1, these BCR cells will have secondary inlet distribution grids located halfway down their beds, and piping within the cells will be such that operators may direct influent MIW to be treated into the gravel substrates using either or both of them. As is the case with BBRs operating in areas with harsh winters, operators of the Mine's EEB System will promote the growth of extra biofilm bacteria in the Autumn by injecting inoculant solution,§ extra methanol and fertilizers into both sets of the BCR cells' inlet distributors before cold weather sets in.

* Optionally, prior to Closure, the electrical requirements of some or all of these can be provided via electrical lines from the transformer station on the Process Area.

† The situation will be similar to that of a corner of a bermed BBR shown in Figure 3.1 in Section 3.2 of Chapter 3, albeit with much thicker substrates and no aeration tubing.

‡ HDPE or PVC liners could also be considered but bituminous liners will be better able to manage the weight of the gravel substrates.

§ Inoculant solution will be prepared in a small fermentation tank located in one of the Process Area's heated buildings and charged initially with methanol, fertilizers and imported anaerobic bacteria (i.e., bioaugmentation will be used). Once Mining is well underway, genomic samples of the microbes in at least one of the BCRs will be taken regularly to allow the fermentation tank to prepare a biostimulated inoculant instead, one that will allow enhanced biodegradation in the BCRs (biocontrol, see Section 1.11 in Chapter 1).

Treated effluent from the bottoms of the second cell of each train of BCR cells will be piped or flow in ditches down the Canyon to enter one or more HSSF Aerobic Wetland* cells constructed just above the River bank at the bottom of the Canyon. These downstream, rectilinear polishing cells (2:1 aspect ratio) will each contain layers of pulp and paper mill biosolids† 0.5 m thick over a 0.5 m thick layer of limestone gravel. These Aerobic Wetland cells will be surrounded by berms made of non-acid generating crushed waste rock, have 3:1 inner and outer slopes, and be lined with HDPE liners over sand bases. Normal water depth in them over their substrates will be 10 cm. The exact sizes and number of Aerobic Wetland cells will depend on the volumes of MIW to be treated and its anticipated, "worse-case" flow rate. Their purpose will be to allow the surface re-aeration of the treated MIW effluent from the upstream (and much upslope) BCR cells, promoting the oxidation/hydrolysis of any residual metal(loid)s in the treated MIW and allowing them to precipitate and settle out.

Also located nearby alongside the River, the Created Wetland‡ will be constructed with a large forebay into which stormwater from the SWCP will be discharged. Effluent from the Aerobic Wetlands will also be discharged into this Created Wetland. Overflow from the Created Wetland will enter the River.

The Aerobic Wetland cells and the Created Wetland will be vegetated with known, relatively zinc- and lead-resistant wetland, plants native to the area (possibly Reeds).

Accordingly, the EEB System will consist of the PWP, ALDs (later), BCR trains, each involving two cells, Aerobic Wetland cells, and a Created Stormwater Wetland. It will be designed for gravity flow, minimal attention and sustainable operation. It will remove metal(loid)s by sulfate reduction and the other mechanisms addressed in this book.§

Longer term, at some point well after Closure as the acidity and metal(loid)s concentrations in the MIW naturally decrease to a minimal level, methanol injection into the BCR cells can be ended and the system allowed to evolve into a fully passive, wetland-based Natural WWT system.

The major state-of-the-art advancement represented by Case 2 will be the use of downflow, semi-passive BCR cells with gravel substrates to treat large volumes of acidic MIW at a northern base metal mine all the way from the start of mining to Closure and beyond it to eventually "walk away" site abandonment. Figure 9.3 illustrates the concept of treatment systems for the treatment of declining ARD Concentrations over time.

---

* Aerobic Wetlands are described in Section B.6 of Appendix B.

† This assumes that adequate volumes of these biosolids are either available locally or can be imported from a convenient source. If not, wood chips or some other form of locally available organic substrate can be used instead.

‡ Created Wetlands are described in Section A.6 of Appendix A

§ See Equations 5.10 and 5.11 in Section 5.5 of Chapter 5 for the removals of lead and zinc by sulfate reduction.

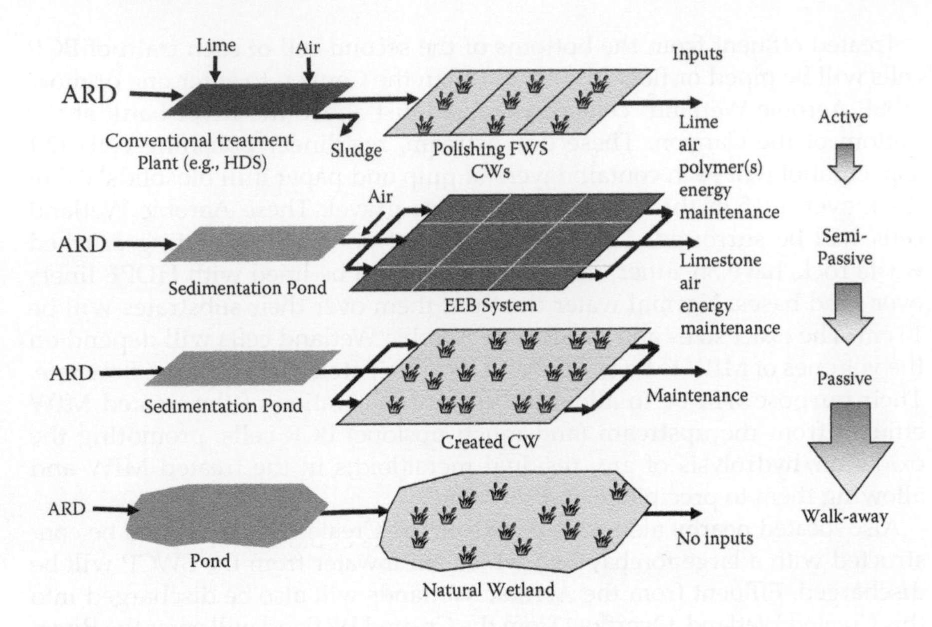

FIGURE 9.3
Concept of the treatment of declining ARD concentrations over time.

As with Case 1, another Case 2 advantage will be the EEB System's use of biocontrol while it is operating to identify and promote optimal biofilm microbial communities in the trains of BCR cells.

In summary, Case 2 is for a relatively simple, gravity flow, semi-passive treatment BCR-based EEB System to manage acidic MIW at a remote northern lead–zinc mine. For it, the KISS principle will apply. At it, the EEB System will treat collected MIW (from ore and waste rock piles, process water from the Mill, and water from a tailings pond) both during the Mine's operation (where it will operate in conjunction with preliminary liming in a PWP) and after Closure (when ALDs will take over the neutralization function of the liming). Depending on the amount of MIW to be treated, two or three trains of large VSSF downflow BCR cells with gravel substrates will clean up the MIW, and methanol will be used as their active medium. BCR effluent will be tertiary-treated in Aerobic Wetland cells located on the River bank and overflow into the stormwater systems' Created (Stormwater) Wetland from which it will flow into the receiving waters, the River.

9.3 Recommended R&D and Treatability Testing

As was mentioned throughout this book, although the basic kinds of EEBs (e.g., BBRs and BCRs) are fully developed and proven, there are a number of ways that their performance might be enhanced still further and this section suggests several areas of R&D that might be (and in some cases, already is being or is planned to be) carried out to do so.

1. The use of genomic techniques to assess what is happening to the microbial populations in the biofilms in EEBs has already proved fruitful (see Section 1.11 in Chapter 1), but further research is needed to better assess how genomic testing can be used to define conditions not only during normal operations but also when these conditions change due to changing environmental conditions, stresses or other situations. The goal should be the integration of genomics into Reaction Kinetic methods. A treatability test project, the *EW Test Project* (see Section 4.5 in Chapter 4), has already indicated that conventional assumptions about microbiological conditions in SAPS Bioreactors (and, to a lesser extent, BCRs) may need to be reconsidered and genomic testing may be valuable, if not essential, in preparing for the design of these kinds of bioreactors in future. It is recommended that the *EW Test Project* be continued for another phase to further assess this.
2. As was suggested in Section 1.11.5 of Chapter 1, a more detailed evaluation is needed of magnetotactic bacteria and the generation of black fines in situations where glycols are being used. It is recommended that R&D be carried out to do so.
3. Bench- and pilot-scale treatability tests need to be carried out for WWs containing traces of so-called "exotics" (flame retardants, endocrine disruptors, biocides, PCBs, nanoparticles, cosmetic chemicals, etc.) to determine how these contaminants are removed in EEB Systems involving aerobic and/or anaerobic EEBs. It is recommended that programmes of R&D and treatability testing be carried out to do so.
4. Certain plants have been identified as hyperaccumulators for select species (e.g., As, Ni, and Se, see Section C.6 in Appendix C). While the trend with aggregate-substrate EEBs (e.g., BBRs and semi-passive treatment BCRs) has been to use thicker substrates with either no vegetation at all or non-involved plants (e.g., terrestrial grasses on the surfaces of airport BBR cells), there may be an opportunity to design thinner (0.4–0.7 m thick) gravel-bed BBRs and to vegetate them with hyperaccumulators whose roots would be induced to grow down completely through the gravel beds, and to direct through them WWs that have passed through a contaminated site or waste accumulation, allowing the removal of a select CoC (or CoCs) by phytoextraction. Periodic plant harvesting would be required to recover the plants which then could be sent to a secure landfill or ashed and the resultant "bio-ore" sent to a metal refinery for recovery. For example, runoff and percolate from an old tailings pile at a closed nickel mine might be directed through a BBR cell vegetated

with certain known *Alyssum* cultivars that hyperaccumulate nickel. There is an opportunity to carry out bench- and pilot-scale testing to develop new EEB ecotechnologies to recover a variety of pollutants in this manner and such is recommended.

5. Salt-contaminated runoff is all too common in many places (e.g., from the ditches alongside streets and highways where road salt is used in winter). There are known plants that take up salt into their aboveground tissues, and even extrude salt on their leaves, if exposed to excess salt in the water in their root zones. In a manner similar to that for the hyperaccumulating species mentioned in the previous point, it is recommended that R&D be carried out to develop EEB Systems capable of removing salt from runoff streams.
6. Legacy phosphorus contamination in sediments at the bottoms of lakes, rivers and other bodies of fresh water continues to be a problem as storms and other sources of turbulence can stir them up, releasing the CoC into the water bodies. Where particular accumulations of such contamination have been identified (e.g., in the contaminated sediments that are common around the discharge pipes from municipal WWTPs), a method is needed to clean them up. The concept of *Hypolimnetic Withdrawal* has been proposed to do so. This concept envisages vacuuming up the contaminated sediments and transporting them to onshore locations for treatment in EEB Systems there.

   A proof-of-concept Hypolimnetic Withdrawal project has already been proposed by Stantec, ETDC and CAWT, and it is recommended that it be implemented after securing governmental support. The project would be a multi-year one (three or more) that would involve initial indoor pilot-scale testing at CAWT's greenhouse, followed by demonstration-scale testing using a series of existing outdoor ponds available to CAWT at the back of Fleming College's property at its Frost Campus in Lindsay, ON. The project would first involve collecting an amount of suitable phosphorus-contaminated sediment near the discharge pipe from an existing older municipal WWTP at Keswick, ON, Canada into Lake Simcoe (a large freshwater lake in southern Ontario). This WWTP is operated by a likely project participant, the Ontario Clean Water Agency (OCWA). Another likely project participant, the Lake Simcoe and Region Conservation Authority (LSRCA), has a suitable boat available. This collected sediment (probably as a slurry) would be transported to CAWT for pilot-scale testing of its treatability using sedimentation, BBR, BCR, Reed Bed Wetland and Pulse Flow Wetland cells (see Sections B.4 and B.5 in Appendix B) in a process configuration to be determined.

   Once a suitable configuration was established during the pilot-scale testing, the project would proceed to larger outdoor demonstration-scale testing that would involve: (a) isolating a target area of contaminated sediments near the discharge pipe from the Keswick WWTP using a floating plastic curtain (to prevent the spreading of contamination when the sediments are removed); (b) vacuuming up suitable amounts of the sediments and pumping them (using pumps on a barge) via floating pipes to tanker trucks on shores nearby; (c) transporting the vacuumed-up slurry for deposit into a suitable impoundment at CAWT into which the injection of a suitable precipitant (e.g., ferric chloride, see Equation 2.18 in Section 2.5.5 in Chapter 2) would allow the settling out of the bulk of the grit and suspended solids; (d) pumping decant liquid/wetland cell effluents through one or more EEB cells to remove any dissolved CoCs amenable to biodegradation and (e) impounding the EEB cell(s) effluent and adding ferrate to it to remove any remaining phosphorus (Section 2.6.2 in Chapter 2).

7. As was indicated in Section 7.6 of Chapter 7 (Engineered Substrates), there are substrate materials that will sorb and/or cause the precipitation of high concentrations of phosphorus, fluorides and other CoCs from streams such as the low pH pore water that continues to leach from gypsum stacks remaining at old phosphorus fertilizer production facilities. Since phosphorus pollution is a growing problem, more R&D and treatability testing is required to determine how more economic and versatile EEB Systems involving primary and/or tertiary cells containing such materials can be used as semi-passive alternatives to the existing practice of operating costly active treatment lime plants for long periods after the production facilities have closed. Such is recommended.
8. As was noted in Section 3.2 of Chapter 3, upflow SSF CWs are not recommended as it has been maintained (Kadlec 2001) that such a flow regime will lead to unstable hydraulics. This raised the question as to whether the same might be true in upflow EEBs. While the highly turbulent conditions in BBRs (due to aeration) might preclude such flow instabilities in them, they might be a problem in some upflow BCRs if internal hydraulic design is not such that this is precluded. It is recommended that, possibly as part of another of the R&D and treatability test projects proposed above, a detailed evaluation of hydraulics in upflow EEBs be carried out.
9. As was mentioned in Chapter 3, it is recommended that further research be carried out at an airport to try out, under demonstration-scale conditions with a BBR test cell, various new kinds of aeration tubing and aeration system morphologies to see if kinds/layouts can be identified that will be much more resistant to allowing organic and/or inorganic materials to build up in them, and design amendments defined that will allow them to be regularly flushed out while on- or off-line.

10. The treatment of WWs from petroleum and petrochemical industry facilities (process waters, cooling water, blowdown, aqueous sludges and site run-off) is an area of some potential for treatment in EEB Systems and it is recommended that R&D be carried out to define treatability for such WWs. One related area is the treatment of hydraulic fracking water and this should be a priority area for EEB R&D and treatability testing.
11. As is outlined in Section 2.6.3 of Chapter 2 and in Appendix E, Artisanal and Small Gold Mining has been identified as an area for potential EEB treatment, and it is recommended that both R&D on the basic chemistry of mercury in EEBs such as BCRs and treatability testing be carried out to test the performance of EEB Systems to treat WW streams at ASGM areas.
12. Usually when air is compressed by a blower, the air is in contact with lubrication oil in the blower and some of this oil may end up in the aeration tubing if such blowers are used to supply air to BBRs. Such may be responsible for some cases of organics build-up in BBR aeration tubing and such material can result in back-pressure that can cause the blowers to trip. While such organic build-up problems are easy to rectify (flushing of the aeration tubing with hydrogen peroxide solution), it is desirable to carry out studies to assess how aeration tubing might be improved or its systems modified to eliminate such problems. Additionally, there are kinds of blowers available that are used in the food industry for which the air being compressed is not allowed to contact the lubrication oil. Such blowers are more expensive than ordinary ones but the difference is not too large and their potential use for blowers supplying air to BBRs should also be included in the studies. It is recommended that R&D and treatability testing be carried out to assess these matters.

Appendix A: Natural Wetlands

A.1 Natural Wetlands and EEBs

To completely understand EEBs, it is necessary to define just what a "wetland" is and is not, what kinds of wetlands there are and how they relate to CWs and EEBs.

Wetlands can be categorized either as being *Natural Wetlands* or *Artificial Wetlands*, the latter being ones deliberately built by people. Artificial wetlands can be further subcategorized into *Created Wetlands* and *Constructed Wetlands* (Hammer 1997). Created and constructed wetlands are those wetlands designed and built for specific purposes. This chapter introduces natural and Created Wetlands, kinds of "Natural" Wetlands built by people for purposes other than WW management.*

A.2 Kinds of Natural Wetlands

Natural Wetlands can be defined as those shallow water areas wherein, at least periodically, the land supports predominantly hydrophytes (water-loving plants) whose substrates are predominantly undrained soils or, where the substrates are non-soil, are saturated with water or covered by shallow water at some time during the growing season each year. Flooding-intolerant vegetation is absent from them. Water in Natural Wetlands tends to be low in dissolved oxygen.

Natural Wetlands are ubiquitous. They are found in flat, protected and tidally inundated areas. They are also found next to freshwater streams, lakes and floodplains and in surface depressions everywhere. They can be called marshes, potholes, sloughs, fens, wet meadows, playas and vernal pools when the bulk of their vegetation is not woody (i.e., herbaceous), or forested wetlands (e.g., swamps) when it is. Natural Wetlands can occur as ponds in low-lying, upland areas fed by springs and/or precipitation, and in arctic areas as bogs where permafrost dominates their ecosystems.

Natural Wetlands are extremely biologically diverse. This is because of their sensitivity to the water level. Seasonal and annual variations can dramatically alter vegetation, microbial communities and wildlife in and around a wetland. These alterations stress the ecosystem but also provide a series of ecological

* Constructed wetlands are addressed in Appendix B.

niches which support a variety of flora and fauna. Some Natural Wetlands have commercial and utilitarian functions. These include aesthetic, educational and other social values. They also can be sources of crops such as lumber, wild rice, fur-bearing animals, fish and shellfish. The productivity of many wetlands exceeds most fertile farm fields as they receive, hold and recycle nutrients continually washed into them from higher, drier ground (Hammer 1997).

Wetland vegetation in Natural Wetlands may be *free-floating* (e.g., duckweed), *rooted floating* (e.g., pennywort), *submergent* (e.g., pondweed) or *emergent* (e.g., Cattails), and there may be shrubs (e.g., dogwood) and trees (e.g., willow) present. The submerged parts of plants in Natural Wetlands provide physical support for microbial biofilms, and a physical–chemical environment for biofilm and other microbes. They also filter WWs passing through them; transfer gases to and from their root zones; sorb metals and metalloids in their root zones; uptake nutrients and other compounds into their tissues and provide sources of carbon for various processes.

Natural Wetlands are found in surface depressions (*palustrine wetlands*), and alongside streams (*riverine wetlands*), lakes (*lacustrine wetlands*) and by the sea (*estuarine* and *marine wetlands*) where they often provide the interfaces between fully aquatic and terrestrial ecosystems. Waters in Natural Wetlands are generally less than 2 m deep (and often very much shallower), and may stand/flow both on the surface and sub-surface in/via soils and substrates. Regular to erratic drying cycles may occur in all or part of the Natural Wetlands. Water level fluctuations are normal in them, and morphologies usually are complex, with many flow channels, backwaters and other heterogeneous areas.

Around the world, there are many names for different kinds of Natural Wetlands (e.g., *tidal marshes, muskeg, billabongs, carrs, sloughs, moors, mires* and *mangrove swamps,* to name only a few). In the United States, Natural Wetlands are categorized using a hierarchical classification that involves five large systems which are progressively divided into 10 subsystems, 55 classes and 121 subclasses which provide a framework for a *National Wetlands Inventory* (Hammer 1991). In Canada, the situation is not so complex and the usual designations for Natural Wetlands are *marshes, swamps, bog, fens* and *shallow open water wetlands.*

Bogs are peat-covered low/no flow areas dominated by *Sphagnum moss* and other ombrotrophic (isolated from water sources other than precipitation) and acidophilic (thriving under acidic conditions) vegetation. Waters in bogs are low in pH, calcium, magnesium and nutrients (nitrogen and phosphorus compounds). While moss and similar herbaceous plants are the most prevalent vegetation, where conditions allow, some shrubs, low bushes and trees such as white cedar, tamarack and black spruce may grow in them too.

Fens are another kind of *peatland* typified by high water tables and slow internal drainage. Like bogs, herbaceous plants are the dominant vegetation in them (e.g., leather leaf), although a few shrubs (e.g., bog rosemary) and trees (e.g., willows) also may be present. Waters in fens tend to be mineral rich.

Marshes are areas permanently or periodically inundated and dominated by stands of emergent herbaceous vegetation such as Cattails and Reeds.

Waters in them are neutral in pH, relatively high in dissolved oxygen and nutrients, and are generally moving. Open water areas in them may also contain floating plants (e.g., duckweed) and submergent plants (e.g., wild celery). As with other herbaceous, vegetation-dominated Natural Wetlands, a few bushes and trees may also grow in marshes.

Swamps are wooded Natural Wetlands with waterlogged sub-surface areas in which shrubs, bushes (e.g., willow and dogwood), and trees (e.g., white cedar) are the dominant vegetation, although herbaceous plants are usually present among them. Waters in swamps tend to be standing or slowly moving most of the time.

Shallow Open Water Wetlands are ponds (*potholes, sloughs and depressional basins*) of standing or flowing water transitional between lakes and marshes. They often have emergent herbaceous plants growing on their peripheries and in shallower areas, and floating and submergent vegetation in the open water.

Natural Wetlands are usually irregularly shaped and highly associated with the catchment that they form part of. Their area may be hard to accurately define as they may involve various different parts (e.g., marshy areas, deeper water in pond areas and stream edges), each with different ecologies. In some jurisdictions, Natural Wetland delineation (the determination of wetland boundaries) has legal ramifications as activities that destroy or greatly alter them may require mitigation and compensation. Consultants are often hired to delineate them in such cases and use a variety of vegetative, morphological, soil type, physiological and hydrological indicators to do so.

Over 45% of all Natural Wetlands lie above 45° North Latitude (Hammer 1997), and these are largely tundra, muskeg, taiga and coastal marsh wetlands.

Many areas are experiencing continuing Natural Wetlands losses or major alterations due to draining, gradual filling, anthropological stormwater incursions and hydrological alterations. For example, less than 30% of the original, pre-settlement Natural Wetlands in southern Ontario are left.

A.3 Wetland Functions

Wetlands are lands that are seasonally or permanently inundated by shallow surface or groundwater at a frequency or duration sufficient under normal conditions to support a prevalence of vegetation typically adapted for life under saturated soil conditions. They are highly autotrophic, complex and dynamic ecosystems intermediate between uplands and deep water aquatic systems.

In wetlands, the presence of abundant water results in the formation of *hydric soils* (those which are saturated with water and are anaerobic in nature) and favors the dominance of either hydrophytic, or *water-tolerant* rooted and floating plants.

Algae and aquatic plants in wetlands release oxygen into water by photosynthesis as a by-product of their growth. Wetland plants also "leak" oxygen into their root zones. This increases the dissolved oxygen content in the water and in the soil in the vicinity of plant roots, thereby allowing aerobic microbial reactions to occur more readily, supplementing the anaerobic ones normal in hydric soils.

The following are the major ecological functions of Natural Wetlands (Hammer 1997):

- Controlling and storing surface waters
- Recharging and discharging groundwater
- Aiding in flood control
- Protecting shorelines from erosion
- Supporting and initiating complex food chains
- Providing habitat and vegetated areas
- Providing refugia and corridors for wildlife movement
- Trapping sediments which might clog watercourses
- Maintaining and improving water quality
- Immobilizing certain contaminants and nutrients

Accordingly, Natural Wetlands can be used as receiving waters and for active, passive or semi-passive WWT facilities such as EEB Systems. They treat contaminants which enter them as sewage and other WWs, polluted waters from industrial operations, spills, contaminated groundwaters and/or as surface stormwater runoffs from non-point pollution sources (e.g., mines, agricultural areas, and urban streets). The impacts of such discharges on the wetlands can be highly variable, but in spite of often rapid changes in the quantity and quality of waters passing through them, they often function well in polishing effluents from other kinds of WWT facilities.

However, it is important to note that the addition of a WW, treated or otherwise, to a Natural Wetland will dramatically alter its ecology and biology. Temperature, flow regime, pH, water levels, plant growth/speciation, etc. will change. Nutrient-deficient, standing-water ones such as bogs may be converted into flowing systems and certain plants in them may proliferate in the new, positively stressed conditions that favor their growth.

A.4 Natural Wetlands Used for WWT

Both Natural Wetlands and constructed wetlands used to treat WW are referred to as Treatment Wetlands (Moshiri 1993). Natural Wetlands will remove some of a variety of materials from any water passing through them, including suspended and dissolved solids, some soluble salts and other compounds, pesticides and other biocides, undesirable microorganisms, heavy metals, spilled fuels, oils and greases and a host of other organic compounds. A wetland's vegetation also will absorb and assimilate nutrients (nitrogen, phosphorus and potassium compounds) from water. The result is that many harmful and/or undesirable contaminants in the water are greatly reduced in quantity and much of them do not move further downstream from a wetland. This, in turn, reduces or eliminates many adverse environmental impacts. Wetlands therefore act as water treatment facilities.

Natural Wetlands have been used for millennia to dispose of sewage (Kadlec & Knight 1996). The first recorded deliberately designed use of a Natural Wetland for municipal sewage treatment began at a freshwater wetland at Great Meadows, MA on the Concord River in the United States in 1912. In Canada, from 1919 on, sewage effluent from the Dundas, ON Canada WWTP has been directed into the Cootes Paradise Natural Wetland at Hamilton, ON at the west end of Lake Ontario. Both fresh and saltwater marshes in Florida have been (since 1939), and still are, used for municipal sewage treatment. However, increasingly, regulators are frowning on their use for WWT, but they are still often used by small Arctic communities for sewage treatment.

Natural Wetlands are also used for WWT in some southerly areas if they already exist convenient to a WW source. For example, the Town of Houghton Lake in Michigan, USA has been using a local fen for sewage disposal since 1978 and this much studied and monitored system was said to operate well (Kadlec & Knight 1996).

Owing to their heterogeneous natures, where Natural Wetlands are used for WWT, very much larger areas are required for them to ensure adequate treatment than are needed for CWs.

A.5 Tundra Wetlands

The permanent settlement and formation of organized communities in the Canadian Arctic have led to a need to manage and treat WW there. Peatlands of various sorts (bogs, fens) are the dominant kinds of Natural Wetlands near these communities. An important northern kind of Arctic natural WWT wetland using peatlands is the *Tundra Wetland*, a kind of bog/pond mixed wetland (Higgins & Liner 2007; Higgins 2008).

Figure A.1 shows part of the Tundra Wetland that operates at the hamlet of Coral Harbour, Nunavut, Canada on Southampton Island in the northern part of Hudson Bay.

FIGURE A.1
A Tundra Wetland at Coral Harbour.

Figure A.2 shows another Tundra Wetland, that in the community of Chesterfield Inlet, NWT, Canada.

FIGURE A.2
A Tundra Wetland at Chesterfield Inlet.

Tundra Wetlands may be viewed as almost the natural analogues of an older CW system morphology: *marsh-pond-marsh* combinations (see the next chapter), and consist of combinations of boggy areas and small ponds. The boggy areas of Tundra Wetlands are spongy accumulations of living and dead Sphagnum moss, lichens and other vegetation, the dead plants usually only partly decomposed. Water flow through these areas is partially subsurface and partly via surface channels. The other aspect of Tundra Wetlands is the inclusion in them of numerous shallow ponds that have no drainage to groundwater in the short summers due to underlying permafrost. Frost heaving during winter creates ridges and depressions with unique polygon configurations. In summer in the north, long days lead to the proliferation of algae in Tundra Wetland ponds, and photosynthesis leads to highly oxic conditions in them.

Tundra Wetlands treat the sewage of many Arctic communities in summer, with the WWs being impounded (and freezing) during colder periods. For Arctic communities in Northern Canada, Greenland and Alaska, Tundra Wetlands and, for larger ones, HSSF CWs (see the next appendix), are being used for WWT, and there is a potential to convert many of them to EEB Systems if enhanced treatment is ever needed. A detailed evaluation of wetlands use in the Arctic prepared by CAWT staff and associates, *Wastewater Treatment: Wetlands Use in Arctic Regions*, is found in an on-line volume of the *Encyclopedia of Environment Management* (Yates et al. 2013).

A.6 Created Wetlands

Created Wetlands are those "Natural" Wetlands built for purposes other than WWT (e.g., recreation, habitat creation, education, water storage, and/or mitigation). Created Wetlands are usually sited in natural settings and most are designed to be artificial marshes, although they can also be designed as swamps or ponds. It is the intention of their designers that most Created Wetlands evolve over time into Natural Wetlands.

From design, engineering and construction points of view, Created Wetlands are much simpler than CWs or EEBs but tend to be much more complex ecologically. Similar to Created Wetlands are Restored Natural Wetlands, ones that have been modified, changed or used for other purposes, then restored.

Mitigation Wetlands are kinds of Created Wetlands designed to replace Natural Wetlands damaged or destroyed by human development. Mitigation Wetlands may be created by restoring back to functional wetlands earlier wetland sites at different locations; by establishing a new functional wetland from a nearby low-quality upland or other non-wetland sites and/or by enhancing or otherwise improving the functions and values of existing local Natural Wetlands without altering their habitat types. Generally, the construction of Mitigation Wetlands is mandated by regulatory or other legislated bodies (e.g., conservation authorities) to replace Natural Wetlands negatively impacted, and such replacement is usually at some factor of the size of the Natural Wetland involved (typically 1.5 times the size) as determined by a process called *wetland delineation*. Ideally, Mitigation Wetlands should be constructed in the same watershed as the destroyed/adversely impacted Natural Wetlands being mitigated.

Unlike Natural Wetlands, subject to site topography and constraints, Created Wetlands cells tend to be somewhat regular in shape, although often not as regular as those of CWs and EEBs which usually have rectilinear morphologies. Hydraulics are less important with Created Wetlands than it is with constructed wetlands, as is the monitoring of water and other parameters in them. The construction of Created Wetlands is often undertaken all or in part by volunteers. Created Wetlands are usually planted with biodiverse local wetland vegetation. Created Wetlands often have associated with them control structures, overflow piping, culverts, rip rap channels and level control structures such as Hickenbottom® drains.

In the United States, Section 404 of the *Clean Water Act* defines wetland mitigation and the mitigation sequence: first avoid and then minimize adverse impacts on aquatic ecosystems, then use compensatory mitigation to offset to the extent appropriate and practical unavoidable impacts. Some have established "Mitigation Banks" to provide developers with Mitigation Wetlands. These banks are property designed expressly for providing compensation for Natural Wetlands losses from permitted activities.

In some cases, wetland plant-vegetated EEBs that resemble constructed wetlands (e.g., some BBRs), that have secure, long-term water sources and that are designed to evolve first into CWs and then over longer terms into natural treatment wetlands may be considered as appropriate for wetlands mitigation.

In certain jurisdictions in the United States, the delineation, restoration and/or mitigation of Natural Wetlands involves a large number of federal and state laws and regulations. This colors perceptions in these areas of anything referred to as a "wetland," reinforcing the idea of not using that term in any context when proposing or referring to EEB Systems to avoid unnecessary regulatory burden.

Appendix B: Constructed Wetlands

B.1 Stormwater Wetlands

Constructed Wetlands can be used either for *controlling stormwater quantity* (*Stormwater Wetlands*) or for *managing wastewater quality* (*Constructed Treatment Wetlands* for pollution control*). As with Natural Wetlands, Stormwater Wetlands (SW WLs) and CWs can be used as receiving waters for effluents from EEB Systems. In addition, CWs can be used as either primary of tertiary treatment cells in EEB Systems.

SW WLs usually have intermittent and varying influent flows,† often involve single cells, and are more likely to be irregular in shape. SW WL basins and ponds can be vegetated with biodiverse local plants. Various kinds of SW WLs are possible including *shallow marsh wetlands, pocket wetlands, extended detention wetlands* and *Pond Wetland versions.* Many also regard stormwater wet and dry ponds as kinds of SW WLs.

Figure B.1 shows part of an SW WL at an ash landfill site in Portland, Maine, USA. This SW WL consists of an inlet pond and a marsh area. It not only receives stormwater runoff from the entire landfill site, but also is the receiving waters for effluent from a five-cell CW system treating leachate from an ash landfill cell.

FIGURE B.1
An SW WL at a landfill site.

* Generally, the acronym "CW" is used to refer only to Constructed Treatment Wetlands, and Stormwater Wetlands are not usually referred to as CWs. Some erroneously refer to CWs as "engineered wetlands."

† In contrast, influent flows to CWs are usually continuous within narrow ranges and CW systems often involve multiple rectilinear cells, see Section B.2.

SW WLs often consist of an *inlet forebay* (a deep pool to remove particles and sediment); a *marshy area* in which wetland plants are growing (to filter out suspended solids and often divided into low and high marsh sectors); a *deep pool* (a further settling area in some kinds of SW WLs); an *outlet micro-pool;* some sort of *inlet and outlet control structures* to manage water flows and often a *wet meadow plant-vegetated riparian buffer area* around the basins which may serve as an extended detention area. Figure B.2 (Schueler 1992) illustrates a typical SW WL layout.

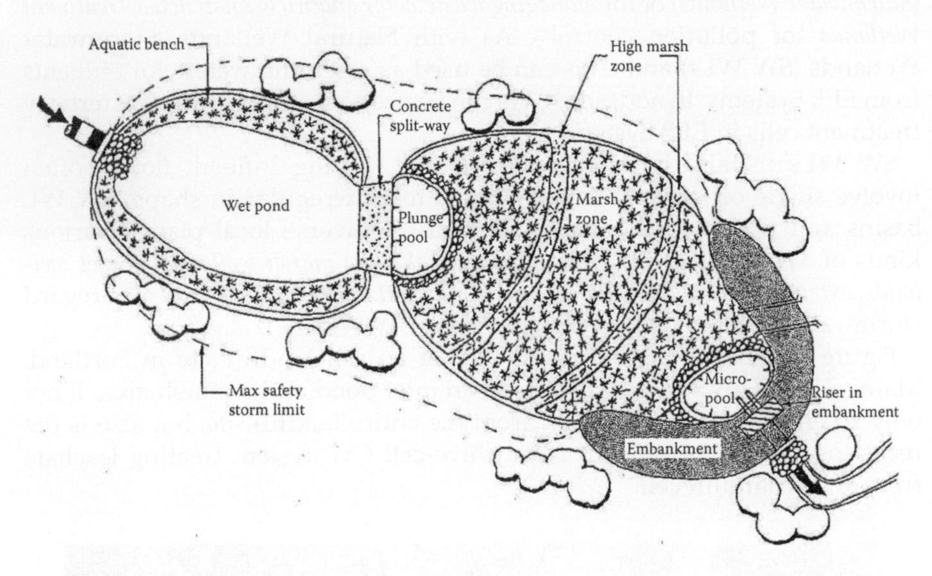

FIGURE B.2
A typical SW WL morphology.

Small amounts of contaminant removal occur in SW WLs but where the removal of larger concentrations of specific contaminants from a stormwater is desired (e.g., the treatment of glycols in contaminated runoff at airports), impoundment and downstream treatment in a CW or EEB can be used instead.* Typically, SW WLs can remove up to 75% of suspended solids in waters passing through them, 30%–40% of the total phosphorus (mostly that associated with suspended solids), 20%–30% of total nitrogen (again, mostly that associated with suspended solids), up to 15% of organic carbon in the water being treated and varying levels of metals and pathogens (again depending mostly on how much of these are associated with removable solids) (Field et al. 1993).

* Nevertheless, there are kinds of EEB Systems called engineered Stormwater Wetlands (ESWs) that are designed to treat the periodic and varying flows of stormwater runoff (Higgins et al. 2010c, see Section 3.10 in Chapter 3).

Two problems with many kinds of SW WLs are that

1. Contaminants deposited in them during one storm event may be remobilized during a subsequent event.
2. The deposited CoCs may be converted to more soluble and mobile forms between storms.

B.2 Constructed Treatment Wetlands

CWs usually consist of a number of individual rectangular (or at least artificially shaped roughly rectilinear) cells (artificially constructed basins) connected in series or parallel and surrounded by berms (dykes) of earth, rock or other materials. They contain structures (distributors, weirs and piping) to ensure good hydraulic dispersion, level and rate control and collection. Additional open water areas, integral ponds and forebays may be involved, depending on the type and application. With some CWs, vegetation may be specifically chosen for effectiveness with certain pollutants. The vegetation of CWs with Reeds (*Phragmites* spp.) and Cattails (*Typha* sp.) is common. Figure 3.27 in Chapter 3 shows the layout of a very large CW at an airport consisting of 12 square cells arranged in six trains of two cells each.

CW cells are usually associated in "CW Systems" with a variety of ancillaries (e.g., surge ponds, ditching, piping, pumps, lagoons, cascades, and land WWT technology such as overland flow swales, Reed et al. 1995). Conventional wastewater treatment methods (e.g., oil and grease removal units, sand filters and grit removal equipment) may also be associated with CW Systems.

CW technology is now well proven. There are textbooks on the subject (Reed et al. 1995; Vymazal 2001; Kadlec & Wallace 2008) and the systems can be designed and engineered with as much confidence as can the other types of biological units widely used in conventional WWTPs (e.g., ones involving activated sludge units). The technology for using CWs to treat municipal, agricultural and industrial wastewater treatment is now mature and there are now thousands of them operating worldwide treating sewage, MIWs, sludges, stormwaters, industrial effluents, leachates, agricultural runoff and many other streams. Their numbers are increasing almost exponentially. Depending on the type of WW being treated in CWs, BOD and COD removals in them can be as high as 90%, while nitrogen removals are usually in the 30%–50% range and phosphorus removals in the 20%–60% range, especially in colder water (Behrends et al. 1996; Arias & Brix 2004).

In Europe, CWs often serve as the main treatment facilities for treating sewage from small communities. In North America, they have often been used as polishing units downstream from lagoon WWT systems and conventional active WWT facilities.

Many early CWs failed to achieve their designers' goals as at first layouts were primitive and proper engineering design principles were rarely followed. CW design evolved through several stages to rectify such limitations through the *kinds of wetland basin used* (e.g., from ponds and failed attempts to create artificial bogs to open water, marsh type CWs); in *morphology* (e.g., from small facilities with one or few long, irregularly shaped basins to the current multiple train, low aspect ratio systems involving one or more rectilinear cells); in the *volumes of water that they could handle*; in *sizing methods used* (i.e., from early empirical relationships based on hydraulic and/or contaminant loadings to modern chemical engineering methods based on Reaction Kinetics [see Section 8.3 in Chapter 8]) and in *engineering design* (from *ad hoc* designs to the use of formal civil engineering techniques) (Kadlec & Wallace 2008).

Table B.1 compares CW systems with Conventional WWTPs.

TABLE B.1
Comparison of CW and Conventional WWT Systems

	CW Systems	Conventional WWTP Systems
Intensiveness'	Land	Energy
Energy source	Renewable	Non-renewable
Major inputs	Solar energy	Fossil fuel energy
	Soils and substrates	Electricity
	Vegetation, other biomass	Chemicals
	Microorganisms	Aeration air
Structures	Largely earthen	Concrete, Steel, Plastic
Operation	Simple	Complex
Susceptibility to hydraulic shock	Low	Often high
Susceptibility to chemical shock	Lower	Higher
Residue production	None	Sludge

These comparisons mostly apply to EEB Systems as well. The three seminal texts on WWT in CWs are: *Natural Systems for Waste Management and Treatment* (Reed et al. 1995); *Treatment Wetlands* (Kadlec & Knight 1996) and *Treatment Wetlands, Volume 2* (Kadlec & Wallace 2008), and the development of EEBs presented in this book follows on their lead.

B.3 Types of CWs

Three types of CWs may be defined

- Pond
- Free-water surface
- Subsurface flow

Pond Wetlands are simple vegetated or partially vegetated basins, free-water surface (FWS) CWs are artificial marshes, and the WWs being treated in SSF CWs flow beneath the surface of permeable substrates such as gravel. The following sections provide more detail on these types.

B.3.1 Pond Wetlands

Pond Wetlands are usually simple, shallow pools (0.1–1.0 m in water depth), often vegetated with emergent wetland vegetation (e.g., Cattails) around their peripheries (10%–30% of area) and having some portion of their surface consisting of open water in which submergent and/or floating wetland vegetation is growing. They are most commonly used in conjunction with other types of CW cells (e.g., as re-aeration basins between FWS cells in the common *marsh-pond-marsh system*), or as components of some kinds of SW WL systems.

Pond Wetlands may provide quiescent areas where sediments and some of the suspended solids in an influent WW can settle out. Hence, Pond Wetlands can be good methods for dealing with any grit, suspended solids, BOD, oil and grease, as well as any pesticides and herbicides, fertilizers, heavy metals and other organics that become associated with them in many WWs.

Figure B.3 shows a Pond Wetland that is part of a CW system treating agricultural WW (e.g., milkhouse water, runoff from cattle holding areas and yards, effluent from an anaerobic lagoon treating liquid manure) (Hurd et al. 1999).

FIGURE B.3
A Pond Wetland.

Some Pond Wetlands are fully and densely vegetated with floating plants such as duckweed and water hyacinth, with their roots hanging down into the water column. Some categorize these "Floating Aquatic Plant" wetlands as separate kinds of CWs.

Pond Wetlands differ from WWT lagoons in that they are almost always deliberately vegetated with wetland plants and most lagoons are not.* In addition, Pond Wetlands are usually much shallower, and hence they tend to be more aerobic than often deeper WWT lagoons.

Where sediment loads are high, Pond Wetland cells can be provided with floating rafts of wetland plants and these can be used in front of downstream FWS or SSF wetland cells. Floating plant rafts in Pond Wetlands allow settled sediments to be periodically easily removed and water levels to be varied without adversely impacting plant hydrology. Floating reed beds are used in an Aerated Lagoon that forms part of what was originally the SSF CW system at London Heathrow Airport in the United Kingdom for treating glycol-contaminated streams (see Figure 3.29 in Chapter 3).†

B.3.2 Free-Water Surface CWs

Constructed Wetlands of the *FWS* type (sometimes alternatively called *Surface Flow* wetlands) are ones where water flows in the open among emergent wetland plants. In FWS CWs, the submerged portions of wetland plants, as well as soil and detritus, act as substrates for biofilm attachment, and the microorganisms in these biofilms (largely bacteria) are responsible for much of the pollution removal.

FWS CWs can be used for primary and secondary municipal and industrial WWT, but most commonly are used for polishing (tertiary) treatment, often after mechanical WWTPs or EEBs. FWS Constructed Wetlands are the most common type of constructed wetland in North America and in tropical countries.

FWS CWs are similar in appearance to natural marshes. They often have open water areas in them with submergent and floating vegetation as well. When vegetation in them is dense, they can be largely anaerobic in nature (except near their water surfaces) but nevertheless still contain aerobic zones and microenvironments. Figure B.4 (Kadlec & Wallace 2008) illustrates an FWS CW cell.

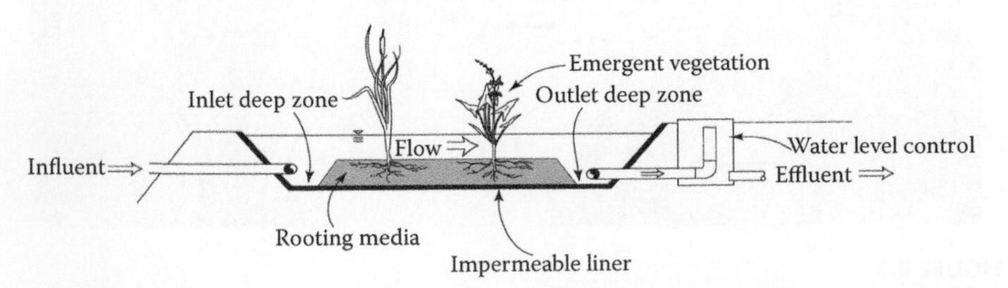

FIGURE B.4
An FWS constructed wetland.

* This does not prevent opportunistic wetland vegetation from colonizing the edges of poorly maintained municipal WWT lagoons!

† It has now been upgraded to an EEB System, see Section 4.9 in Chapter 4.

Figure B.5 shows a demonstration-scale, Cattail-vegetated, FWS CW cell at the Alfred Test Unit in Ontario, Canada.

FIGURE B.5
An FWS CW demo cell.

A typical FWS CW System designed by Jacques Whitford has successfully operated since 1998 at the *Region Waste Services* ash landfill in Portland, MA, USA treating ammonia- and iron-contaminated leachate in a Pond Wetland cell and four FWS CW cells. Figure B.6 shows a view of this CW system including one of the FWS cells in the foreground and the inlet Pond Wetland cell for removing iron from the feedstock leachate and underdrain water in the background (this CW System discharges into the SW WL shown in Figure B.1).

FIGURE B.6
A CW system.

Since decaying vegetation can maximize anaerobic conditions in FWS CWs, it is best to incorporate un-vegetated deeper, open water areas in them to balance out aerobic conditions due to increased atmospheric diffusion (perpendicular open areas also tend to smooth out flow through wetland cells).

As was indicated in Section 3.7 of Chapter 3, the secondary cells of some of the six two-cell trains of the now upgraded EEB System at EIA have been converted to FWS CW cells.* FWS CW cells also often form the last polishing cells of Engineered SW WL systems (see Section 3.10 in Chapter 3) and EEB Systems treating MIWs and landfill leachates.

B.3.3 SSF Wetlands

Sub-Surface Flow Constructed Wetlands (SSF CWs) are sometimes variously referred to as *Vegetated Submerged Bed, Gravel Bed, Rock Reed Filter* or *Root Zone* wetlands. With these kinds of CWs, the wastewater being treated flows under the surface of permeable materials consisting of one or more beds of gravel, sand, rock, wood chips or other granular material. Most SSF CWs are vegetated and wetland plants that grow out of the substrate surfaces of the wetland cells. It is even possible to walk dry-shod on their normally dry surfaces if one can get in among the normally dense vegetation.

SSF CWs are used where the WW being treated is noxious or odorous; where sludges such as biosolids from conventional WWT facilities need to be dewatered; where either very high or very low levels of dissolved metals and other contaminants need to be removed; where a higher degree of freeze protection is desired; where space (i.e., "footprint" area) is an important consideration; where the attraction of waterfowl may be undesirable (e.g., near airports); where ample economic supplies of substrate material are readily available and/or where operation in an engineered mode is desired.

SSF Constructed Wetlands are usually smaller in area than FWS CWs for the same levels of pollutant removal, and can tolerate higher loadings. They are, however, generally more expensive, costing several times the cost of an FWS CW of equivalent size. The higher capital cost is due to the higher costs of designing and building them, especially for the cost of providing their substrates. Figure B.7 shows a pair of SSF CW cells treating domestic sewage in Europe.

There are two types of flow in SSF CWs (and aerated EEBs): *horizontally fed* (HSSF) in which the WW flows horizontally through the substrate parallel to the cell substrate surfaces; and those that are fed vertically by applying WW either to their surfaces (downflow *vertically fed*, VSSF). It then percolates down to buried effluent collection pipes below the substrate) (Table 1.3 in Chapter 1 compares flows in both modes).

* Normally, pond or FWS wetlands are not recommended at airports as they can attract birds and lead to bird strike concerns, but at EIA the treatment system is so far away from the runways that the AoA, the ERAA, has decided that having some FWS CW cells there is not a concern.

FIGURE B.7
SSF wetland cells.

Figures B.8 and B.9 (Wallace & Knight 2006) illustrate HSSF and VSSF CWs. It is noted that while HSSF CW cells have to contain relatively homogeneous beds of substrate (to avoid channelling), no such restriction exists for VSSF CWs which can, if desired, involve multiple layers of different kinds of substrate materials as is illustrated above. Most VSSF CWs contain only one kind of substrate material but some such as Reed-Bed Wetlands ([RBW] see below) benefit from multiple layers.

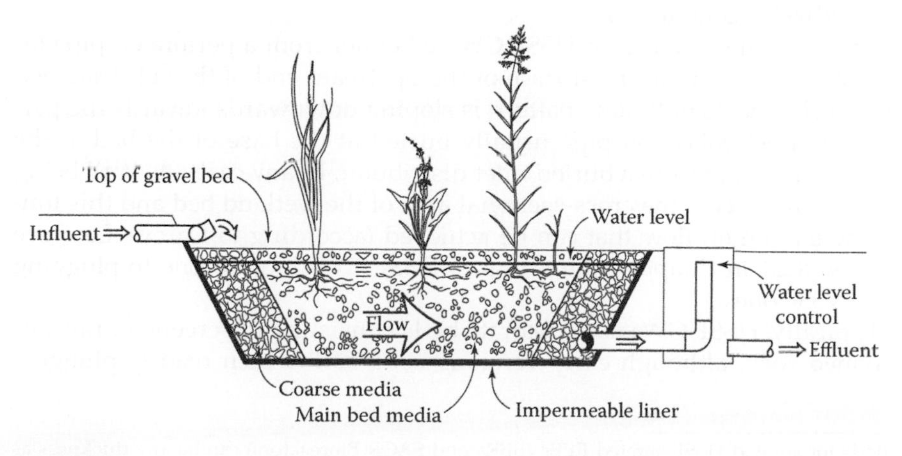

FIGURE B.8
An HSSF CW.

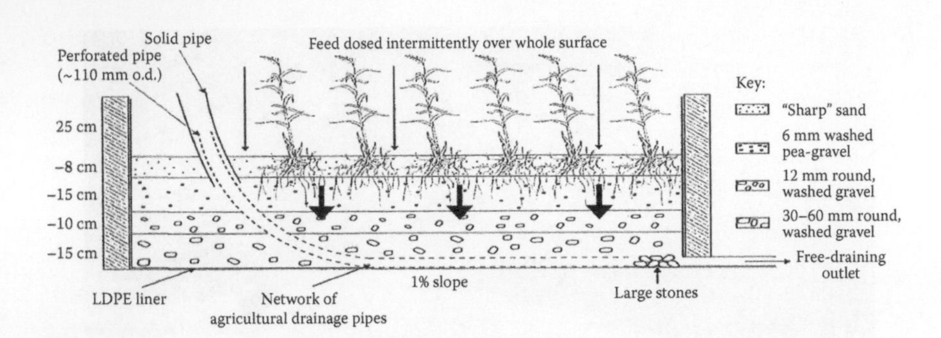

FIGURE B.9
A VSSF CW cell.

Some SSF CWs in colder regions operate in an FWS mode in summer and in a (more weather resistant) SSF mode during the rest of the year. SSF CWs are much more complex to design, engineer and build correctly so that proper hydraulic control is maintained and desired performance achieved.

B.3.4 Horizontal SSF Wetlands

The most common type of SSF CW is the HSSF variety (see Figure B.8). In HSSF CWs, the WW being treated enters the substrate bed at one end of the cell and flows generally horizontally across the cell parallel to the cell substrate surface. Bed thickness for HSSF CWs is usually from 0.5 to 0.8 m. This is the maximum depth to which the roots of the oxygen-transferring wetland plant roots can penetrate, and, were beds any thicker, some of the WW would flow below the plant roots, missing needed aerobic removal processes for oxidizable contaminants.*

The influent can enter an HSSF CW cell either from a perforated pipe (or similar distributor) on the surface of the upstream end of the substrate bed (in which case its bulk flow pattern is sloping downwards towards the perforated outlet collection pipe usually buried at the base of the bed at the other end), or out from a buried inlet distributor. In any case, the WW being treated only "sees" the cross-sectional area of the wetland bed and this limits the maximum flow that can be achieved (according to Darcy's Law, see Section 8.2.1 in Chapter 8) and makes HSSF CWs more prone to plugging than VSSF ones.

Typically, HSSF CWs use substrate beds consisting of screened gravel or crushed rock, although early versions used soil (which readily plugged,

* Beds for aerated HSSF aerated EEBs (BBRs and SAGR Bioreactors) can be any thickness as blowers supply the oxygen needed from below the beds; generally, the beds for them are 1.5–3.0 m thick.

Kickuth 1989), and some existing types use shales, peat, biosolids and compost as all or part of their substrates. Since most SSF CWs have no open water (water levels are usually maintained just below the substrate surface), atmospheric diffusion into them is even lower than that into Pond Wetlands and FWS CWs (0–3 g-O_2/m^2) (EPA 2000a,b) that is usually small compared to the oxygen demand exerted by even a dilute WW or stormwater.

With poorly designed HSSF Constructed Wetlands, substrates can plug or clog, resulting in the surfacing of wastewater flows. While this will lead to reduced performance, it is not always a complete disaster as the failure mode of an HSSF CW is an FWS CW.

B.3.5 Vertical SSF Wetlands

VSSF CWs are variously referred to as *Vertical Flow Constructed Wetlands (VFCWs), Reed-Bed Filters (RBFs), Vertical Filter Beds* (VFBs), and *Infiltration Beds.* VSSF Wetlands treating sewage from small communities are becoming popular for the primary and secondary treatment of sewage from small communities (<500 persons equivalent, Pe) in France, where they often operate in a reciprocating flow mode (two or three intermittently dosed primary cells in parallel followed by two secondary cells in parallel). VSSF CWs are also often used in a continuous flow mode where WW flows are so high that hydraulic considerations preclude the HSSF morphology.

Downflow VSSF Wetlands can be operated in two modes: saturated bed and freely draining bed. *Pulse flow VSSF CWs* (see below) are an example of the latter and are operated by dosing a cell with a WW (usually sewage), then allowing the water to drain down through the bed while another cell is dosed. This draws oxygen into the bed, improving aerobic processes such as nitrification. RBWs are another free drainage example.

B.4 Pulse Flow Wetlands

Pulse Flow Wetlands are another kind of SSF wetland where the influent is dosed into a cell of an ordinary wetland vegetated with wetland plants for a period, after which inflow is stopped and dosing moves on to another cell.* Water in the initially dosed cell is then allowed to drain down into the substrate, drawing air into it.

* It is a moot point whether a Pulse Flow Wetland should be categorized as a CW or an EEB. In this book, the former is used but the differentiation is arbitrary, as it is for RBWs and the aerobic and Anaerobic Wetlands described in the following sections.

Pulse Flow Wetlands are also variously referred to as *fill and drain wetlands, reciprocating bed, tidal flow* and *cyclic wetlands*, and are kinds of high freeboard, downflow VSSF CWs where the substrate beds are periodically flooded, then drained, resulting in partially unsaturated beds. Applied batches of wastewater percolate down through the substrates, and air is drawn into the beds as the water levels drop. This allows aerobic reactions such as nitrification to occur at rates up to double that possible in ordinary SSF CWs, with removal rates above 95%. For example, with an ammonia-contaminated WW, during the flood stage, if sufficient alkalinity is present, NH_4^+ ions adsorb to negatively charged biofilm located between the substrate particles, and aerobic microbes (aerobes) in the biofilm oxidize them in steps to nitrate (see Section 2.4 in Chapter 2). During a subsequent flood stage, these NO_3^- ions are desorbed back into the bulk WW where denitrification will occur if there is sufficient carbon present. Enhanced (>90%) removals of several other contaminants (BOD, TSS, and pathogens) also occur in Pulse Flow CWs (Behrends et al. 1996; Lemon et al. 1996; Matthys et al. 2000). Pulse Flow CWs are usually vegetated, often with Reeds or Cattails.

There are two basic ways to dose a Pulse Flow Wetland: (1) by flowing or pumping sequential batches of WW onto the surface of the cell (i.e., dose that cell, then direct the WW to a different cell while the first cell drains) or (2) more passively by the use of a dosing siphon attached to the outlet of a single cell (Vinci & Schmidt 2002). No controls or mechanical/electrical equipment are needed with a dosing siphon (see Section C.5 in Appendix C). In operation, WW to be treated in a fill and drain SSF wetland cell equipped with a dosing siphon on its outlet piping is allowed to build up in the headspace above the substrate surface and then is discharged through the wetland bed when it opens.

B.5 Reed-Bed Wetlands

A kind of downflow VSSF CW used for municipal biosolids (sewage sludge) dewatering in both North America and Europe is the *RBW*, named after the usual plant with which it is vegetated, Reeds (RBWs are sometimes also referred to as *Biosolids Dewatering Wetlands* or *Biosolids Stabilization Wetlands*).

Most municipal WWTPs are based on the intense biological sewage treatment of sewage in aerated tanks (i.e., the activated sludge process), and they generate large amounts of excess biosolids (sewage sludge made up of recalcitrant organics and dead microbes, called *waste activated sludge* [WAS]). That part of the biosolids not recycled to feed which must be removed from the process circuit must be stabilized (to kill pathogens), conditioned (to allow agglomeration during subsequent dewatering), then dewatered (to reduce the solids content from anywhere from 0.5% to 2% solids of the raw WAS (it is a black liquid at this point) to the by-product 20%–25% solids "cake" which has to be then disposed of at no little expense.

Indeed, up to 2/3 of the OPEX and a not insignificant part of the CAPEX for such active WWTPs is due to biosolids management (Aerated Lagoon-based WWT facilities also produce biosolids requiring similar management). If they have the space, small conventional WWTPs at rural communities sometimes have the option of avoiding the expense of fossil-fuel-based biosolids management by instead treating their un-dewatered biosolids in RBWs using solar and gravitational energy for power.

RBW cells are high freeboard ones which usually consist of a surface layer of sand underlain by progressive layers of pea gravel and coarser gravels. In biosolids management service, several RBW cells are periodically dosed (flooded above their surfaces) in rotation with 0.3–0.5 m of liquid biosolids from the WWTP applied to dewater, dry and ultimately convert these residues into a compost-like material on the cell surface. Each dosing cycle leaves a thin layer of dewatered biosolids on the cell surface, and the roots of the prolifically growing Reeds penetrate the layers and grow densely in them. The reed plants maintain and facilitate liquid percolation, transpire water and utilize nutrients from the biosolids.

RBWs for biosolids management are considerably more economic and simpler to use than many other biosolids management methods, especially for smaller, more rural communities. RBWs do not eliminate the need for solids disposal. In them, the biosolids are converted to a compost-like material which is suffused by plant roots and this must be removed (usually after 2–5 years). This compost can be used for land application or some similar organic soil use. Percolate from RBW cells is usually recycled to the WWTP, and must be injected into its influent sewage feedstock carefully to avoid disrupting the process as these streams have very high concentrations of contaminants.

Winter freezing of RBW beds hastens the biosolids degradation process. As much as 40% of the deposited biosolids can be degraded by the plants but this volume reduction is partially balanced by the reed roots in the resultant compost. Nevertheless, disposing of the compost from RBWs is far easier and more acceptable than handling the dewatered biosolids from WWTPs that do not use them. Generally, vegetation from CWs is not harvested, but RBWs are an exception and the Reeds in them are usually harvested or burnt once a year. Figure B.10 shows a number of RBW cells alongside a small (8000 Pe) activated-sludge-based WWTP in Skovby, Denmark (Kadlec & Wallace 2008).

In addition to biosolids, septage and other liquid sludges can also be dewatered in RBWs but in such cases extensive primary treatment (e.g., feedstock preparation by filtration) has to occur before the cells are dosed and some

FIGURE B.10
RBWs at a US WWTP.

sort of provision must be made to deal with the even stronger percolates (e.g., treatment in an EEB System).

The problem with RBWs is one of diminishing returns. The several RBW cells required take up a lot of space and the sites of many rural WWTPs do not have enough. In addition, as the size of a WWTP increases, so does the amount of biosolids it produces and with this the size and number of RBW cells needed if this option is to be considered.

B.6 Aerobic Constructed Wetlands

The removal of many contaminants requires aerobic conditions (e.g., ammonia nitrification), and in many cases such removals require the provision of oxygen to aerobic microbes living symbiotically or opportunistically in the root zones of wetland plants (as ammonia nitrification is a bacterially mediated aerobic process). This can be difficult in the normally anaerobic environment below the water surface of CWs. As was discussed above, in them oxygen may come from that dissolved in the influent wastewater, from surface re-aeration, from "leakage" from the plant roots/rootlets in the root zone (such is the nature of many wetland plants), via the venturi effect through the hollow wetland plant stems (the reason aerobic reactions can continue in a CW even in winter), and/or artificially. All of these methods other than the artificial ones can only provide limited amounts of oxygen, and thus restrict ordinary CWs requiring the removal of oxidizable contaminants to low loading conditions (i.e., such wetlands will be large in size, have high water residence times and, in any case, will only achieve limited removals of the oxidizable components).

Accordingly, most ordinary CWs typically consist of cells in which aerobic and anaerobic processes take place simultaneously at different locations in the wetland at much lower efficiencies than would be possible in a cell where one type of process was optimized. However, there are methods to do just this—optimize one kind of condition (aerobic conditions or anaerobic conditions) and the multiple cells of a wetland system offer opportunities to do so in separate cells. A wetland involving enhanced oxic or aerobic conditions is known as an *Aerobic Wetland,* while one involving enhanced anoxic or anaerobic conditions is known as an *Anaerobic Wetland.* A multi-cell treatment wetland system or EEB System can contain one or more of either type of cell.

In the past in Eastern US, coal mines and Aerobic Wetlands were used to treat ARD which contained net alkalinity with the purpose of neutralizing metal acidity (Skousen 1998). Until very recently, sizings were based on the loading method (see Section 8.2.3 in Chapter 8), with a hydraulic loading rate of 200 m^3/ha.day and a seven-day retention time being recommended for some (Wile et al. 1985). Others suggested sizing them for contaminant loadings of 10 g/m^2.day for Fe removal and up to 2 g/m^2.day for Mn removal if sufficient alkalinity was present (Hedlin et al. 1994) and they could tolerate up to 3.5 g of acidity/m^2/day (Kadlec & Knight 1996). Sizings were considered valid only for relatively low ARD flow rates, and were applicable for the treatment of certain kinds of coal mine drainage. Generally, Aerobic Wetlands were designed to have 0.03–0.1 m of organic substrate covered over with a thin layer of water.

Now Aerobic Wetland cells can be designed using modern methods and such can be used for polishing as tertiary treatment components in EEB Systems

B.7 Anaerobic Constructed Wetlands

Anaerobic Wetlands were first used by the mining industry to remove iron (and, to a lesser extent, manganese) from small streams and seeps as sulfides in a reducing environment. In the substrate beds of such early Anaerobic Wetlands, which were made up of organic material such as compost mixed with limestone, these were from 0.1 to 0.3 m thick, and it was intended that in them any ferric iron (Fe^{3+}) in the MIW being treated would be reduced to ferrous iron (Fe^{2+}), which, along with any other ferrous iron present, would react with the H_2S produced by the sulfate-reducing bacteria to form an insoluble precipitate in the bed.

$$Fe^{2+} + H_2S \rightarrow FeS_2\downarrow + 2H^+ \tag{B.1}$$

Over time, the insoluble ferrous sulfide would build up in the bed. With such Anaerobic Wetlands, which mostly were located at or near coal mines in the eastern US, acidic minewater was applied at the surface and flowed along and down through the substrate beds (i.e., in a mixed HSSF/VSSF mode) to collection distributors at the end of the beds.

The required size of these early SSF CWs was a function of metal loading, pH and alkalinity. These simple Anaerobic Wetlands (sometimes referred to as *Compost Wetlands*) proved effective for metals removals only at low feedwater flow rates and only if the water was net acidic. They are rarely used now for MIW treatment as they have been superseded by more efficient designs. However, versions of them, sized using reactor-based models and modern design methods, are still used and these are called anaerobic CWs.

Anaerobic SSF CWs, now designed using modern methods and ancillaries, are not limited to treating ARD and can be used to remove not only dissolved metals (and not just iron), but also other contaminants (e.g., organics and nitrates) from a variety of WWs. They are designed as downflow VSSF cells and may be either vegetated or un-vegetated, with mixed organic/limestone substrates from 0.1 to 0.5 m thick.

B.8 Wetland Vegetation

The most common types of emergent vegetation used in FWS CWs in Europe and North America are Cattails, Bulrushes (*Scirpus* spp.) and Reeds, although a variety of other wetland vegetation can be used as well, such as Reed Canary Grass (*Phalaris arundinacea*) and Managrass (*Glyceria maxima*). In all but the smallest CWs, monocultures are usually used rather than biodiverse vegetation since, under the stressed conditions of a CW, the more "aggressive," stress-resistant wetland plants (e.g., Reeds and Cattails) will quickly displace others anyway if they are present.

In some cases, SSF CW cells are operated with no vegetation at all, especially for the early cells of a CW train, ones where the cells operate anaerobically, and/or in cases where relatively easy to degrade pollutants (e.g., low-molecular-weight alcohols) are the main contaminants being treated. Nevertheless, in most CWs, plants can and do enhance treatment and vegetated CWs usually allow better performance than non-vegetated ones.

While wetland plants can and do take up nutrients such as nitrogen and phosphorus, amounts are limited, and the nutrients are released to the water again when the plants die and decay.

By their natures, wetland plants are efficient at gas transfer to and from their root zones (*rhizospheres*), which are the ideal habitat for a variety of symbiotic and opportunistic bacteria and other microbes. Where the gas is oxygen, "leakage" from the rootlets of the plants allows the formation of aerobic microenvironments where aerobic microbes can interact with oxidizable contaminants in WWs. The root zones of wetland plants will readily sorb dissolved metals and other contaminants (e.g., metalloids and certain salts) from wastewaters passing through them, and such can serve as an important metals removal method.* However, generally, except for metabolically important metals, sorbed metals and metalloids on wetland plant roots are not translocated from roots into above-ground plant tissues, and, in any case, many sorbed metals are released back into the water when the plant dies.†

Ordinary SSF CWs provide reasonable removals of some kinds of BOD, suspended solids and pathogens from WWs being treated in them. However, they require relatively large surface areas to achieve even moderate levels, so, if the flow rate of a WW being treated is high and/or the concentrations of contaminants are elevated, they may have to be so large to obtain desired effluent contamination levels that they become infeasible and/or uneconomic. In addition, as is mentioned above, a CW will only remove part of a WW's nutrients.

B.9 Advantages and Disadvantages of Constructed Wetlands

The use of CWs for treating or polishing WWs has a number of advantages, including that they

- Provide effective and reliable wastewater treatment
- Are relatively inexpensive to construct
- Are relatively economical to operate and have low labour requirements
- Are easy to maintain and have low-energy requirements
- Are able to accept varying quantities and concentrations of pollutants
- Are relatively tolerant of fluctuating hydrologic conditions
- Provide indirect aesthetic benefits (e.g., habitat, green space, and recreation)

The use of CWs for WWT is not a panacea; there are disadvantages to their use as well. These include that Constructed Wetlands

- Require large land areas
- Are ecologically and hydrologically complex

* There is even a phytoremediation method, rhizofiltration—see Section C.6 of Appendix C—which exploits this mechanism by growing plants hydroponically to allow the removal of the sorbed metals by root harvesting.

† There are, however, certain phytoremediating plants that can phytoextract certain metals and salts out of the root zones into roots, leaves and shoots, but this is rare under normal conditions.

- Can lead to pest problems (e.g., mosquitoes in pond and FWS ones)
- May not prove practical in some situations where local conditions (topography, drainage, soils, etc.) are not suitable
- May require some time before optimum efficiency is achieved
- Do not have many years of experience to draw on
- May be unfamiliar to regulatory authorities who may not have precedents
- Many early ones were poorly designed, leading to negative perceptions
- May operate at lower efficiencies during winter

Nevertheless, constructed wetland technology has evolved rapidly and engineers are now able to design wetland and other natural WWT systems with as much confidence as to their operability and pollutant removal levels as with comparable conventional (mechanical) WWT systems. There are now constructed wetland systems operating in all States, Provinces and Territories from the mouth of the MacKenzie River in the Arctic to tropical areas.

B.10 Polishing Wetlands

CWs are often used as part of EEB Systems. For example, as is mentioned in Section 3.7 of Chapter 3 regarding the EEB System at EIA, when the first cells of the two two cell trains of the old CW system at that airport were converted to BBR cells from CW cells, the gravel substrate to do so was excavated from the second cells of the respective trains, leaving these second cells as FWS CW cells.* Also, Engineered SW WLs often have an FWS CW as the final step in their process train (see Section 3.10 in Chapter 3).

Indeed, the use of CW cells as the final polishing (tertiary) step of an EEB System may be highly desirable in some cases, as these passive treatment cells are relatively economical to build, require only minimal ongoing attention and maintenance, and can be useful to remove small amounts of residuals from upstream treatment in EEB cells. For example, in cases where passive treatment anaerobic EEB cells (e.g., BCRs) are used to treat MIWs, their carbonaceous substrates may leach small amounts of BOD and other COCs, and a downstream CW can be ideal to clean up these residuals. Similarly, where they are located convenient to the site of the rest of an EEB System, Natural Wetlands can also be used as Polishing Wetlands, and it may be moot whether they are described as the final steps of EEB Systems or receiving waters.

* However, it is generally not a good idea to have FWS CWs at airports as they can provide habitat for waterfowl and represent a potential bird strike hazard.

Appendix C: Associated Technologies

C.1 Technologies Used in Association with EEB Systems

There are a number of WWT technologies/methods that are often used as part of or in association with EEB Systems. These include

- EEB System Impoundments
- Limestone treatment systems
- Aerated Lagoons
- Dosing siphons
- Phytoremediation
- Permeable reactive barriers
- PROPASYS

The first four of these (EEB System Impoundments, limestone treatment systems, Aerated Lagoons and dosing siphons) have been/are being used as parts of EEB Systems. The latter three (phytoremediation, permeable reactive barriers (PRBs) and PROPASYS-based methods) are often used either to help prepare WW streams for later treatment in an EEB System or, as part of an *Integrated Treatment System,* used to deal with minor streams around a site that are not large enough/conveniently located to allow/justify treatment in an EEB System but are still important enough to require management.

These methods/processes are addressed in more detail in this appendix.

C.2 EEB System Impoundments

As part of active, semi-passive and passive WWT systems, many WWs being treated contain entrained particulates (grit) as well as suspended solids and coagulated materials which can be removed by allowing them to settle out by gravity in an impoundment. In addition, a WW being treated in an EEB System may also contain suspended, colloidal and/or dissolved CoCs that can be induced to settle by mechanical, chemical or biological means. Such

EEB System Impoundments* may form part of the primary treatment component of EEB Systems, or, in some cases, part of 3° treatment. Depending on their main functions and uses, these components of EEB Systems may also be described as *Sedimentation Ponds, Settling Ponds, Dosing Ponds/Tanks, Balancing Ponds/Tanks, Tailings Ponds, Facultative Lagoons* and *Oxidation Ponds.* Two or more kinds of such impoundments may form parts of an EEB Systems and streams may be recycled from them or pumped/allowed to flow among them.

CoC removals in EEB System Impoundments may be dominated by a specific mechanical, chemical or biological process, but in them other processes will occur as well. For example, in Sedimentation Ponds, as their name suggests, sedimentation processes (gravity settling of grit and suspended solids) dominate, but flocculation/precipitation, sorption, redox reactions and biodegradation processes, as well as those associated with any plant growth, morbidity and decay, also occur in them.

Balancing ponds/tanks do not just accumulate and even out flow into downstream EEB Systems; they may allow the diversion of excess flow away from them and the recycle into them of some of an EEB System's treated WW if the streams being collected are too strong for ready treatment in downstream EEB Cells. Where balancing ponds/tanks are used, they can be configured as above-ground metal vessels depending on space at the specific airport and height restrictions, but, as was mentioned above, they can also be configured as buried tanks and vaults.

EEB System Impoundments used by the mining industry can act as settling ponds, Sedimentation Ponds and/or dosing ponds in which certain WW species such as the ferrous iron in an influent MIW naturally oxidize to ferric iron and then hydrolyse to ferric (oxy)hydroxides, which in turn will then settle as sludge at the bottom of the impoundment (see Equations 2.20 and 2.21 in Chapter 2). The removal of dissolved, colloidal and suspended contaminants in EEB System Impoundments at some mine sites may be greatly enhanced by adding appropriate chemicals such as lime and/or other materials such as slurried clay, bentonite or active silica (Na_4SiO_4), either by injection into the impoundments or their influents, spraying them onto their surfaces or by simply hand dispersing them on water surfaces. In some cases, the removal of ferrous iron in such an impoundment may be further enhanced by providing some aeration in part of it,† either by using submerged aeration grids, or by using floating aerators, mixers or aspirators.

* The term "impoundment" here is not limited to ponds and lagoons as these may also be configured in an EEB System as tanks, vaults and other kinds of vessels. For example, the main stormwater collection impoundment at BNIA is a very large 1.2 MM USG concrete vault that is mostly buried in the ground with only it top meter of height protruding aboveground. GCSW that collects in it is pumped to the airport's EEB System for treatment or, if measured to already meet regulatory criteria, discharged to the airport's receiving waters, a small stream.

† But not at such levels that the entire impoundment could be described as an Aerated Lagoon, see Section C.4 below.

Where high sulfate MIWs are being treated, EEB System Impoundments can be used downstream of BCRs (i.e., as 3° treatment cells) to allow the removal by oxidation back to sulfates of any potentially hazardous, excess hydrogen sulfide formed in the bioreactors and persisting in their effluents*.

Phosphorus is often removed from the waters in municipal WW lagoons and ponds by the continuous injection or surface application of alum, which results in an aluminum hydroxide floc (see Equation 2.20 in Chapter 2).

Depending on the nature of the WW and the particular situation, many chemicals can be used to enhance the removals of CoCs such as phosphorus from EEB System Impoundments of various kinds where the treatment of a variety of WWs is involved. These include ferrous sulfate ($FeSO_4$), ferric sulfate ($Fe_2[SO_4]_3$), ferric chloride ($FeCl_3$), calcium chloride ($CaCl_2$), magnesium chloride ($MgCl_2$), and sodium aluminum oxide ($NaAlO_2$). Where low pH wastewaters are involved (e.g., ARD), alkalinity is often added to neutralize it by adding slaked lime (CaO), hydrated lime ($Ca[OH]_2$), limestone ($CaCO_3$), dolomite ($Ca/MgCO_3$), caustic soda (NaOH), sodium carbonate (Na_2CO_3), magnesium oxide (MgO), and many other chemicals (see Section 2.5.4 in Chapter 2).

Liming (using slaked or hydrated lime) is the most common and is usually used to neutralize acidic, iron-rich WWs. It is often carried out as part of active WWT facilities (in which case it is usually carried out in tanks or in ponds by injecting a lime slurry into the WW). This results in a highly amorphous sludge of calcium sulfate (gypsum) and iron (oxy)hydroxides containing 1%–5% solids. Where WWs are acidic, high in iron content and contain sulfates (as do many MIWs), liming will raise pH and precipitate out the iron. The resulting yellowish-red ferric oxyhydroxide floc (ochre) will also co-precipitate other dissolved metals in the WW. The principal reactions of liming with hydrated lime can be simply expressed as

$$Ca(OH)_2 + Me^{2+}/Me^{3+} + 3H_2SO_4 \rightarrow Me(OH)_2/Me(OH)_3\downarrow + CaSO_4 + H_2O \tag{C.1}$$

where Me represents any bi- or tri-valent metal in the WW, including iron. In addition to the ochre sludge, if the pH rises above about five, any aluminum ions in a WW will hydrolyse and create an aluminum hydroxide floc as well.

Not only will such hydroxides remove a large portion of the dissolved metals in a WW, but their settling flocs will also carry down part of any other contaminants associated with suspended solids in the WW.

* See Section 5.5 of Chapter 5. For the same reason, pilot-scale BCR cells usually are enclosed and are fitted with vents, see Figure 1.4 in Chapter 1.

C.3 Limestone Treatment Systems

Where a low pH wastewater (e.g., an ARD) needs to be treated in an EEB System, limestone treatment can be used for neutralization as part of the system's primary WWT step instead of liming. The purpose of limestone treatment is to provide alkalinity to WWs by passing them through or over beds of limestone (calcium carbonate) (Brown et al. 2002) and the following equations illustrate the limestone neutralization process:

$$CaCO_{3(s)} + 2H^{+} \rightarrow Ca^{2+} + H_2CO_3 \quad (C.2)$$

$$CaCO_{3(s)} + H_2CO_3 \rightarrow Ca^{2+} + 2HCO^{3-} \quad (C.3)$$

$$CaCO_{3(s)} + H_2O \rightarrow Ca^{2+} + HCO^{3-} + OH^{-} \quad (C.4)$$

$$CaCO_{3(s)} + SO_4^{2-} + H_2O \rightarrow CaSO_4 + HCO^{3-} + OH^{-} \quad (C.5)$$

$$CaCO_3 + Fe^{2+}(\text{or } Mn^{2+} \text{ or } Zn^{2+}) \rightarrow FeCO_3(\text{or } MnCO_3 \text{ or } ZnCO_3) \quad (C.6)$$

These reactions will raise the pH of an acidic WW to the 6–7+ range. Other than the addition of alkalinity, ideally no treatment should occur in a limestone treatment system.

$CaCO_3$ dissolution in a limestone treatment system is inversely proportional to pH, Ca^{2+} and HCO^{3-} activity, and directly proportional to the partial pressure of CO_2 (P_{CO2}). Usually, calcite (agricultural limestone) is involved but occasionally dolomite (calcium magnesium carbonate) is used instead.

Effluent from limestone treatment can be directed to a Sedimentation Pond where the ferrous iron will oxidize to ferric iron, hydrolyse and precipitate. Alternately, it may be directed into a BCR where it will precipitate as insoluble materials such as sulfides.

There are five basic kinds of semi-passive limestone treatment processes

- Open limestone channels
- Anoxic limestone drains
- Oxic limestone drains
- Limestone ponds
- Limestone leach beds

Any of these systems may be used for primary treatment in an EEB System.

Open limestone channels (OLCs) are the simplest form of limestone treatment and involve placing in or lining open trenches with crushed limestone or limestone cobbles. These can then be used to convey acidic WWs such as ARD from its source to the rest of an EEB System. In OLCs, once the ferrous iron in the ARD is exposed to air, it oxidizes and hydrolyses, and the surfaces of the limestone particles may encrust (armor) with ochre (see Equation C.1), gradually limiting the limestone's neutralization potential.

Figure C.1 (Higgins et al. 2006a) illustrates the beginning of the accumulation of ochre on the limestone particles of a limestone treatment system.

FIGURE C.1
Limestone armoring.

With OLCs, this negative effect can be minimized by ensuring that they are located on steep slopes (>20%) and that water velocities in them are at least in the 0.1–0.5 m/min range to keep the precipitates in suspension while limestone dissolution occurs (Younger 2000).

The most common kinds of limestone treatment systems are called *Anoxic Limestone Drains* (ALDs). In them, acidic WWs are passed through beds of calcite or dolomite from which air is excluded by burial or some other method. ALDs have been widely used as Stand-Alone treatment methods at power plants, mines and other industries that produce low pH, metal-contaminated WWs requiring treatment.

The amount (mass M) of limestone required for an ALD (or for the limestone layer in a SAPS Bioreactor, see Section 4.3 in Chapter 4) can be calculated using the following formula (Hedlin et al. 1994):

$$M = (Q.\rho.\tau_L)/V_v + (Q.C.T)/P \qquad \text{(C.7)}$$

where Q is the design maximum ARD flow rate, ρ is the limestone bulk density, τ_L is the residence time in the limestone bed, V_v is the bed void volume, C is the predicted alkalinity of the effluent, T is the desired bioreactor lifetime before the limestone may need replacing (usually set at 10 years or more) and P is the limestone purity (usually 80%–95%).

ALDs work best when neutralizing ARDs containing <1 mg DO/L, <1 mg Fe^{3+}/L, <1 mg Al^{3+}/L and <2000 mg SO_4^{2-}/L.

The presence of moderate-to-high concentrations of dissolved aluminum in a WW will lead to the formation of gelatinous aluminum hydroxide floc between an ALD's limestone particles as soon as the WW's pH is raised, exacerbating the armoring and plugging that the ochre sludge may cause. Various methods have been proposed to limit the buildup of potentially plugging iron and aluminum hydroxides in ALDs, and these, depending on particular WW chemistries and flow rates, have varying degrees of effectiveness.

In order to maintain WWs at low dissolved oxygen levels, ALDs are built into the ground and covered over (buried) with impermeable soil (e.g., clay) and/or geotextiles to exclude further oxygen ingress from the air.

It has been found that incorporating sandstone into the limestone bed of an ALD can substantially offset the armoring of its limestone substrate bed (Sasowsky et al. 2000). When this was the situation with an ALD, the majority of the sulfides were precipitated on the sandstone.

Where it is impossible to exclude oxygen from a WW, another kind of limestone drain, an *Oxic Limestone Drain* (OLD), may be used instead. With OLDs, the use of some sort of mechanism such as pumps or dosing siphons are included to promote the periodic flushing of the drains and to limit ochre armoring. With OLDs, there is no attempt to prevent ferric iron formation in the drain as the pH is raised. Unlike ALDs, OLDs need not be buried (but they still may be).

Related to limestone drains are Limestone Leach Beds in which separate, uncontaminated streams of water are passed through beds of limestone or dolomite to make them alkaline, after which they are mixed with low pH WW streams to neutralize them.

C.4 Aerated Lagoons

Aerated Lagoons can form part of the secondary WWT component of EEB Systems where they are positioned before EEB cells. In EEB Systems involving SAGR Bioreactors (see Section 3.9 in Chapter 3), the system is designed to have the Aerated Lagoons remove suspended solids and BOD, leaving the SAGR Bioreactor cells for nitrification service only. In other EEB Systems, the Aerated Lagoon complements downstream EEB cells.

Aeration equipment for an Aerated Lagoon must be sized to continuously support a minimum of 2.0 mg/L of oxygen in the WW passing through it. In order to determine the size of equipment that can establish these concentrations, bacterial oxygen demands must be calculated and balanced against the ability of the equipment to introduce oxygen into the WW. In general, aeration by diffusers is more efficient for water depths of more than 3 m, and aeration by floating aspirators is more efficient at depths less than 3 m, but this may not apply for very large lagoons (Rich 1999).

Stoichiometrically, 4.6 kg of oxygen is required for bacteria to oxidize 1.0 kg of ammonia. Given the daily oxygen requirements, aeration equipment suppliers can size equipment for site-specific temperature and elevations. The oxygen transfer efficiency for aeration equipment is subject to local constraints and varies for the different types of aeration equipment. Floating aerators are typically sized for a field transfer efficiency of 0.5 kg/HP-hr.

Submerged diffusers that have air supplied by blowers typically have a transfer efficiency of approximately 5.2% per meter of submerged depth.

Aerated Lagoons are typically classified as either complete mix or partial mix. Complete mix is defined as providing a uniform suspension of solids (bacterial biomass) throughout the mixed volume. Partial mix is defined as providing uniform concentrations for soluble items in the mixed volume. To achieve nitrification, complete mix conditions are required. However, *Nitrifier* bacteria are relatively light compared to other heterotrophic bacteria and, for this reason, less energy is required to keep a uniform concentration of them suspended. In general, if a lagoon is mixing limited, mechanical mixing is more efficient than mixing using submerged diffusers.

C.5 Dosing Siphons

As is discussed in Section 4.3 of Chapter 4, the purpose of the kind of EEB ABR known as a SAPS Bioreactor is to convert ferric iron in an ARD to ferrous iron, and then to raise its pH to circum-neutral levels in a limestone layer after its organic substrate. The WW, which will then be rich in ferrous hydroxide, will flow out of the EEB cell either into a Sedimentation Pond or CW where the ferrous hydroxide will hydrolyse to ferric hydroxide. Alternately, it can be sent to a BCR for further treatment (see below).

An ordinary SAPS Bioreactor will work well as long as the MIW being neutralized does not contain any appreciable amounts of dissolved aluminum. If it does, when the pH is raised in the limestone bed, aluminum hydroxide will precipitate out. While aluminum hydroxide will not cause armoring as will ochre, it can build up in the void spaces between the limestone particles, reducing throughput and eventually plugging the limestone bed. To prevent this from being a problem, there is a version of a SAPS Bioreactor available called an Aluminator™. With this kind of SAPS Bioreactor, the bed is manually flushed every few hours to remove accumulating $Al(OH)_3$ sludge.

There is another, less intensive way to remove both accumulating ochre and aluminum hydroxide sludge from a SAPS Bioreactor. That is to attach a Dosing

Siphon to its outlet. Wikipedia defines a siphon as a device that moves a liquid upward above the surface of a reservoir with no pump but powered by gravity as it discharges at a lower level than the surface of the reservoir from which it came. Dosing siphons are kinds of automatic siphons that convert small continuous flows into large intermittent flows and move finite volumes of liquid (doses) at flow rates from a few m^3/min to many hundreds of m^3/min.

Figure C.2 illustrates one kind, a bell dosing siphon.*

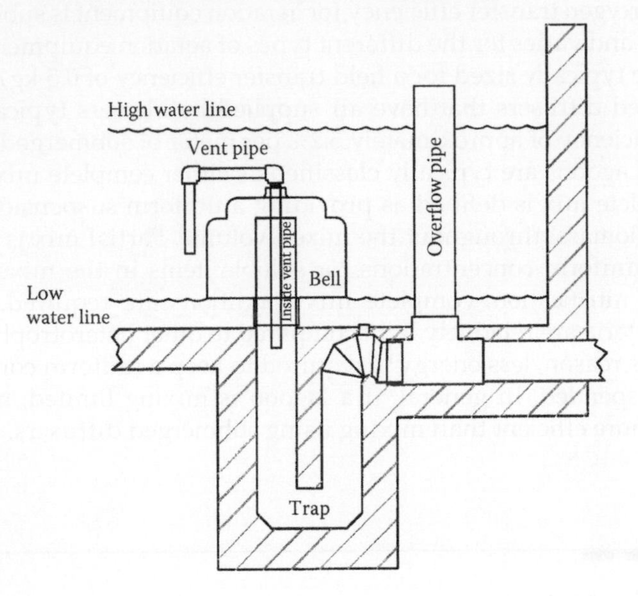

FIGURE C.2
A dosing siphon.

Dosing siphons have no moving parts, require no power sources, require little maintenance and can be made of corrosion-resistant materials well able to stand up to ARDs. They have been used for a long time at municipal WWTPs to dose trickling and sand filters, but in EEB Systems they can be associated with SAPS Bioreactors to flush neutralized ARD into downstream Sedimentation Ponds. Figure C.3 illustrates how a dosing siphon would be connected to the outlet of a SAPS Bioreactor in an EEB System.

C.6 Phytoremediation

Phytoremediation is that type of bioremediation where the degradation and/or removal of pollutants from a contaminated medium is caused, mediated

* Details on the operations of dosing siphons may be found on Wikipedia.

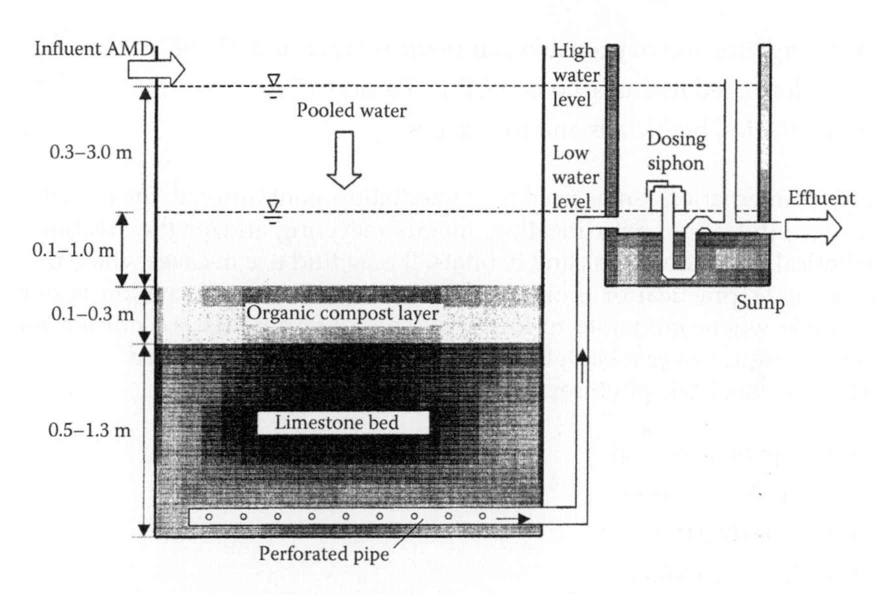

FIGURE C.3
A dosing siphon with a SAPS Bioreactor.

and/or assisted by plants and as such EEB Systems involve many of its mechanisms. Phytoremediation can be classified as a set of economic, generally on-site, *in situ* techniques for the degradation, removal and/or stabilization of various pollutants from soils, sediments, sludges, WWs, groundwaters and indoor air. It can be a soil remediation methodology, a waste management method or a way to carry out WWT. Phytoremediation can be used to clean up both organic and inorganic contaminants including heavy metals, pollutants that can be difficult to handle by other methods. It can be carried out either terrestrially or aquatically and, for the latter, can even involve hydroponics.

The following are some of the kinds of contaminants, the clean-ups of which are, or can be made to be, amenable to phytoremediation.

- Metals
    - Heavy Metals (Pb, Cr, Zn, Cu, Hg, Cd, and Ni)
    - Radionuclides (3H, fission wastes, and transuranic wastes)
    - Other inorganics
    - Metalloids (As, B, and Se)
- Nutrients (PO_4, NH_3, NO_3, and K)
- Salts (chlorides, sulfates, and fluorides)
- Organics
    - Petroleum hydrocarbons (VOCs, oil and grease, BTEXs, and oils)
    - Heavy hydrocarbons and aromatics (TPHs and PAHs)

- Energetics and oxygenated compounds (TNT and MTBE)
- Chlorinated hydrocarbons (PCP, PCBs, and TCE)
- Pesticides, herbicides and fungicides

Phytoremediation can be used for remediation, containment, the reduction of water infiltration, detoxification, metals recycling and/or the creation of aesthetically pleasing areas and habitats. It may find use in cases where there are no other practical or economic alternatives. Phytoremediation is most applicable where moderate to low levels of contaminants are present and where clean-ups over relatively long timescales are not precluded.

There are six basic phytoremediation mechanisms as follows:

- Phytovolatilization
- Phytodegradation
- Phytoextraction
- Rhizodegradation
- Phytostabilization
- Phytofiltration

Phytovolatilization relies on the phytoremediating plants to transpire out of plants growing in soils and sediments volatile contaminants such as light organic compounds (e.g., gasoline and some light chlorinated compounds) and certain inorganic contaminants (e.g., mercury and selenium compounds). Such volatile materials, when present in a contaminated medium, may (and usually will) volatilize over time anyway, and phytovolatilization will only accelerate the process.

Phytodegradation, sometimes also called *phytotransformation*, refers to the uptake (translocation) by plants into their saps of lighter, partially hydrophilic organic compounds and their subsequent metabolization within the phytoremediating vegetation. When such organic pollutants are taken up into the plants, they are either metabolized into plant tissue or mineralized all the way to carbon dioxide. With phytodegradation, contaminants generally are destroyed, and there usually is no need for plant harvesting.

The ability of plants to take up and metabolize metals from water in their root zones (*rhizospheres*) is well known. Indeed, many metals are essential nutrients (e.g., iron, potassium, copper, zinc, nickel, manganese, and magnesium) and plants have evolved sophisticated mechanisms to translocate them into their tissues. Where such nutrients are present in a contaminated medium (e.g., a soil) in large amounts, or when toxic concentrations of metals are involved, certain plants have evolved mechanisms to deal with such situations as well, by either taking some of them up into their tissues where they are detoxified, isolating them in vacuoles and/or by adsorbing them on/into the surfaces of their roots and rootlets where they are sequestered. Some of these plants (*accumulators*) are able to take up metal contaminants in soils through their roots into their shoots, stems, leaves and other above-ground tissues.

Some accumulators, owing to acclimatization, for defence purposes and/or because of earlier development, are able to take up very large amounts of metal contaminants through their roots into above-ground tissues and these are called *hyperaccumulators*.

Phytoextraction is the utilization of special varieties of plants to extract, transport and concentrate inorganic contaminants (metals, metalloids, salts and nutrients) from soil porewater into the plants' above-ground tissues. These may then be transported within the plant and may be concentrated in above-ground biomass (leaves and stems) in cell walls or cellular vacuoles. One form of phytoextraction uses hyperaccumulators which are acclimatized to severely metal-contaminated soils and which sometimes can accumulate as metal in plant tissue up to many percent of the plant's dry weight. Harvesting of plants is required to remove phytoextracted inorganics from a site.

A plant-soil system's ability to phytoextract can be *enhanced* by irrigation, by over-fertilization, by the application of certain mobilization chemicals (e.g., chelators) or other amendments and/or by sophisticated vegetation establishment methods, all of which increase inorganics mobility (and hence potential bioavailability).

Rhizodegradation, sometimes called *Enhanced Rhizosphere Biodegradation*, is the plant-mediated microbial breakdown (oxidation) of organic contaminants in plant root systems. With it, soil microbes that live symbiotically in plant root zones first oxidize contaminants to biomass (i.e., more microbes) and lower molecular-weight compounds and later (in some cases) to more still biomass, carbon dioxide and water (mineralization). The root zones of the plants supply an environment favorable for microbial attachment, growth and proliferation. As with phytodegradation, organic pollutants are destroyed by rhizodegradation and plant harvesting usually is not required.

Phytostabilization, sometimes referred to as *in situ inactivation combined with re-vegetation*, is the opposite of phytoextraction or phytodegradation. Its purpose is to convert to less toxic forms and/or decrease the bioavailability of metallic and/or high-molecular-weight organic contaminants in soils and sediments, thereby preventing their entry into groundwater and/or food chains. This is accomplished by the mixing (usually by tilling) into soil of one or more *ameliorants* (special kinds of amendments) which more or less permanently adsorb, precipitate, change the valence state of and/or otherwise inactivate and immobilize soil contaminants. Vegetation is then planted on the amended soil to protect it from erosion, reduce water infiltration, further precipitate and sequester the pollutants and create better conditions for microbial activity in the rhizospheres (which in turn further helps to sequester contaminants and render them biologically unavailable). A variety of alkaline and exchangeable materials are used as phytostabilization ameliorants, including limestone, phosphate minerals, hydrous oxides, zeolites and industrial by-products such as fly ash, compost, biosolids and iron shot. Phytostabilization does not remove contaminants from the soil and is a waste management rather than soil remediation method.

Phytofiltration is the use of plant root systems to adsorb, concentrate and even precipitate metals and organic contaminants from a WW stream. It can be carried out in aquatic WWT systems (e.g., constructed wetlands), or hydroponically in ponds or vessels using terrestrial plants (in which case it is known as rhizofiltration). It is a powerful new way to remove radionuclides (e.g., isotopes of uranium, cesium, strontium, technetium and plutonium) and toxic metals (e.g., mercury) to very low levels (e.g., ppb), both in wetlands and hydroponically.

EEB Systems can involve or be associated with the following kinds of phytoremediation methodologies as follows:

- Phytoremediation caps
- Phytoremediation barriers
- Biological pump and treat systems
- Phytoirrigation systems
- Phytoremediation rafts
- Phytoremediation covers
- Vegetative Covers
- Evapotranspiration covers

Phytoremediation Caps are used to remediate shallow accumulations of organic-contaminated soil. Generally, trees are planted over a soil-capped waste disposal site and their roots grow down into the waste. There they remediate it by phytodegradation, phytovolatilization and rhizodegradation mechanisms. Phytoremediation caps can be used over waste accumulations whose seepages are treated in down-gradient EEB Systems.

Phytoremediation Barriers consist of groves of trees, often hybrid poplars, placed in staggered rows perpendicular to the flow direction of a shallow, contaminated groundwater plume. Their purpose is to mitigate/stop the flow of the contaminated groundwater. The *phreatophytic* trees have high water needs and draw the groundwater into their root balls where it is used by the trees and transpired into the air. Contaminants in the water drawn into the root balls are remediated by various phytoremediation mechanisms. Phytoremediation barriers are kinds of WWT methods that can be used to treat minor flows of contaminated groundwater at a project site where more substantial flows are treated in an EEB System.

Figure C.4 illustrates a phytoremediation barrier.

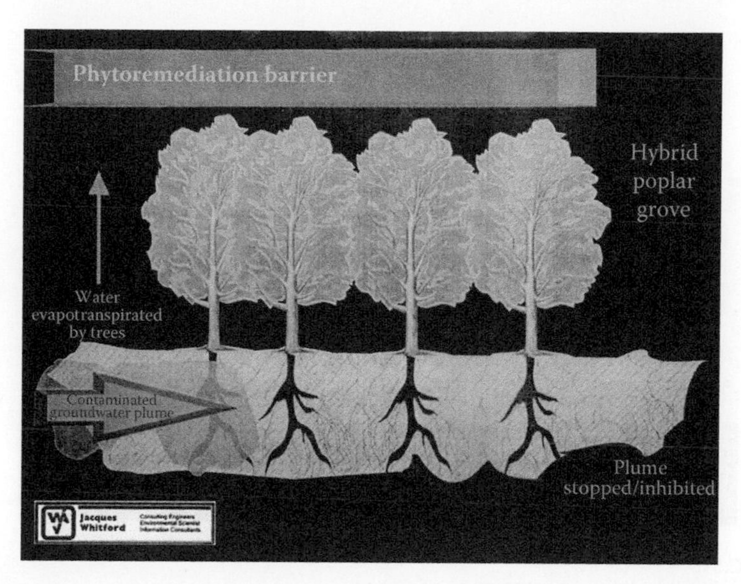

FIGURE C.4
A hybrid poplar phytoremediation barrier.

Biological Pump and Treat Systems are plantations of trees, often hybrid poplars, willows and/or alders, planted over contaminated soil. They function almost the same as phytoremediation barriers but their purpose is to remove groundwater and porewater from soil on contaminated sites such as brownfields, while allowing the contained contaminants to be phytoremediated. These systems share aspects of remediation and WWT methods.

Phytoirrigation systems are another name for spray or drip irrigation land WWT systems. With them, contaminated water such as food plant process waters and effluents from other WW systems are used to provide irrigation water for vegetated fields and/or copses of trees. Generally, the water for the irrigation is pumped into the spray or drip irrigation systems. Its source may be surface water or groundwater from a sump or well. The irrigated water is removed by evapotranspiration and the contained contaminants are phytoremediated in the soils around the vegetation by the various phytoremediation mechanisms. Like biological pump and treat systems, these systems share aspects of remediation and WWT methods.

Phytoremediation Rafts are floating artificial rafts in which wetland plants are growing hydroponically suspended on flat geotextile meshes. They are used for the pretreatment of WWs and to cover the surfaces of surge ponds in front of other WWT systems (e.g., CWs) where open water surfaces that might attract waterfowl are undesirable (e.g., at airports). Phytoremediation rafts are used in the surge pond upstream of the BBR cells of the EEB System at London Heathrow Airport (see Section 3.8 in Chapter 3).

Figure C.5 shows some of them.

FIGURE C.5
Phytoremediation rafts in the EEB System at LHA.

Phytoremediation Covers, sometimes called *phytostabilization covers or caps,* are especially intended for covering waste accumulations containing potentially toxic metals. With them, the purpose is to minimize as far as possible the infiltration into the covered material of any precipitation falling on a contaminated, usually metal-contaminated, site. In addition, they involve layers of phytostabilizing ameliorants to treat any upwelling contaminated water from the waste accumulation and are bottomed by an impervious layer such as clay to greatly inhibit infiltration. The soil cover materials in phytoremediation covers may contain fillers, fertilizers, bulking agents and organic wastes. Vegetation for phytoremediation covers is usually chosen to have exudates that help immobilize contaminants, complementing the phytostabilization action.

Phytoremediation cover systems often include active aspects (drains, ditches and channels) to divert "clean" rainwater falling on the site away from the cover into Stormwater Wetlands, and the greatly reduced leachate may flow into CWs. Like biological pump and treat systems, these systems share aspects of remediation and WWT methods.

Vegetative Covers are simple kinds of phytoremediation covers often used to vegetate and reclaim the surfaces of waste accumulations such as tailing and waste rock piles at mine sites and the filled cells at landfills. Such waste accumulations can be capped with dry covers that in turn may be further categorized into *Barrier Covers* and *Alternative Covers.* The former are those that use materials of low hydraulic permeability to minimize infiltration, thereby providing physical barriers between the environment and underlying waste (e.g., geotextile covers, clay liners and multilayer earthen covers). The latter (Alternative Covers) have less reliance on minimizing percolation (as it will often get treated downstream anyway), instead using hydrological methods that rely in part on the properties of the cover materials to store water until part of it is either transpired through vegetation growing on the cover and/or evaporated from the cover surface (EPA 2003).

Barrier (impermeable) kinds of dry covers (e.g., clay covers) are designed to divert most of the meteoric water falling on them, but even so there always will be some infiltration through them and usually also some groundwater passing through the covered material. This approach may still necessitate some sort of provision for long-term water treatment such as an EEB System, although the volumes to be treated will be very much less than those required for a water cover, a way which is often used by the mining industry to close out acid-generating tailings and waste rock areas.

Alternative Covers will be intermediate between water covers and barrier-type dry covers as to the amount of excess water that will need to be treated with them, as with them there is less concern about eliminating infiltration, and there will be times when percolation into and through them will be higher. The kinds of vegetation that are planted on barrier-type dry covers are usually selected not to have deep roots which might penetrate the cover material and provide routes for the influx of water into the underlying waste. This practically limits cover vegetation to kinds of grasses, and also necessitates long-term maintenance programmes to prevent deeper rooting vegetation such as trees and bushes from colonizing the cover surface. This is a special problem for the closure of mines located in forested areas where trees are the natural local vegetation (as is the case at many mines). With Alternative Covers, the penetration of roots through the cover material, and even into the underlying wastes, is not a problem. Alternative Covers can be combined with WWT systems, if required, to deal with that infiltration and down-gradient seepage which may occur. Accordingly, with Alternative Covers (e.g., Vegetative Covers and ET Covers), woody plants native to an area can be used.

Alternative Covers are usually more economical to install and maintain than are barrier kinds of dry covers. The main considerations for selecting an alternative cover are climate; underlying mine material to becovered (i.e., its physical and chemical properties); the availability and cost of suitable cover materials; the physical, chemical, and agronomic properties of these cover materials and the kind(s) of plants with which it is intended to initially vegetate the cover (since natural succession will inevitably change the kinds of plants growing on a cover, this aspect too must be considered when designing an alternative cover).

Evapotranspiration Covers (ET Covers, sometimes called *store and release covers*) are special kinds of Alternative Covers which combine the permeable properties of the organic parts of the cover materials used in simpler Vegetative Covers (e.g., compost, wood wastes and biosolids) with the phytoremediating properties of the vegetation planted in them (in this case the ability to evapotranspiration back into the air a significant part of the meteoric water falling on them for at least part of every year). While Vegetative Covers are only sometimes combined with the collection of meteoric water,

ET Covers are almost always associated with this collection and its downgradient treatment, often using an EEB System. In addition, ET Cover systems are often associated with the provision of some way to divert stormwater from areas adjacent to the area being covered (to minimize the amounts of meteoric water flowing onto the cover), and can be sloped to allow clean water to flow off them into an adjacent stormwater system once the cover reaches field capacity during precipitation events (and hence no more water can infiltrate).

ET Covers use one or more layers of organic and other materials, and may be further categorized into monolayer and capillary versions. Monolayer (sometimes referred to as *monolithic or monofill*) covers involve a single layer of soil or other growth medium (e.g., compost, wood waste and excavated topsoils) in which the plants are growing. A capillary-type ET cover consists of a layer of finer-grained material similar to that of a monolithic cover overlying a layer of coarser material such as sand or gravel. The differences between the unsaturated hydraulic properties between the two layers minimize percolation into the lower layer (EPA 2003).

With either monolayer or capillary covers, under the saturated conditions of large precipitation or snowmelt events, there will be some infiltration into the underlying material (and the waste), but the cover material quickly will reach field capacity (i.e., saturate), allowing the uncontaminated excess water to flow off the cover for separate management (e.g., passage through a stormwater system).

Figure C.6 illustrates an ET cover.

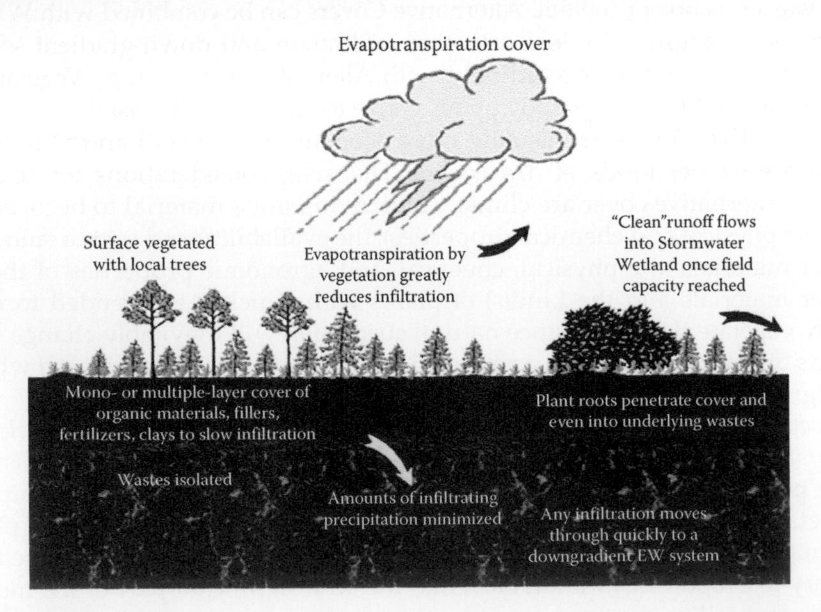

FIGURE C.6
An evapotranspiration cover.

C.7 Permeable Reactive Barriers

PRBs are covered sub-surface trenches filled with active materials that intercept contaminated groundwater plumes. These plumes must move through them as they flow under their natural gradients. Contaminants are immobilized or destroyed *in situ* by redox processes, co-precipitation, adsorption and/or biological processes. In the case of the latter, the PRB active media are often of the same kinds as those used in EEBs. As contaminants move through the active media of a PRB, reactions occur that transform them to less harmful (i.e., nontoxic) forms (Blowes et al. 2003).

There are three common morphologies for PRBs: (1) porous wall, (2) trench and gate and (3) funnel and gate. For the porous wall variety, plumes can pass through any part of the entire barrier wall continuously, while for the other two varieties, groundwater plumes are routed by impermeable wall sections to permeable openings in which the (re)active media are placed. PRBs may be designed to be permanent, semi-permanent or replaceable. Various materials can be used as the reactive media of PRBs depending on the contaminants involved and the particular situation including peat, biosolids, compost, wood wastes, bauxsol (the seawater-neutralized red mud from bauxite manufacture) and zero-valent iron (ZVI).

There are three kinds of PRBs: degradation barriers in which contaminants are degraded *in situ* by chemical or biological means, precipitation barriers in which contaminants react chemically and/or biologically to form insoluble products which are retained in the barrier material, and sorption barriers in which contaminants are sorbed and/or biosorbed and/or chelated on the surfaces of barrier materials. The active parts of some PRBs function as de facto anaerobic bioreactors, and their operations are similar to those used in EEB Systems. Also, like phytoremediation barriers, PRBs are used in conjunction with EEB Systems treating more substantial flows at contaminated sites.

Figure C.7 (Blowes et al. 2003) shows the morphology of a PRB.

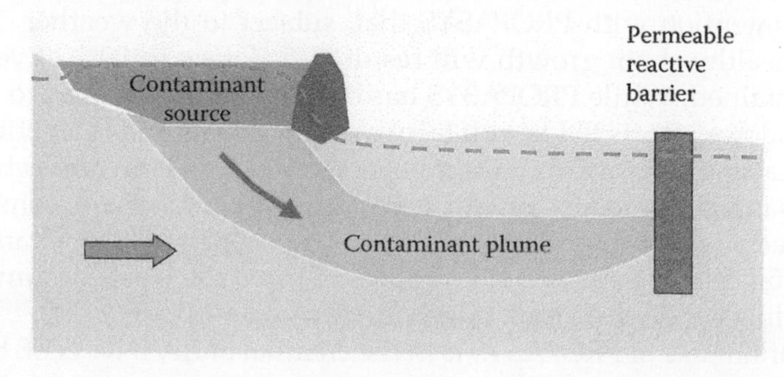

FIGURE C.7
Permeable reactive barrier.

C.8 Propasys

PROPASYS (from *PROPAgation SYStem*) is the name of a series of innovative products for soil remediation and vegetation establishment developed by SESI and offered by ETDC. PROPASYS is a rapid vegetation establishment method that can be used to reclaim stressed environments. PROPASYS is manufactured by blending together *Vegetated Growth Substrates* (humus-like artificial soil materials), fertilizers and nutrients, biostimulants, and *propagules* (pre-germinated vegetative mixtures).

PROPASYS formulations are complex, living ecosystems in themselves. They provide all the nutrient requirements needed for plant growth, and have high moisture retentions. Since it is its own soil, PROPASYS can be applied to very deficient rocky, barren or sterile areas such as brownfield sites. The use of PROPASYS on a site being remediated will result in vigorous plant growth in relatively short periods, and can allow the covering over of areas that would otherwise be very difficult to vegetate. Also, PROPASYS is less impacted by site conditions or adverse weather following application than are other application methods such as by the mechanical spreading of cover soils with seeds or hydroseeding.

The *Vegetated Growth Substrates* of PROPASYS are blended from waste materials (e.g., wood wastes, sawdust, composts and slags), pollution control by-products (e.g., WWTP by-product sludges) and other materials (e.g., ground limestone). Vegetated Growth Substrates are custom-designed and their compositions will vary depending on the local availability of suitable components, the application site situation, environmental conditions and several other factors. Sometimes Vegetated Growth Substrates can allow the productive use of low-value materials that might otherwise require disposal.

The *propagules* of PROPASYS are bioengineered mixtures of various plant species including seeds, roots, shoots, rhizomes and vegetative cuttings. Propagules can be designed to create many kinds of plant ecosystems.

With PROPASYS, ecological compatibility can be assured, vegetation local to the site being remediated used, and biodiversity preprogrammed.

It is the intention with PROPASYS that, subject to the weather, 15 cm of vigorous, healthy shoot growth will result in 5–7 days in 80% of vegetative species contained. While PROPASYS has been successfully used to vegetate lawns, meadows, sports fields, golf fairways and greens and recreation areas, its ecological engineering markets involve its use to reclaim and rehabilitate stressed, damaged and/or contaminated sites. Examples are construction sites, eroded areas, brownfields, rights-of-way, spillway and channel bottoms, landfill cell surfaces and the surfaces of mine site waste accumulations such as tailings and waste rock piles.

An important use of PROPASYS is in the creation of LIVING WEB blankets. These are pre-grown, biodiverse vegetation systems made up by impregnating a quantity of PROPASYS into a jute or coir erosion control blanket spread out on a large sheet of plastic located at some convenient, open flat site (e.g., an old parking lot, an inactive industrial site or an unused road). Water and fertilizers are applied and the plants pre-grown until they are well started. Once the vegetation on the sheet has grown enough, the blanket is cut up into mats (sheets) that are rolled up and transported to the site where they are to be used. LIVING WEB mats can be rolled down the steepest slopes, and fixed in place to result in instant vegetation in a manner that cannot be duplicated by other methods. Where a suitable shallow water preparation site is available, sub-aqueous LIVING WEB mats vegetated with growing biodiverse wetland vegetation also can be prepared. These mats can then be used to cover over and encapsulate contaminated sediments on parts of the bottoms of Natural Wetlands, particularly shallow water areas along shorelines where wave action often precludes establishing wetland vegetation by any other methods.

The most important use of LIVING WEB in association with EEB Systems is in covering over the sides of waste rock and tailings piles at mine sites, thereby reducing the amounts of ARD, NMD or alkaline MPWs produced when precipitation and/or groundwater percolates through the waste accumulations.* The reduced and attenuated amounts of MIWs thus generated can then be routed to EEB Systems handling MIWs at the site or used for the management of such streams as part of integrated treatment systems.

Figures C.9 through C.11 show the preparation, transportation and installation of LIVING WEB mats on an otherwise very-difficult-to-vegetate steep slope above a stream that was behind some private residences, whose backyards were being eroded away, potentially endangering the houses. Figure C.8 shows the slope.

FIGURE C.8
A steep slope requiring remediation.

* For waste accumulations generating ARD, the soil mix of the PROPASYS used in preparing LIVING WEB may include ground limestone in its mix to provide some neutralization potential as well.

Figure C.9 shows the preparation of LIVING WEB mats at a distant, unused flat industrial site.

FIGURE C.9
The preparation of LIVING WEB mats.

Once prepared, the LIVING WEB mats were transported to the project site (Figure C.10).

FIGURE C.10
Transporting LIVING WEB mats.

At the project site, the LIVING WEB mats were installed by rolling them down the slope to be remediated and pinning them in place (in this case—not shown—with "live" wooden stakes that would themselves later propagate to form bushy plants; Figure C.11).

FIGURE C.11
Installing a LIVING WEB mat.

Figure C.12 shows the slope after the LIVING WEB mats were installed.

The remediation of the sides of tailings and waste rock accumulations at mine sites can be accomplished in much the same manner as that shown for the above relatively simple residential project.

FIGURE C.12
A steep slope remediated.

Appendix D: Glycols and Airports

D.1 Glycol Freezing Point Depression Chemicals

As was introduced in Section 2.3.2 of Chapter 2, among the largest and most important uses of Eco-Engineered Bioreactor Systems are those used at airports, and these systems rely on BBRs (see Section 3.2 in Chapter 3) for the secondary treatment of wastewater streams contaminated with glycols resulting from the de-icing and anti-icing of aircraft in cold weather. Accordingly, defining the properties of these glycol-contaminated streams is essential in considering the use of EEB Systems at airports.

Glycols are dihydric alcohols used for suppressing the freezing point of aqueous solutions. All are fully miscible in water and nonflammable. Table D.1 (Higgins & MacLean 1999; ACRP 2012) compares some of the physical properties of the three glycols used for aircraft de-icing and anti-icing: ethylene glycol (EG), propylene glycol (PG), and diethylene glycol (DEG).

TABLE D.1

Properties of Glycols

	EG	PG	DEG
Formula	H H | | HO–C–C–OH | | H H	H OH H | | | H–C–C–C–OH | | | H H H	H H H H | | | | HO–C–C–O–C–C–OH | | | | H H H H
Molecular weight (g/mol)	62.3	76.1	106.1
Density (g/mL)	1.115	1.038	1.118
Freezing point (°C)	−13	−59	−80[a]
Boiling point (°C)	197	189	244
Vapour pressure (mm Hg)	0.05	<0.01	nd
Oxygen demands (g O_2 per g glycol)			
ThOC	1.29	1.69	1.51
COD	1.29	1.69	1.51
BOD_5	0.47	1.08	0.10

[a] Sets to glass below −60°C.

The oxygen demands are most relevant to the treatment of glycols in BBRs.*

* The derivation of that for EG is presented in Section 2.3.2 in Chapter 2.

The total aerobic degradation of pure ethylene glycol (also see Equation 2.6) in a BBR can be represented by

$$HOCH_2CH_2OH + 2.5\,O_2 \rightarrow 2\,CO_2 + 3\,H_2O \tag{D.1}$$

while that of pure propylene glycol* can be represented by

$$CH_3CH(OH)CH_2OH + 4\,O_2 \rightarrow 3\,CO_2 + 4\,H_2O \tag{D.2}$$

Such degradations may be biotic (microbial) and/or abiotic.

Since the ThOC is practically equal to the COD for ethylene glycol, the latter can be calculated as

$$\begin{aligned} COD_{EG} = {} & (1.29\,g\,O_2/g\ EG \times 1.115\,g\ EG/cc\ EG \times 1000\,cc/L \times 1\,mol\,O_2/32.0\,g\,O_2 \\ & \times 1\,mol\ EG/2.5\,mol\,O_2) \times (62.3\,g\ EG/mol\ EG \times 1000\,mg/g) \sim 1{,}120{,}000\,mg/L \end{aligned} \tag{D.3}$$

The cBOD concentration of pure EG is 0.47 g O_2/g EG, or 36% of the THOC (~403,000 mg cBOD per L).

Similarly, using Equation D.2, for pure PG the ThOC is 4.0 mol O_2 × 32 g O_2/mol O_2 divided by 76 mol PG/g PG = 1.69 g O_2/g PG. Also, the PG concentration times 3 × 12 (the moles of carbon as part of the CO_2 product in this case) divided by the molecular weight of PG (76) gives the TOC concentration for PG, 0.47 times its concentration in mg/L.

Again, since ThOC is practically equal to the COD (see Table D.1), for pure PG this parameter can be calculated as

$$\begin{aligned} COD_{PG} = {} & (1.69\,g\,O_2/g\,PG \times 1.038\,g\ PG/cc\ PG \times 1000\,cc/L \times 1\,mol\,O_2/32.0\,g\,O_2 \\ & \times 1\,mol\ PG/4.0\,mol\,O_2) \times (76.1\,g\,PG/mol\,PG \times 1000\,mg/g) \sim 1{,}043{,}000\,mg/L \end{aligned} \tag{D.4}$$

The cBOD concentration of pure PG is 1.08 g O_2/g PG, or 64% of the THOC (~667,000 mg cBOD per L).

As was mentioned, the aforementioned calculated theoretical COD values for pure glycols are relatively low compared to those often measured for ADF concentrates in commercial analytical laboratories.† Besides the glycol,

* PG has two stereoisomers: 1,2 propylene glycol and 1,3 propylene glycol. PG-based ADAFs are made from racemic mixtures.

† Usually, for 88%–92% EG and 88% PG ADF concentrates, laboratory-measured CODs tend to be >1.3 MM mg/L unless dilutions are measured using weight not volume.

the remaining content (e.g., 8% for most EG ADF concentrates) is water, impurities and ADPAC chemicals (circa 1%). With adjustments to the calculation methods and the use of advanced TOC meters calibrated for varying TOC and COD (and BOD) concentrations, much more accurate COD measurements are now possible of the high values of them found with ADF and AAFs.

Even though the CODs of the two pure glycols are similar (1,120,000 and 1,043,000 mg COD per L for EG and PG, respectively), owing to its higher molecular weight, the cBOD of pure PG is higher than that of EG (36% of COD or 403,000 mg cBOD per L). Accordingly, PG-based ADAFs should be harder to treat than EG-based ones, and this is the case. So, all other things being equal (and they are not!), it should be more advantageous to use EG-based fluids for de-icing as the degradation by-products will be lower in molecular weight and hence easier to manage.

While Equations D.1 and D.2 represent the complete oxidation of the glycols, in some cases only partial transformations occur and, as was addressed in Section 2.3.2 in Chapter 2 (see Equation 2.7), there are intermediate degradation steps that form VFAs. The main aerobic thermal biodegradation volatile fatty acids from aqueous EG biodegradation are glycolic acid, oxalic acid and formic acid (Dwyer and Tiedje 1983; Clifton et al. 1985; Rossiter et al. 1985). Glycolic acid is the first metabolite of EG and is responsible for most of its toxicity. Much higher levels of glycolic acid will form if copper is present, while the presence of aluminum will make the thermo-oxidative degradation of EG more temperature sensitive and the VFAs will involve higher levels of formic acid.

Biodegradation is unlikely in concentrated glycol solutions* (i.e., those with glycol concentrations above ~30%) but can occur in more diluted ones. That biodegradation is occurring/has occurred in a glycol-contaminated wastewater stream will be signalled by lower than anticipated pHs. An impact of this is that, in addition to the nitrogen compounds (usually urea) and phosphorus-rich material (usually phosphate fertilizer) that need to be added to the feedstock of the usually nutrient-deficient glycol-contaminated airport wastewater streams being treated in BBRs, caustic solution may need to be added as well to keep influent pH in the circum-neutral range. Since glycols and VFAs will both report as organics in laboratory or instrument measurements, BOD or COD or TOC can be misleading when only used to measure glycol concentrations in dilute streams but this does not matter if they are treated in BREW Bioreactors as these will transform ("burn") the VFAs as readily as the glycols.

* This is because, at elevated glycol concentrations, the osmotic pressure will be so high as to lyse any bacteria that might contaminate the fluids. It is only when the glycol concentration drops when the fluids are diluted after use during de-icing and anti-icing that osmotic pressure will drop far enough so that bacteria can survive in the solutions and biodegradation can commence.

The main aerobic biodegradation product of aqueous PG solutions is the VFA, lactic acid ($CH_3CH[OH]COOH$), with other VFAs, formic, acetic and oxalic acids, as minor products.

If copper is present, glycolic and lactic acids can form copper (II) metal complexes that will buffer the solutions and form precipitates over time.

Anaerobic glycol biodegradation will involve one or more biological fermentation steps followed by hydrogen oxidation and the cleavage of acetate (CH_3COO^-) to eventually form CO_2 and methane (Switzenbaum et al. 2001). The exact mechanisms involved and their extent will depend on specific conditions including the microbial consortia present.

The complete anaerobic biodegradation of pure ethylene glycol in an anaerobic bioreactor can be represented by

$$HOCH_2CH_2OH + H_2O \rightarrow CH_4 \uparrow + HCO_3^- + 3H^+ \quad (D.5)$$

Intermediate products in this case are the aforementioned acetate and ethanol (CH_3CH_2OH).

The total anaerobic biodegradation of pure 1,2 propylene glycol can be represented by

$$CH_3CH(OH)CH_2OH + H_2O \rightarrow 2CH_4 \uparrow + HCO_3^- + H^+ \quad (D.6)$$

In this case, intermediate products include propionate ($CH_3CH_2COO^-$), n-propanol ($CH_3CH_2CH_2OH$) and acetate. Unlike aerobic biodegradation, which can occur in wastewaters at temperatures as low as 0.1°C,* anaerobic bacteria are usually ineffective below about 15°C (EPA 2002), and accordingly, where anaerobic biodegradation is to be considered, wastewaters must be preheated.

Expanding on the aerobic degradation reaction of Equation D.1, and assuming that microbial biomass may be represented by $C_5H_7O_2N$, the aerobic microbial biodegradation of ethylene glycol such as that which occurs on/in the substrate biofilms of a BBR can be given by (Stell 2010)

$$C_2H_6O_2 + O_2 + 0.3HCO_3^- + 0.3NH_4^+ \rightarrow 0.3C_5H_7O_2N + 0.8CO_2 + 2.7H_2O \quad (D.7)$$

Equation D.7 emphasizes the role of a nutrient (here illustrated by a nitrogen compound, ammonia) in the biodegradation process and illustrates the importance, when treating a glycol-contaminated wastewater in an EEB, of

* This is exploited in BBRs and SAGR Bioreactors by designing them in a manner so that the growth of large numbers of aerobic bacteria in their substrate beds can be carried out before cold weather commences. Although these bacteria will not reproduce well once the water cools as winter approaches, those that are already there will continue to metabolize contaminants such as organics and ammonia right down to the freezing point.

the presence of adequate amounts of the nutrients required to maintain biofilm vitality (and not just N, but also P, K, and micronutrients). (Figure 1.11 of Section 1.11.4 in Chapter 1 illustrates what might happen if nutrients are deficient!)

As was mentioned earlier, some wastewaters (e.g., those from airports) may be nutrient-deficient and for these, nutrients can be added to the influents to the BBR cells.* Nutrient requirements for a BBR can be based on estimated composition of bacterial biomass (Grady et al. 2011). To do so, influent concentration as COD is multiplied by an assumed bacterial yield (mg biomass/mg COD). The biomass is multiplied by an empirically derived elemental fraction to calculate nutrient requirements (C/N/P is usually set at 100/5/1).

BBR cells that "rest" for part of a year (e.g., those treating glycol-contaminated streams at airports outside of cold weather de-icing periods) can tolerate higher loadings, as any excess, only somewhat-recalcitrant organic material that may be deposited in the substrate in the de-icing season, may biodegrade during the rest periods.

As may be seen from Equation D.5, for every mole of EG microbially transformed (see the footnote on page 5 of Chapter 1) in a BBR, one mole of oxygen and 0.3 mole of alkalinity is required, while 0.3 moles of new microbial cells (bacterial biomass) are formed.

The equivalent equation for the aerobic biodegradation of PG is given by

$$C_3H_8O_2 + 1.6\ O_2 + 0.48\,HCO_3^- + 0.48\,NH_4^+ \rightarrow 0.48\,C_5H_7O_2N + 1.08\,CO_2 + 3.52\,H_2O \tag{D.8}$$

As was mentioned, propylene glycol has a higher BOD than ethylene glycol and if the residues of its use as an ADAF at an airport are discharged to receiving waters, it will persist longer and will consume more oxygen in degrading biotically (see Equations D.5 and D.6), so PG might be imagined as having a higher environmental impact than EG.

However, even at low concentrations, traces of EG (or rather its biodegradation product, glycolic acid) in the residues of de-icing fluids discharged to receiving waters can be toxic to aquatic and mammalian organisms. In the United States, ethylene glycol is classified as a hazardous pollutant under the Clean Air Act and a hazardous substance under the Comprehensive Environmental Response, Compensation and Liability Act (CERCLA), while PG is much less toxic (although it is a teratogen) and is even used in cosmetics, medical products and foods.

Owing to such regulatory complications, most US airports use PG-based de-icers and anti-icers. This is not the case in Canada where most (but not all) airports use EG-based ADAFs, not because of less concern with

* If they were deficient, the bacteria in the cells would begin generating "slime" (see Section 1.10.4 in Chapter 1) that might plug BBRs and/or lead to greatly off-spec effluents.

environmental impacts but because EG is more appropriate for the usually colder weather there and the fact that it biodegrades much more readily than PG.

EG is manufactured in Canada, while PG is not. Both EG and PG are manufactured in the United States. In Canada, EG is a listed chemical on the National Pollutant Release Inventory, while PG is not. The same is true for the US Toxic Release Inventory.

D.2 Glycol Management at Airports

Pure glycols such as EG and PG are not used for de-icing and anti-icing at airports. Instead, as was introduced in Section 2.3.2 of Chapter 2, aircraft de-icing fluids (ADFs) and aircraft anti-icing fluids (AAFs) (collectively ADAFs) are used and these contain the glycol plus water, small amounts of impurities residual from their manufacture and the ADPACs that allow them to meet performance-based standards established by the Society of Automotive Engineers (SAE) and Aerospace Material Specification (AMS).*

Aircraft de-icing fluids (ADFs) are mobile fluids that are used to provide short-term protection from icing. ADFs have little adhesion and a large proportion of them are quickly shed as Spent Glycols at their point of application. The remaining ADFs drop off during taxiing and take-off, and much of these may end up in glycol-contaminated stormwater.

Aircraft anti-icing fluids (AAFs) are gel-like ("pseudoplastic") fluids applied to aircraft after snow removal. These non-Newtonian fluids contain a polymeric thickening agent as well as the same kinds of ADPAC chemicals and other materials contained in ADFs. AAFs tend to stick to aircraft surfaces, thereby inhibiting subsequent ice formation while the aircraft is stationary and later during taxiing and take-off. They are used where an aircraft is to be left parked for a period (e.g., overnight) during icing conditions. The ADPACs in AAFs used to amount to a couple of percent in total volume but as their manufacturers (Manufacturers) continue to reformulate them, they now usually constitute less than 1%.

Generally, about four times as much diluted ADFs are sprayed at airports as are AAFs. The ADFs are often delivered to customers at airports as 88%–92% concentrates. These ADF concentrates are diluted with hot water at 55–80°C by sprayers and others carrying out de-icing to the

* AMS 1428 addresses ADFs, while AMS 1429 addresses AAFs. Pure freezing point depressing chemicals such as EG and PG would not be able to meet SAE standards without having such additives in them (Higgins & MacLean 1999; ACRP 2008, 2013).

30%–60% glycol range before use. The AAFs delivered to customers at airports generally contain 40%–60% glycol, and are usually used as received or "neat" without dilution, although hot water may be used to remove accumulated snow and ice on fuselages before the application of anti-icing fluids begins.

In a few cases, those applying the glycol solutions (particularly major airlines carrying out their own de-icing operations) may continuously alter the amount of dilution of their ADFs according to prevailing conditions (e.g., temperature) so that the freezing point temperature of the resulting solution is always below ambient temperatures (load following). In such cases, ADFs containing as low as 30% glycols may be sprayed when temperatures are warmer (but still low enough to cause icing).

The US Transport Research Board (TRB), an arm of the American National Research Council, usually cosponsored by the US Federal Aviation Administration (FAA), has published a series of reports and other documents under the Airport Cooperative Research Program (ACRP). These documents (ACRP 2008, 2009, 2012, 2013 and others not referenced herein), available on the web at www.TRB.org, address many aspects of the use of ADAFs at airports including formulations, toxicity, guidelines for use, monitoring and residuals treatment.

Glycol-contaminated wastewater streams at airports* are generated in colder weather from residual streams resulting from the spray application on aircraft of ADAFs at departure gates (at-gate de-icing), at nearby or remote dedicated de-icing pads or at central de-icing facilities (CDFs) where de-icing is carried out after aircraft leave terminal gates.

As was introduced in Section 2.3.2 of Chapter 2, the procedure for application of ADAFs involves spraying these glycol-based fluids (ADFs and/or AAFs) onto aircraft skins first to remove any existing accumulation of snow and ice on the fuselages, and thereafter to prevent any further ice buildup.† Much of the ADAF, plus most of the removed snow and ice (and any dirt from the skin), falls as Spent Glycols to the tarmac under or beside the aircraft, but significant fractions also end up in glycol-contaminated stormwater runoff and ploughed snow.

The de-icing season in most parts of North America begins in October–November and may extend to April–May of the next year. Accumulated glycol-contaminated residual WW streams at airports may be treated in EEB Systems during the de-icing season, and after it ends.

Figure D.1 shows an aircraft being deiced at a terminal gate.

* As was introduced in Section 2.3.2 of Chapter 2, glycol-contaminated wastewater streams at airports include Spent Glycols, glycol-contaminated stormwater runoff (GCSW) and snowmelt.

† One ADAF Manufacturer is quite eclectic about what its ADAFs remove, mentioning water, snow, ice, slush, frost, hoarfrost, rime, glaze and snow pellets.

FIGURE D.1
Aircraft glycol de-icing.

A medium-sized airport might use ~1000 m^3 of ADAFs during a de-icing season, while a major airport may use several thousand m^3 (Richter 2003). Glycol-contaminated residual WW streams at airports will contain most of these fluids as well as much of the precipitation (snow and ice) that enters their collection systems during the de-icing season plus most of the dilution water added to ADFs plus any other streams that may enter the collection systems (e.g., washwaters, local surface, and/or ground-water entering collection systems). As a result, the quantities of glycol-contaminated WW that need to be treated in EEB Systems (or otherwise require management) at airports can range from a few hundred m^3 per day to several thousand m^3 per day.*

Regardless of how efficiently de-icing and anti-icing is carried out, a good part of the ADAFs used do not end up in the Spent Glycols collected in the areas where de-icing and anti-icing occur. Some drift or are windblown away as aerosols. Some run off through cracks or channels on pads, aprons, taxiways and runways. Blasts from jets and propellers blow some off de-icing areas onto adjacent grassed areas. Some adhere to the aircraft skin (especially AAFs which are designed to do so) and later run or blow off elsewhere on airport property as aircraft taxi and take off. A small amount evaporates, although the relatively high boiling points and low vapour pressures of the glycols indicate that evaporation can be expected to be only a small source of loss when they are used.

ADAFs dripping on taxiways and runways from queuing aircraft waiting to take off can result in glycol concentrations of the order of 100 mg/L beside these pavements and up to several thousand mg per L in stormwater ditches and storm sewers at airports.

---

* The design basis influent flow rate for glycol-contaminated WW being treated at BNIA in its four very large BBRs is 4620 m^3/day (820 m^3/day of Spent Glycols and 3800 m^3/day of GCSW).

The management of Spent Glycols, GCSW and glycol-contaminated snowmelt varies from airport to airport. It is estimated (Hartwell et al. 1995; Higgins & MacLean 1999) that, in the past on average, 80% of the sprayed fluids were lost in runoff, spray drift, jet blast and wind shear. At many smaller airports, the proportion was (and unfortunately still is for some) 100%. At many airports, Spent Glycol collection at terminal gates and at nearby or more remote dedicated de-icing pads by GRVs and/or into dedicated sewers can collect up to half of the ADAFs used (but usually less). At other airports, sophisticated CDFs are used to ensure the collection and removal of a larger proportion of the sprayed ADAFs as Spent Glycols (sometimes up to 65% or more but, owing to the frozen nature of the soils at many northern airports during the periods of ADAF use, the most likely receptors of the rest will be snow and surface runoff). However, at still others, a large proportion of the sprayed ADAFs simply flow or blow into nearby ditches, aprons or grassed areas and/or accumulate nearby as contaminated snow. These eventually end up in GCSW. Increasingly, regulatory authorities are insisting that glycol management at airports address the treatment or recycling of all glycol-contaminated streams, not just Spent Glycols.

ADFs and AAFs are categorized in four "types." Type I ADFs are unthickened, mobile Newtonian fluids used almost exclusively by most airports for the de-icing of aircraft (Dow 2004, 2009). In addition to additives and water, most Type I fluids contain a single glycol (EG or PG), but some in Europe and elsewhere involve (or have involved) mixtures of EG (35%–95%), PG (10%–90%) and DEG in varying proportions. Type I de-icing fluids are dyed orange.*

Types II, III and IV AAFs are higher viscosity, thickened fluids. Type II glycols (straw yellow in color) have been mostly phased out but are still used at some small airports. They provide about 30 min of freeze protection, while the Type IV fluids now being mostly used for anti-icing provide about twice as much protection. Type III AAFs are dyed bright yellow. Type IV anti-icing fluids are dyed green.

As was mentioned, the glycol-contaminated residual wastewater streams at airports do not contain sufficient nitrogen (or phosphorus)—more complex versions of Equations D.5 and D.6 can be shown including P in the term for bacteria biomass†) to provide the needed nutrient for the biofilm bacteria, and nutrients have to be added to the feedstocks of BBRs treating such streams. However, with the reformulation of some ADAFs by their Manufacturers (see Section D.2), this may no longer be completely the case

* The dyes are selected so that they degrade quickly in dilute solutions when exposed to air and UV light. For example, although one can see large piles of orange-colored snow accumulated beside the de-icing pads at EIA during the winter, once the snow melts during the freshet, the resulting water in the ditches leading to the Gun Club Storage Pond is clear (see Section 3.7 in Chapter 3).

† An equivalent microbial biomass formula including phosphorus might be $C_{60}H_{87}O_{23}N_{12}P$ (Liner 2013).

as some of the phosphorus (and perhaps some of the nitrogen) needed may now be supplied by chemicals in the ADAF's proprietary ADPACs.

Manufacturers sell ADAFs to those parties that see to their applications at airports (AOAs, airlines, the military, air cargo firms) and/or to firms contracted to carry out glycol spraying operations at airports (the Sprayers).

The additive packages in ADAFs (the ADPACs) include chemicals such as surfactants, other wetting agents (e.g., 1,4 dioxane), defoamers, corrosion inhibitors (e.g., sodium nitrite, benzoates, borax, silicates), pH buffers, antioxidants, hydrophobic oils, anti-precipitation agents, thickeners (in AAFs), foam suppressors, fire inhibitors, and dyes. In addition, contaminants in ADAFs can include small amounts of their polymeric forms (e.g., polyethylene glycol in the case of EG, especially EG's dimers and trimers) as well as urea, ethylene oxide and acetaldehyde (Hartwell et al. 1995; Higgins & MacLean 1999; ACRP 2008).

The 2008 ACRP report (Document 3: Formulations for Aircraft and Airfield De-Icing and Anti-Icing: Aquatic Toxicity and Biochemical Oxygen Demand) states that various ADPACs in ADAFs used at airports have contained 25 kinds of freezing point depressants, 21 surfactants, 11 corrosion inhibitors, 13 thickening agents (in AAFs), nine pH modifiers, five dyes, four oils and four antioxidants/antimicrobial chemicals. It goes on to state that ADAF Manufacturers constantly modify their proprietary ADPAC formulations to improve performance, address environmental considerations and reduce costs. This ACRP Document 3 notes that prior to about 2006, most ADPACs contained enough components with antimicrobial properties that, although they increased environmental impacts, inhibited degradation until large amounts of dilution and exposure to air occurred after use. However, post-2006, changes in ADPAC compositions to address toxicity concerns have reduced or eliminated many antimicrobial components, resulting in glycol-contaminated WW streams (Spent Glycols, GCSW) that may be more prone to biodegradation than they did before and contain larger amounts of nitrogen and phosphorus that can impact the environment in different ways.

As the exclusion of oxygen will inhibit degradation, concentrated ADAFs as received by airports and stored in closed tankage before use will be relatively resistant to biodegradation* even if stored between one de-icing season and the subsequent one. However, once ADAFs are used for de-icing/anti-icing and are exposed to air as diluted Spent Glycol or glycol-contaminated stormwater streams at airports, degradation may be quick.

Airports often monitor de-icing and anti-icing fluids' quality by BOD tests but, as was mentioned, these will not distinguish between the remaining glycols and their VFA by-products as all will report as BOD. (BRIX

* As was mentioned, this is not surprising as the concentrated glycol in ADAFs to be used for spraying (usually ~50%) would lyse any microbes that might contaminate these fluids. It would only be at some level of dilution after their use in de-icing and anti-icing that microbes could survive in resulting, more dilute glycol-contaminated residual streams and begin to biodegrade them.

refractometer tests are used at many airports to monitor ADAF quality but these methods too are of little value for measuring the glycol content of more dilute streams.)

Prior to about 2006,* product literature (MSD Sheets) for ethylene glycol-based and propylene glycol-based ADAFs indicated that these fluids had a "shelf life" of two years. More recent product literature from Manufacturers indicates that a significant proportion (~80% or more) of such fluids will biodegrade after a short period during warm weather (but only a small amount will biodegrade in the same period if the temperature is low).

In the past, two of the most important constituents of the ADPACs, the ones most mentioned by the ACRP (ACRP 2008) to be associated with the toxicity where glycol-contaminated streams were discharged to the environment, were alkyl phenol ethoxylate surfactants (ethoxylated surfactants (0.4%–0.5% of the volume of ADAFs) and benzotriazole-derived corrosion inhibitors such as benzotriazole and tolyliazoles (BDCIs). With AAFs, neutral, anionic and/or cationic polymers have been used as thickening agents. Neutral and anionic polymers are usually not toxic, but some feel that they could cause the suffocation of aquatic organisms in the receiving waters to which airport runoff may be discharged if they were present in high concentrations (Hartwell et al. 1995; Zitomer 2001). Cationic polymers can be highly toxic, and some of them (e.g., polyamines) also lead to the formation of ammonia in runoff.

Most ADPACs contain flammability inhibitors. These compounds are very toxic, but NASA insists on their inclusion. A variety of wetting agents (surfactants or detergents) are used in the antifreezes to keep the other additives in solution. Some of these too can be highly toxic to aquatic organisms. The degradation of already toxic components in ADPACs was suspected of leading to even more toxic breakdown products. As a result, ADFs and AAFs were more toxic than the pure glycols that were the largest ADAF components, and in most cases where toxicity was a concern, it was the additives that were the cause (ACRP 2008). Experiments were carried out where toxicity was seen to occur in receiving waters by airports following storm events where ADAFs were used and showed that as little as 1%–2% of glycol-contaminated stormwater runoff could debilitate or kill aquatic organisms (ACRP 2013). It was also shown that, owing to the nature of their additives, Type IV AAFs were much more toxic than were Type I ADFs (by up to two orders of magnitude), and presumably the same was true of Type II and Type III AAFs.

According to the ACRP (ACRP 2008), while pure EG is toxic (LC_{50} to aquatic organisms) only when its concentration reaches 18,500 mg/L in a body of

* One Manufacturer's publication indicates that reformulation to address the toxicity of its ADPACs was commenced as early as 2001. However, it is not known to what extent, with which products or in what market areas these changes occurred and the 2006 date suggested by the ACRP seems a more relevant one.

water, EG-based Type I ADF is toxic at approximately 10,000 mg/L and Type IV AAF at 2500 mg/L. Accordingly, prudence dictated that any glycol-rich streams at airports had to be regarded as dangerous to the local environment if discharged into receiving waters.*

As a result, the ADAF Manufacturers were under pressure to replace the ethoxylated surfactants, BDCIs and other potentially toxic chemicals (e.g., pH modifiers, dyes) in their ADPACs with less toxic alternatives. They did so.

Whether or not the ethoxylated surfactants and/or BDCIs (and their breakdown products) were solely responsible for all of the toxicity of specific ADFs and AAFs in the past is not relevant as they were perceived as such, and are mentioned in the literature as the worst offenders. According to the ACRP (2008), most glycol Manufacturers reformulated their Type IV AAFs in time for the 2006–2007 de-icing season, and some of their Type I ADFs in time for the 2007–2008 season. The exact compositions of the ADPACs either before or after reformulation to reduce toxicity have never been disclosed by the Manufacturers (they are judged to be business confidential and proprietary), but since the glycol-contaminated streams at airports resulting from the use of the fluids after doing so appear to be more biodegradable and contain much more phosphorus than they did before, some of the reformulations changes can be speculated.

Potassium phosphate is mentioned by the ACRP (2013) as a potential alternative corrosion inhibitor to replace all or part of the BDCIs in ADFs and AAFs. Lauryl alcohol phosphoric acid ester ethoxylate is mentioned as a potential alternative phosphorus-containing surfactant. In addition, disodium phosphate and other phosphates (at up to 0.25% by volume) have been mentioned as potential alternative pH modifiers for use in reformulated ADF and AAF ADPACs. It is not known whether ADAF Manufacturers have included these specific phosphorus-rich chemicals in their reformulated ADPACs but these or similar phosphorus-containing chemicals must have been introduced to the ADPACs as analyses of these fluids now find hundreds to thousands of mg TP per L in them.†

Since some of the ADPAC chemicals that had antimicrobial properties (e.g., the ethoxylated surfactants, BDCIs and pH modifiers) have now been replaced with less toxic ones containing higher proportions of nutrient chemicals (e.g., P, N), it is not surprising that the residual streams from ADAF use are now more biodegradable than they were before. The unintentional result

* Many airports still collect and treat/recycle on- or off-site only the relatively concentrated Spent Glycols that fall under aircraft being deiced but take no action regarding that part of the sprayed ADAFs that do not end up in Spent Glycols (and, as was mentioned above, these can be >50% of the ADAFs used even where an airport has sophisticated Spent Glycol collection methods such as CDFs). The rest end up in GCSW and, along with pavement de-icers and other contaminants, enter the local environment.

† An EG-based, 92% concentrate Type I ADF may now contain in the order of 1000 mg TP per L as well as increased amounts of TN. When used as a de-icer, the glycol-contaminated residual streams that result may contain amounts of phosphorus that may exceed discharge criteria for it (1.0 mg TP per L in Canada where required, see Table D.2).

of this may be problems for those involved in glycol management at airports (AOAs, Sprayers, air cargo firms, the military, firms cleaning up Spent Glycols at de-icing pads and Recyclers). This is not only because glycol-contaminated WW streams entering the environment from airports may now exceed regulatory requirements for phosphorus, but also because facilities for the storage, handling, treatment and transportation of these glycol-contaminated streams may now be more subject to sludge and VFA buildup.* As a result, AOAs and other parties involved in glycol management at airports now may face increased costs for maintenance, monitoring and operations due to the increased biodegradability of the ADAFs they use.†

D.3 Regulatory Considerations

Controlling and treating glycol-contaminated streams from airports is a matter of increasing environmental concern and is now having a major effect on airport design and operations. In Canada, most of the land at major airports is now owned by a Canadian Government Ministry, Transport Canada, although the actual operation of the airports is usually delegated to publicly controlled local AOAs.‡ The overall responsibility for wastewater treatment at these airports rests with the Airport Operating Authorities, although they and/or various other parties (e.g., airlines, Sprayers, Recyclers) may be involved in applying the de-icing fluids, collecting Spent Glycols and other glycol-contaminated streams, and managing used glycols.

Transport Canada expects AOAs to comply with some reporting standards. Chapter 13 of TP14052 (January 1995) of its publication on Aviation Safety outlines environmental and related aspects (e.g., impacts, standards and guidelines) that airports should adhere to, presents an Airport Quality Manual, defines expected reporting, reviews the effects of ADAFs on the environment and specifies that AOAs must prepare and keep up-to-date Glycol Management and Mitigation Plans (GMPs) for their airports. A GMP must detail: (1) information on de-icing activities, equipment and sites at an airport, (2) the facilities involved for ADAF storage and handling, (3) how relevant operator training will be carried out, (4) how glycol-contaminated WW streams will be contained and disposed, (5) information on safety matters, (6) how ADAF inventories will be monitored and controlled, and (7) how the

* A recent need to clean out sludge from YYT's CDF diversion chamber might have been due to the formulation changes.

† EEB Systems can, of course, be designed to remove the enhanced amounts of phosphorus and nitrogen from such streams before they are discharged to the environment.

‡ Smaller Canadian airports are often owned or operated by local communities or provincial/territorial ministries (e.g., in Ontario, its Ministry of Transport).

airport will report on ADAF use. A GMP must include or complement the airport Emergency Response Plan.

Chapter 13 of TP14052 states bluntly that "…the discharge of untreated runoff containing aircraft de-icing fluids into receiving waters creates an unacceptable pollution problem and a hazard to aquatic life." GMPs that only address the management of Spent Glycols and ignore the impacts of glycol-contaminated stormwater from around airport property may not meet the requirements under the Canadian Environmental Protection Act (CEPA 1994), but up to this point there has been little federal enforcement on this aspect.

In Canada, the following requirements under CEPA are based on EPS 1-EC-76-1, April 1976, Guidelines for Effluent Quality and Wastewater Treatment at Federal Establishments, and apply to wastewaters from all airports owned or operated by the federal government.

TABLE D.2
Guidelines for Effluent Quality at Airports

BOD	20 mg/L
Glycols	100 mg/L
TSS	25 mg/L
NO_3-N	13 mg/L
Oil and grease	15 mg/L
TP	1.0 mg/L
Phenols	20 μg/L
pH	6–9
FC	400/100 mL
Temperature	Not >3°C+ for receiving waters

The 100 mg glycol per L guideline is incompatible with the 20 mg BOD per L one (i.e., any stream containing 100 mg/L of EG, for example, will probably have ~80–90 mg BOD per L). GMPs often specify only the 100 mg glycol per L criterion, leaving other criteria for a glycol-contaminated stream leaving airport property to be specified by provincial/territorial authorities that may require BOD discharge criteria of 20–40 mg/L.

The Canadian Council of Ministers of the Environment (CCME) has prepared water quality guidelines relevant to Canadian conditions and these specify (2011) maximum concentrations in receiving waters of 3 mg/L for EG, 31 mg/L for DEG and 74 mg/L for PG.*

As was mentioned above, in addition to meeting these federal guidelines, glycol-contaminated streams that flow off airport property will have to meet quality guidelines and criteria set by Provincial or Territorial

* These, of course, are for concentrations in receiving waters, not for glycol-contaminated effluents entering them!

Environment Ministries, and these may be more stringent than those above and may include parameters not included in Table D.2 (MacDonald et al. 1992). Hitherto, Canadian Federal and Provincial/Territorial quality criteria for receiving waters around airports in Canada have not included tighter criteria for nitrogen (either as TN or NOx-N—the current federal criterion is 13 mg NO_3 per L) but new regulations imposing more stringent criteria for these are said to be imminent.*

In the United States, the Environmental Protection Agency (EPA) has set regulatory guidelines (EPA 2000a) for the management of Spent Glycols and glycol-contaminated stormwater at airports.

Section 402 of the Clean Water Act, the National Pollutant Discharge Elimination System (NPDES), defines the system for permitting point source discharges into the waters of the United States. This includes wastewaters from municipal, commercial and industrial sources as well as stormwaters contaminated with pollutants (ACRP 2013). The US EPA has authorized regulatory agencies in most US states to enforce this and this requires dischargers such as airports to obtain NPDES permits or a state equivalent. In addition to direct discharges to receiving waters, US airports may discharge into municipal sewers and other off-site treatment facilities.

NPDES and equivalent permits require dischargers to implement Best Management Practices (BMPs) to minimize pollution, and to prepare and implement Stormwater Pollution Prevention Plans. For stormwater discharges from airports, benchmark NPDES limits usually are 30 mg BOD per L, 120 mg COD per L and pHs in the 6–9 range. These are stringent limits but are ones easily achieved by BBR-based EEB Systems.

The EPA has issued technology-based effluent limitation guidelines and new source performance standards† (the Guidelines) to control the discharge of glycol-contaminated wastewater from airports in the United States. The Guidelines apply to both aircraft de-icing and anti-icing, as well as pavement de-icing/anti-icing. They specify that existing and new primary airports with 1000 or more annual jet departures are either not to use urea for surface de-icing or alternatively to meet numeric effluent limitations for ammonia (2.14 mg ammonia per L). New airports with more than 10,000 annual departures located in certain cold climate zones are required to collect/treat/recycle at least 60% of the glycols used and meet numeric discharge criteria for COD in streams discharging into the "waters of the U.S." The Guidelines do not establish uniform national requirements for existing airports but leave them to be addressed by general or site-specific NPDES limits.

* It will be difficult for many glycol management methods other than on-site EEB Systems to meet Table D.2's total phosphorus criterion (1.0 mg TP per L), especially in the light of the aforementioned ADPAC reformulations that greatly increase ADAF phosphorus contents. EEB Systems including DNBRs as well as BBRs could easily meet any future TN or NOx-N criteria.

† Effluent Limitation Guidelines and New Source Performance Standards, http://water.epa.gov/scitec/wastetech/guide/airport/index.cfm

D.4 Pavement De-icers at Airports

De-icing and anti-icing activities at airports are not limited to those associated with aircraft. Airports must also ensure that runways, aprons, taxiways, parking lots, sidewalks and roads remain clean and ice-free during the winter period. In addition to ploughing, this can be accomplished by using a variety of solid and/or liquid pavement (surface) de-icing and anti-icing chemicals. When used as de-icers, they melt down through ice and snow to the pavement surface, where the rich solutions spread out under the ice breaking the bonds between it and the pavement, allowing the ice to be removed more easily. When used as anti-icers, they are applied to the pavement prior to a snow or freezing rain event. Some are better than others at de-icing or anti-icing (Higgins & MacLean 1999).

Pavement de-icing and anti-icing chemicals and their associated materials also lead to runoff contamination, and while aircraft ADFs and AAFs can be largely confined to relatively small areas, surface de-icers and anti-icers are usually applied to much wider areas, so the effective control of contaminated runoff from them is more difficult.

Owing to potential corrosion problems and concern about the ingestion of their crystals into jet engines, the ordinary salt (sodium chloride) or calcium chloride surface de-icers are not used on runways and other places used by aircraft ("Airside" to use airport terminology), although they are will be present in airport runoff if used elsewhere on the airport (e.g., on parking lots, i.e., "groundside" or "landside").

For those airside areas used by aircraft, the following chemicals can be used as surface de-icers:

- Urea
  - Solid forms
  - In solutions
- Glycols
  - Ethylene glycol
  - Propylene glycol
- Acetates
  - Sodium acetate
  - Potassium acetate
- Sodium formate

Urea can be used airside as either a de-icer or an anti-icer. As the latter, it is six times more effective. It is relatively cheap, effective down to −11.5°C, gives good de-icing performance, and is readily available in both solid and liquid forms. The solid is available in pill form (less tendency to be blown about by jet blast) and the liquid is usually available in bulk in a glycol solution. It used to be the most widely used chemical. However, it is readily microbially transformed (mineralized, see Section 2.3.2 in Chapter 2) to ammonia and its use can result in unacceptably high concentrations of ammonia in ditches, surface water impoundments and receiving waters around airports above allowable levels. The hydrolysis of urea can be given by

$$CH_4ON_2 + 2H_2O + H^+ \rightarrow 2NH_4{}^+ + HCO_3{}^- \quad \text{(D.9)}$$

Whether the ammonia will be in its molecular or ionic form depends on pH (see Section 2.4 in Chapter 2). Regardless, ammonia is regarded as highly toxic in receiving waters and regulators impose strict limits on its concentration in streams entering them. As a result, urea's use in this service has been curtailed or dropped at many airports, and only airports where almost all of the Spent Glycols and GCSW can easily be assembled for treatment (e.g., BUF, YLW) can consider going back to urea as a surface de-icing chemical.

Using the same format as was used in Equations D.5 and D.6, Equations 1.9 and 1.10 in Section 1.13 of Chapter 1, the overall ammonia oxidation and microbial cell synthesis can be rewritten as (Metcalfe & Eddy 2004)

$$\begin{aligned} NH_4^+ + 1.83\,O_2 + 1.98\,HCO_3^- \rightarrow\ & 0.021\,C_5H_7O_2N + 0.98\,NO_3^- \\ & + 1.88\,H_2CO_3 + 1.041\,H_2O \quad \text{(D.10)} \end{aligned}$$

where in this case $C_5H_7O_2N$ represents the synthesis of the nitrifying bacteria.

As can be seen, the nitrification reaction consumes oxidation and alkalinity. As was mentioned, nitrification is slow compared to the biodegradation of other de-icing chemicals by hetrotrophic bacteria (Stell 2010) and does not proceed until the concentration of organics has declined to about 20–40 mg/L, a fact that is important in the design of EEB Systems.

The oxygen demand of urea is much higher than that of other pavement de-icers; on an equivalent basis, it is 90% higher than that of sodium formate, 73% higher than that of potassium acetate, 47% higher than that of EG and 31% higher than that of PG (Stell 2010).

In addition to their use in ADAFs, liquid glycols are often used as pavement de-icers as well, either alone or mixed with other de-icers and anti-icers where they also serve as wetting agents for other solid de-icers. At some airports, glycols are used alone for most de-icing/anti-icing applications, and urea is only added to them occasionally when freezing rain is encountered. As with their use as aircraft de-icers and anti-icers, the biggest problem with the use of glycols as pavement de-icers/anti-icers is their high BODs.

Sodium acetate (CH_3-CO-O-Na·3H_2O), which is also available in pill form, has the advantage of not leading to ammonia pollution in GCSW as it contains no nitrogen. It reacts with ice faster than does urea, and only two-thirds as much of it is required to achieve similar results over a range of temperatures. Its BOD is lower too (0.41–0.65 g O_2 per g), but it costs much more, up to four and a half times as much as urea. Sodium acetate is usually used in conjunction with a liquid pre-wetting agent such as EG, PG, or liquid potassium acetate (CH_3-CO-OK).

Sodium formate (NaO-CHO-CHO-ONa), another solid pavement de-icer, also does not result in ammonia pollution. It has been widely used in Europe and is now used in North America at some airports. It can be used at lower temperatures than urea and only about half as much is needed to obtain the same results. Its BOD is relatively low (0.23 g O_2 per g), but it costs about three times as much as urea. It too is usually used with a liquid pre-wetting agent such as potassium acetate (BOD of 0.46 g O_2 per g).

One of the problems with some pavement de-icer formulations (e.g., sodium formate/potassium acetate) is that expensive equipment is required to apply them, and that to be effective they are often applied before the fact (i.e., before a precipitation event occurs) as an anti-icer. This requires preplanning and careful weather prediction (a normal activity at airports anyway), but if the forecast freezing conditions or snow do not occur, the preparation and chemicals may be wasted.

The selection of a particular pavement de-icer/anti-icer will depend on a variety of local factors, including regulations and whether enough of the glycol-contaminated WW streams at an airport can be collected for treatment in WWT facilities such as EEB Systems. All things considered, many airports would like to go back to using urea and glycol mixtures for surface de-icing but are being pressured not to do so unless effective means of dealing with runoff ammonia levels are possible.

The following table, adapted from ACRP's Document 3 (ACRP 2008), compares the laboratory-measured BODs and CODs of aircraft and pavement de-icing materials as delivered to airports (ADAF BOD and COD concentrations are in mg per L and those of surface de-icing chemicals are in mg per kg) (Table D.3).

TABLE D.3

BODs and CODs for Select De-icing Chemicals

	Concentration %	COD	BOD	BOD/COD
Type I EG-based ADF	92	1,220,000	511,000	0.42
Type IV EG-based AAF	64	590,000	249,000	0.40
Type I PG-based ADF	88	1,314,000	916,000	0.70
Type IV PG-based AAF	50	479,000	539,000	0.64
Potassium acetate (liquid)	50	315,000	247,000	0.78
Sodium acetate (solid)	96	700,000	571,000	0.82
Sodium formate (solid)	98	242,000	n/a	–

As may be seen, even without the contributions of residual ADFs and AAFs dripping from the fuselages of aircraft taking off, surface de-icing and anti-icing operations also lead to the contamination of GCSW at airports, emphasizing the need to treat such streams. EEB Systems represent one of the few options able to do so.

D.5 The Measurement of Concentrated Samples

Airport stormwaters often contain large concentrations of glycols and their measurement by standard commercial analytical laboratories can sometimes lead to problems if they are not appreciated. The situation can be illustrated by Table D.4, which provides a comparison of organic measurements for a sample of glycol-contaminated airport stormwater during a de-icing season before and after dilution with distilled water at 1500:1 and the resulting new diluted sample analyzed for the same parameters. It compares the results for these two sets of analyses and compares them with the results that mathematically should be expected if the results of column two (measured by standard methods) were indeed correct and they were prorated based on the dilution.

TABLE D.4
Organics Comparisons (mg/L)

Organic	Measured by Standard Methods	Measured for a Subsample Diluted at 1500:1	Mathematically Expected Results Based on Dilution
Glycol	65,000	167	44
TOC	24,000	890	16
COD	20,000	1400	80
cBOD	70,000	890	47

Assuming that both sets of measurements in Table D.4's columns 2 and 3 were accurate, it can be seen that organic results from diluted sample measurements presented in column 3 (which did not have to be diluted and the results back-calculated) were consistently lower than might be expected.

Column 2's ratio of cBOD to glycol concentration (cBOD/glycol) was 1.08 but that of column 3 was 5.32. However, the ratio of BOD/glycol would be expected to be somewhat lower were the contribution to BOD of ammonia to be considered and these illustrate the problems with trying to use BOD as a criterion. COD measurements can be measured more accurately and analytical laboratory methods are different for when lower and higher values are being measured. COD would be a much more practical parameter to use in measuring the efficiency of treatment in an EEB System.

D.6 Eco-Engineered Bioreactor Systems and Airports

BBR-based EEB Systems are presently in use at some northern airports for treating the glycol-contaminated wastewater streams that result at them from the de-icing of aircraft in cold weather. These include those at BNIA (BUF), Edmonton International Airport (YEG) in Leduc, Alberta, Canada, London Heathrow Airport (LHA) in the United Kingdom, and Long island McArthur Airport (ISP) in Islip, NY, USA. (Those at YEG and LHA represent the upgrade of earlier CW systems.) Projects involving BBR-based EEB Systems are also underway to consider/design systems at several other airports including St. John's International Airport (YYT) in Newfoundland, Canada and Kelowna International Airport (YLW) in British Columbia, Canada.

Airport EEB System projects usually proceed through the same phases as those of other WWT projects (see Section 8.1 of Chapter 8) including: Feasibility Study, Conceptual Design, Detailed Design and Engineering, Tendering, Construction and Commissioning (Higgins et al. 2007a, 2010a). Since every airport has different conditions (e.g., climate, soils, topography, de-icing facilities, operating methods, ADF and AAF types and mixes, logistics), a treatability test (see Chapter 6) should always be carried out as part of the Feasibility Study or Conceptual Design stage of the project.

Loadings to any kind of BBR must be carefully controlled to prevent excessive biomat formation (particularly near cell inlets), but doing so with those used at airports is especially critical as the glycol-contaminated streams involved with them can vary greatly in soluble BOD concentrations over short periods.* If not allowed for, biomat formation may clog the gravel particles in the bed and restrict the hydraulic throughput of the system. There is a relationship between the allowable daily mass flux rate of organics (g-BOD/m^2.day) to a BBR and the aggregate size (d_{10}) used in the beds. For smaller pea gravel with a d_{10}<4 mm, a suggested mass flux rate of 250 g-BOD/m^2.day is recommended.

However, the larger aggregates particles typically used in airport BBRs are suitable for accommodating higher flux rates because of larger void spaces. (Figure 7.2 in Chapter 7 shows the gravel substrate used in the four BBR cells at BNIA.)

* See Section 1.10 in Chapter 1 for a review of biofilms, biomats and slime formation in EEBs.

Appendix E: ASGM

E.1 Artisanal Gold Mining

A potential new market area for EEB Systems may be to help manage wastewaters from sites where artisanal and small gold mining (ASGM) operations are occurring or have taken place. These are located largely in developing counties, and are often associated with wastewaters contaminated with both mercury and cyanide (see Sections 2.3.5 and 2.6.3 in Chapter 2).

Artisanal gold mining refers to extralegal, uncontrolled small-scale mining operations carried out on a very small scale by individuals with little or no training in mining technology (Viega 1997; Viega & Hinton 2002; Veiga et al. 2006; Utomo et al. 2010). An example is the situation in Suriname, South America (Versol 2007) where the practice is usually referred to as "porknocking."* While artisanal mining for other metals (e.g., cobalt) and gems (e.g., jade) occurs, the most common target commodity is gold. Artisanal and other small-scale mining operations (often involving small local rock crushing and other facilities) are defined as "... informal and unregulated systems of ... mining involving operations" which are "... often illegal, unsafe and environmentally and socially destructive" (Utomo et al. 2010). Opportunities to recycle mercury in an artisanal gold mining flow sheet, where they exist, are poorly exploited, and it is estimated that 1–2 g of mercury are lost for every gram of gold produced, amounting to a worldwide loss annually into soil, air and water of over 1000 tonnes of mercury (Veiga et al. 2006).

While some panning for alluvial gold and river dredging does occur, the bulk of artisanal gold operations that exploit placer gold involve hydraulic mining (land dredging) whereby shallow gold-containing placer alluvial or colluvial (soft rock close to the surface) saprolytic gold-bearing ore is dug up (now usually using excavators), placed in piles and dosed with mercury.

This latter step is called whole-ore amalgamation, a practice that is known to be ineffective, yet is still carried out in many places anyway. The collected ore is then slurried into a pit by teams of porknockers using high pressure water from pumps driven by a diesel generator. The slurry is then pumped to elevated wooden sluice boxes for the gravity separation of the contained gold. The floors of the sluice boxes contain riffles, matting and/or copper

* Those who practise porknocking are usually untrained locals or men from neighbouring countries who use relatively unsophisticated methods and equipment to access gold and other valuable materials at what they perceive to be "hot spots" for them at or near active and closed mines or on concessions where exploration is underway in places like Guyana, Suriname, Brazil and Indonesia. The name is said to have been derived from the salted pigs' tails that once formed the staple diets of porknockers in remote jungle areas.

plates covered with mercury to settle out a gold-rich "concentrate" containing mercury, gold-mercury amalgam, larger pieces of gold and other heavier materials. More mercury may be added at this point. The bulk of the slurried ore (tailings) flows out of the sluice box into a nearby "tailings pond" (often a mined out pit) where solids settle out and water decants for reuse. If there are nearby streams or ponds with sufficient water, the tailings (which still contain an appreciable fraction of the gold and much mercury) are simply discharged into them or the adjacent jungle.

After 10–14 days, slurrying is stopped and the "concentrate" collected in the bottom of the sluice box at such a hydraulic mining operation is thoroughly washed; rocks and any gold nuggets are manually removed, and the mats are removed and beaten and/or the copper plates scraped into the bottom of the sluice box to release more gold. More mercury is then added to the sluice box and manually worked into the solids there to amalgamate as much of the remaining settled gold as possible. The collected amalgam is then removed and may be squeezed through a cloth to remove excess mercury (which is usually discarded). The amalgam (about 50% gold) is then heated in a pan with a blowtorch or placed in an open pot on a camp stove to drive off ("burn") most of the mercury out of it, leaving a sponge-like gold "dore" which accounts for about 10%–50% of the gold which was present in the original ore. In Suriname, gold dore is generally sold by artisanal miners to registered and unregistered gold merchants in the country's capital, Parimaribo. It contains 2%–8% mercury, and this leads to significant air pollution in the city when the dore is later melted to drive off the mercury.

E.2 The Environmental and Social Costs of Porknocking

In 1997, it was estimated (Viega 1997) that approximately 15,000 porknockers were active in Suriname producing about 8–12 t/year of gold, and about 8000 of these were illegal entrants from Brazil. Currently, artisanal mining gold production in Suriname may be as high as 20–30 t/year. Almost 20 years ago, Viega (1997) described porknocking in Suriname as "... out of control" and reported that "... uncontrolled gold mining and the uncontrolled use of mercury is having a devastating effect on the environment in the gold mining regions of Suriname" and "... artisanal gold miners use inadequate or inappropriate exploratory ore estimation, mining and ore processing [methods]" and "... there is little or no waste management, reclamation or re-vegetation [involved]" and "... there is insufficient awareness of the threats posed by mercury and mercury vapor."

In addition to causing mercury pollution, porknocking also has many negative environmental impacts including deforestation, the disruption of water courses, damage to sources of drinking water, the silting of streams and rivers, the destruction of wildlife habitat and the removal of the wildlife

(Veiga & Hinton 2002). Artisanal mining also has many negative social impacts including the spread of disease (especially malaria and STDs such as AIDS) and the promotion of violence, gambling, alcoholism and drugs.

Soils in areas of porknocking operations may end up highly polluted with mercury metal and the inorganic mercury ions (Hg^{2+}) that form part of it. Concentrations of 12 mg Hg per kg, with spots containing as high as 200 mg Hg per kg, have been reported from artisanal mining operations in the Brazilian Amazon (Veiga et al. 2006). Porknockers regularly swim in mercury-contaminated tailings ponds and stand in the slurring pits where they are exposed to mercury vapours and where there is dermal exposure to mercury metal and inorganic mercury in the water.

Mercury vapour on its own is a dangerous pollutant that can cause a variety of serious acute and chronic health problems to porknockers exposed to it. Normal city air contains about 0.01 $\mu g/m^3$ of mercury and the safe limit for exposure is 1 μg Hg per m^3, but when burning amalgam, porknockers may be exposed to as high as 60,000 μg Hg per m^3 (Versol 2007; Kadlec & Wallace 2008).

Another serious aspect of mercury pollution is the ready methylation of metallic mercury and inorganic mercury to form even more toxic methyl mercury (CH_3Hg^+) (Cordy et al. 2011). Methylation occurs under anaerobic conditions and is a consequence of the activity of sulfate-reducing bacteria. The risk of methylation occurring is increased when mercury vapour condenses and/or when discarded mercury and mercury metal left in the "tailings" from porknocking sluice boxes is transported away from artisanal mining sites into local soils, sediments in streams, ponds, jungle areas and Natural Wetlands. This risk is due to the propensity of elemental and inorganic mercury to bind readily to dissolved organic carbon in water. Methyl mercury bioaccumulates in the food chain (especially fish) and biomagnifies as it does so. Consequently, this represents a hazard not only to the porknockers (who eat a lot of fish) but also to other sectors of the population in a mining area who are not directly involved with mining. Coupled with widespread and cumulative deforestation that is occurring in the artisanal gold mining areas, the steady and also cumulative increase in mercury pollution due to porknocking is a major environmental disaster in the making.

E.3 Methods to Minimize Mercury Losses

There are ways to minimize mercury losses, although they are rarely practised in most areas where porknocking occurs. In some places where artisanal gold mining occurs, concentrated amalgam is often collected from several sites and burnt in a retort* located at a central amalgamation site. The

* A retort is a vessel in which amalgam is placed and heated to drive off the mercury up a tube to condense in an adjacent cooler chamber.

collected mercury can then be reused. There are several varieties of retorts used and they are easy and economic to build or buy. Without a retort, as much as 50% of the mercury in an amalgam can be lost in burning. However, with effective use of a retort, losses can be cut to 0.05%. The reason that retorts are not used in many areas where porknocking occurs is said to be that there is insufficient infrastructure (e.g., roads) in the gold-producing parts of these areas to allow transporting of amalgams to central sites for burning, and it is easier to burn them where they are mined.

There are available technologies that allow the collection of gold concentrates from artisanal mining operations without using mercury at all (Cordy et al. 2011). Examples are using what are called Cleangold® sluice boxes in place of wooden ones, or replacing sluices entirely with centrifugal concentrators to concentrate the gold-bearing part of the slurried ore. Once the miners have generated a concentrate, there is a worldwide push to teach artisanal gold miners to use a simple and safe cyanidation flow sheet to recover the gold from the concentrate. Education of miners in safe and efficient mining techniques was advocated by the University of British Columbia (UBC) as early as 1995, and has been proven time after time to be a viable option for development of the artisanal mining sector. Education underpins the mission statement of the Training Centres for Artisanal Miners (TCAM) project described in another paper presented at the *1st International Symposium on Sustainable Mining Operations* (Anderson et al. 2014).

However, assuming cyanidization is adopted as the most viable gold-recovery flow sheet from ore concentrate or whole ore, a waste stream will still be apparent. This waste stream will constitute both tailings and process waters. It is suggested that EEB Systems could be viable options to manage environmental risk associated with the wastewaters released from tailings at ASGM locations.

Underpinning the global push for cyanidization to be used at ASGM operations is the reported efficiency of this technique that can be as high as 95% recovery (Hylander et al. 2007). However, gold is not the only metal dissolved from the ore during cyanide leaching. In particular, if mercury is also present within the ore as a consequence of amalgamation at some point prior to cyanidization, or owing to the high level of mercury in the ore (e.g., in Indonesia), mercury will also be leached from the ore during cyanidization. Mercury will form soluble complexes with cyanide, although at a dissolution rate that is slower than for gold (Cordy et al. 2011). Velasquez-Lopez et al. (2011) reported a mercury balance for a small-scale carbon-in-pulp cyanidization circuit of amalgamation ore designed for artisanal miners in Ecuador, and showed that approximately 27% of mercury in the feed material was leached over three days.

The risk then is that a portion of this complexed mercury will be discharged from the process facility with the tailings. If this mercury remains in solution, then there is concern that, in a poorly managed tailings facility, leaching of mercury-contaminated water into a receiving environment may occur. In this scenario, the cumulative amount of mercury discharged into the environment may be equivalent to existing cases where whole-ore is amalgamated.

Krisnayanti et al. (2012) proposed a scenario where tailings at the end of an artisanal or small-scale cyanidization circuit were collected into containment areas. Once these areas are full, they proposed that the tailings could be planted with a suitable species, and phytoextraction used to recover residual metal in the tailings. Gold phytomining in this scenario would most likely be affected using cyanide to induce the uptake of gold into the plants (Anderson et al. 2005). The aim of this scenario would not be to directly make a profit, but to recover value from the final crop that could be used to offset the cost of the tailings containment, and allow for long-term tailings management.

This proposal, however, has one major flaw. There is no mechanism to manage or degrade dissolved contaminants of concern within the aqueous phase of the tailings discharge during the filling of the tailings dam, and after treatment of the vegetated tailings with cyanide to induce uptake of gold. It is unlikely that tailings dams at ASGM locations will be built to international standards, and therefore there is risk for contaminants to leach into the surrounding environment. These contaminants may include cyanide, complexed metal(loid)s and excessive levels of nutrients.

E.4 EEBs and ASGM

It is suggested that a relatively simple kind of Eco-Engineered Bioreactor System could be built at the outlet point of a tailings dam designed for phytoextraction, or at some other convenient location where it could be used to manage any contaminated water from an ASGM site (e.g., site runoff, tailings and waste rock areas seepages, process waters and even local sanitary sewage). In many cases, EEB Systems involving only BBRs as their secondary treatment components would be adequate, but if there were other CoCs that required anaerobic treatment as well (e.g., other dissolved metals, phytoextraction enhancement chemicals), DNBRs and BCRs could be integrated with the EEB System.* In the field, such systems would involve simple surge ponds to collect contaminated water and even outflows. These would be followed by the EEB cells in series, with either a single train or more than one being dictated by local considerations. Aeration air for BBR cells could be supplied

* ASGM EEB Systems would have to be piloted first to determine mercury speciation and fate. An essential part of such treatability testing would be R&D to define the chemistry of metallic, inorganic and organic mercury compounds in such systems and methods for their safe handling (see Section 9.3 in Chapter 9).

by small adjacent blowers, the electricity for which could be supplied from local sources (e.g., the generators at the ASGM facilities or solar panels).

While full-scale EEB Systems in developed countries usually involve geomembrane-lined cells and large quantities of expensive aggregates such as gravel, those at ASGM EEB sites could involve cells lined with local clays and filled with crushed rock mined locally. By these means, very economic EEB Systems could be readily constructed using local labour with a minimum of imported materials (only the blowers, possibly aeration tubing, some pumps and some simple monitoring instruments). Any BCR cells could involve chipped wood from local trees and manure as active media. Effluents from such facilities could be discharged for polishing into small CW cells or local Natural Wetlands if such were conveniently available. Designs would involve gravity flow as far as possible. Such ASGM EEB Systems would readily clean up contaminated waters during the operation of the ASGM sites, greatly improving health and safety of the operators and local populations and preventing the level of environmental pollution of surrounding areas that is the current norm. They would have the further advantage of being able to evolve after the closure and abandonment of the ASGM sites into fully passive systems that would continue to clean up contaminated waters from the former ASGM sites into the future. It is suggested that this described scenario would greatly assist in the sustainable development of ASGM operations.

*Appendix F: Acknowledgements**

In addition to the "Champions" mentioned in the Preface, the Co-Authors would like to acknowledge the support for, contribution to and involvement in the matters presented in this book, *EEB Ecotechnology*, and in the development of EEBs of the following people: Dr. Tereza Dan, Al Leggett, Mickey Liu, Dr. Susan MacFarlane, Warren Martin, Dr. Angus McGrath, Gregory Shaw, Dr. Andrew Sinclair, Sheldon Smith and Dr. Higgins' and Mr. Stiebel's other colleagues at Stantec and its legacy firm, Jacques Whitford; Dr. Mattes' Nature Works colleagues including the late James Hill, Dr. Bill Duncan, Matt Pommer and their colleagues at Teck Cominco; Dr. Gordon Balsch, Heather Broadbent, Stephanie Collins, Jennifer Paul and other staff of Dr. Wootton at CAWT; as well as Dorothy Ashworth of ETDC; Mr Wallace's and Mr Liner's colleagues at NAWE, Richard Wagner, David Austin and Curtis Sparks; Gareth Meal, Donald Smith and Paul Parker at Urban Engineers of NY (Urban); Michael Walters of the Lake Simcoe and Region Conservation Authority (LSRCA); Professor Tom Wildeman of the Colorado School of Mines; Dan Harrington of the South-West Solid Waste Corporation (SWSWC); Lloyd Rozema of Aqua-Treat Technologies; Dr. Jack Adams of the University of Utah & Inotec Inc.; Mark Pizey and Dr. Don Elder of Solid Energy NZ; Dr. Claire Depardieu of Laval University; Wally Sencza of Newmont Canada; Chris Kinsley, Anna Crolla, Claude Weil and Sarah Hurd of Alfred College's OWRC; Prit Kotecha of Suncor; Dr. Jane Pagel of the Ontario Clean Water Agency (OCWA); Dr. David Blowes, Dr. Trevor Charles and other colleagues at the UoW; Jim Gusek of Sovereign Consulting Inc.; Vasudha Seth of Dofasco Steel; Shawn Taylor of Dillon Consulting Ltd.; Joe Yakimchuk at the Kelowna International Airport (YLW); Dr. Gould's former colleagues at CANMET-MMSL, Jonathan Kawaja and Janice Zinck; Tjep Verkuijl, Rubin Wallin and the lab and engineering staff at RGM in Suriname; Don Baxter, Tom Myatt, Jerry Janig and Ellerton Castor of Ontario Graphite Limited (OGL); Mr MacLean's colleague at the ERAA, Lisa Dechaine; Phil Lalande and Wendall Wain of SESI; Ms Minkel's colleagues at the NFTA, John Diebold, Tom Dames, John Schaefer; Don Haycock of Conestoga-Rovers and Associates (CRA); Phil O'Connell, Glenn Mahon, Peter Avery and other staff of the SJAA; Derek Gray and Ken Eastman of the Greater Toronto Airport Authority; Professor Cathy Chin and Atif Vaniyambadi of the Department of Chemical Engineering at the University of Toronto (UoT); Steve Ostafichuk and Luke Daly of Ferrate Treatment Technologies Inc. (FTT); Professors James Longstaffe and Susan Glasauer of SES at the UoG; Dr. Chris Anderson of

* The companies and organizations mentioned were those at which the people listed were associated at the commencement of their involvement with the Co-Authors for the development of the EEB ecotechnology, and the companies and organizations may have since changed, and the people may currently be at different ones.

Massey University in New Zealand; Professor Sven Jørgensen of Copenhagen University; the late Professor John Meech of UBC and Irma Shagla Britton, Claudia Kisielewicz, Molly Pohlig, and Florence Kizza of CRC Press/Taylor & Francis and Arun Kumar T, Project Manager at Nova Techset Pvt Ltd.

The Co-Authors would also like to acknowledge the contributions to and support for the Eco-Engineered Bioreactor ecotechnology of the following organizations.

F.1 Soil Enrichment Systems

SES (Soil Enrichment Systems Inc.) (SESI) was a small, privately owned ecological landscape contracting firm based in Vaughan, Ontario, Canada that was incorporated in 1989 and ceased operations in 2000. SESI sought to operate according to the principles of Ecological Engineering (see Section 1.9 in Chapter 1), and provide unique bioengineered products for rapid vegetation establishment. SESI's main products were PROPASYS and LIVING WEB (see Section C.8 in Appendix C). Following the purchase of half of the equity of SESI in the mid-1990s by CRA, the Lead Author was sent to become President of SESI. In addition to expanding the scope of SESI's bioengineering activities, Dr. Higgins conceived, obtained financing and support for, and managed the *BREW Project* (see Section 1.4 in Chapter 1) that led to the development of what first were called "engineered wetlands" and later evolved into the EEB ecotechnology. When SESI ceased operations, the Lead Author and several of the staff of SESI joined Jacques Whitford, a *BREW Project* participant, and established its Ecological Engineering group.

F.2 Jacques Whitford Environment Limited

Jacques Whitford was a large employee-owned environmental engineering, environmental science and geotechnical engineering firm headquartered in Dartmouth, Nova Scotia, Canada. It was purchased by Stantec at the end of 2008 and integrated into that firm's operations. At the time, Jacques Whitford had almost 800 employees located in offices across Canada (including major ones in Ottawa and Toronto), the United States and offshore.

Jacques Whitford's Ecological Engineering group operated out of its Toronto office and provided services to clients in a variety of areas including the use of: natural systems (e.g., constructed wetlands) for wastewater treatment, engineered wetlands (as EEBs were then called), bioengineering (e.g., Stormwater Wetlands), soil bioremediation (e.g., phytoremediation), industrial soil ecology (e.g., composting), and ecological management (e.g., ecological risk assessments). Jacques Whitford's Ecological Engineering group was involved in a number of wetlands engineering and phytoremediation projects including: a tree phytoremediation project involving the use of over 3000 hybrid poplar and willow trees to clean up hydrocarbon-contaminated soil at the site of a former fuel oil and gasoline loading facility; a major community-based environmental risk assessment which included phytoextraction and phytostabilization of metal-contaminated soils and the treatability testing for, and design and engineering of, the EEB Systems project at BNIA. Jacques Whitford organized and directed the projects at the test facilities at Alfred College and CAWT that were used to test the treatability of several especially recalcitrant wastewaters.

F.3 Stantec Consulting Inc.

Stantec Consulting Inc. is a major Canadian engineering consulting company with offices in the United States, Canada, the Caribbean, Europe, and elsewhere, providing professional services in environmental services, planning, engineering design and construction, architecture, surveying, economics, landscape architecture, project management, and other disciplines. Stantec supports both the public and private sectors by providing services in a number or practice areas, which are grouped into five broad categories: buildings, environment, industrial and project management, transportation, and urban land.

Stantec has extensive experience and capabilities in the field of wastewater treatment using active, passive and semi-passive treatment systems. In the case of the former (mechanical WWTPs), the company has designed, assisted with the tendering of, supervised the construction of and assisted in the start-up and commissioning of many types of WWTPs used for treating municipal and industrial wastewaters. These include facilities for primary WWT (e.g., screening, flotation, coagulation/precipitation, sorption, aeration); secondary WWT (activated sludge plants, e.g., extended aeration, SBRs, membrane filtration facilities, e.g., reverse osmosis, ultrafiltration, anaerobic bioreactors, e.g., contact, sludge blanket and fluidized and fixed film bioreactors, e.g., RBCs and trickle filters); and tertiary WWT facilities (e.g., those for ion exchange, oxidation and nutrient removal).

Stantec was already a leader in the design and engineering of WWTPs when, at the end of 2008, it acquired Jacques Whitford which brought to Stantec comparable experience and expertise in the field of natural WWT, especially in the design and engineering of Eco-Engineered Bioreactor Systems.

More information on Stantec may be found on its website, www.Stantec.com

F.4 North American Wetlands Engineering, LLC

NAWE was a small ecological engineering company founded by Scott Wallace and Curt Sparks and based near St. Paul, MN, USA. The privately owned company provided residential, municipal, industrial ecological and project engineering services as well as site evaluation, environmental assessment, engineering design, permitting construction supervision and operator training services particularly in the area of CW and EEB Systems. NAWE carried out over 250 projects ranging from those for single-family homes to the 6000 m^3/day of contaminated groundwater project at the old Amoco refinery property in Casper, WY (see Section 3.4 in Chapter 3). NAWE first patented as *Forced Bed Aeration*™ what would later become the aerated gravel-bed part of EEB ecotechnology (BBRs).

BREW Project participant, Jacques Whitford, purchased NAWE in April of 2007, and the firm became Jacques Whitford NAWE, then still later Stantec NAWE, when Jacques Whitford in turn was purchased by Stantec at the end of 2008.

F.5 Nature Works Remediation Corporation, Inc.

Nature Works designed, built and operated a field-scale BCR-based EEB System at Teck Cominco's lead–zinc metal refinery and smelter in Trail, BC, Canada (see Section 5.6 in Chapter 5), and has designed and started up full-scale BCRs at the Yankee Girl mine in BC and the Park City mine in Utah. Nature Works has been involved in a number of bench-, pilot, demonstration-, and full-scale field BCR-based EEB System projects on its own and in conjunction with Jacques Whitford/Stantec and ETDC. These include ones for removing As, Cd, Cr, Cu, Mo, Ni, Pb, Sb, Se, and Zn at concentrations ranging from thousands of milligrams per liter down to low milligram per liter levels, and even for certain CoCs to microgram per liter levels. Nature Works has worked with ETDC and Stantec on several other similar systems, including those for Newmont's Golden Giant Mine (Mo-contaminated tailings pond water); a major electrical utility (for wooden utility pole storage yard runoff contaminated with copper chrome arsenate wood preservative); OGL (tailings area runoff) and Vale (acid rock and neutral mine drainages). Nature Works pioneered the use of pulp and paper biosolids to remediate barren rock areas on contaminated areas and steep slopes around the Teck Cominco facilities.

More information on Nature Works may be found on its website, www.nature-works.net

F.6 Centre for Alternative Wastewater Treatment

CAWT carries out applied research in five key areas:

1. The performance of constructed wetlands in cold climates with the objective of developing systems for small or isolated communities and specific industrial applications
2. The development of effective constructed wetland systems for water treatment and reuse for the Canadian aquaculture industry
3. The design of low-cost/minimal maintenance natural water treatment solutions for developing countries
4. To investigate the ability of innovative wetland treatment system designs to remove problematic environmental contaminants such as cyanide, heavy metals, pathogenic bacteria, organic contaminants, and excess nutrients
5. To carry out R&D on on-site wetland treatment systems for individual homes

Since the doors opened for business in 2005, CAWT has secured over $15 million in resources, over 100 industry partnerships and expanded applied research activities into many technologies, including

- Constructed wetlands
- Phytotechnology
- Eco-Engineered Bioreactors
- Anaerobic digesters
- Composting technology
- Gasification technology
- Environmental spill technology
- Bioremediation of heavy metals with bacteria
- Phosphorus removal technologies
- Micronutrient stimulants
- Lagoon circulation technology
- Well design and installation
- Modelling Software

CAWT's test facilities are located in a large greenhouse at the Frost Campus of Fleming College at its School of Environmental and Natural Resources in Lindsay, ON, Canada. In addition to floor space for test units, bench areas, a walk-in refrigerator and a large environmental chamber there, CAWT operates an adjacent two-train outdoor constructed wetland/EEB test facility and has comprehensive high-level wet chemical and advanced analytical laboratories on site.

Details on CAWT, its treatability test facilities and lab capabilities are found in Section 6.3 of Chapter 6. More information on CAWT may be found on its website, www.cawt.ca

F.7 Urban Engineers of NY

Urban Engineers, Inc. is an award-winning design and engineering firm of over 500 employees founded in 1960 and is headquartered in Philadelphia, PA, USA. The company maintains a hands-on approach, focusing on technical excellence, on-time performance, diversity and innovation. Its early work focused on highway and bridge design and it now also provides planning, design, environmental, construction support, project management and training in other market areas such as arts and culture, aviation, energy, railroads, transit, hotels, and ports for public and private clients.

Urban's New York branch, Urban Engineers of NY, D.P.C. partnered with Jacques Whitford to provide civil engineering services for the design and construction of the seminal BBR-based EEB System at BNIA (see Section 3.6 in Chapter 3).

More information on Urban can be found at www.urbanengineers.com

F.8 Niagara Frontier Transportation Authority

The NFTA is a diversified and synergistic public benefit organization set up in 1967. It has over 1550 employees and provides the Buffalo-Niagara area of upstate NY, USA with cost-effective, quality transportation services. It operates two airports, bus and rail systems, and the Buffalo Boat Harbor, and has its own engineering, property, traffic, transit police, and fire departments.

The NFTA operates two airports in the Niagara peninsula: BNIA and Niagara Falls International Airport. Kimberly Minkel, BSc, MBA, the NFTA's current Executive Director, was Director of Health, Safety and Environmental Quality for the NFTA when the project was to design and build what is now called a BBR-based EEB System at BNIA (then referred as to as an EW system) and was instrumental in conceiving, proposing and overseeing this seminal US $13.1 MM project.

More information on the NFTA may be found at www.nfta.com

F.9 Environmental Technologies Development Corporation

ETDC is a licensed Canadian ecological engineering consulting firm in which the Lead Author is President. ETDC provides strategic consulting, training and planning services for environmental solutions. The company carries out the design and engineering of Eco-Engineered Bioreactor Systems and other natural wastewater treatment methods; prepares closure plans; carries out environmental technology and commercialization evaluations; gives courses on a variety of subjects (e.g., EEB ecotechnology, CWs, phytoremediation, chromium remediation) and provides business planning services for various firms.

More information on ETDC may be found at www.etdc.org

Glossary

A: Area (m^2)
AAF: Aircraft anti-icing fluid
Abiotic: Not involving biological processes
ABR: Anaerobic bioreactor, an old name for a BCR, now referring to all anaerobic bioreactors, both EEB ones and Stand-Alone versions
Acid: A chemical substance that can release excess protons
ACEC: American Council of Engineering Companies
ACRP: Airport Cooperative Research Program
Active medium (media): Carbonaceous substance(s) forming part of a bioreactor substrate or added to a semi-passive BCR
Active treatment: Involving mechanical WW treatment (e.g., in a conventional WWT plant)
ADAFs: Aircraft de-icing and anti-icing fluids (ADFs + AAFs)
ADF: Aircraft de-icing fluid
ADPACs: Proprietary "additive packages" of chemicals added to ADFs and AAFs to meet SAE and other requirements
Aeration: The addition of air to water, usually for the purpose of providing higher oxygen concentrations for biological reactions
Aerobe: An aerobic microbe
Aerobic: Requiring free oxygen
ALD: Anoxic limestone drain
Ammonification: The microbial conversion of organic nitrogen into ammonia
AMS: Aerospace material specification
Anaerobe: A microbe that does not use O_2 to obtain energy, one that cannot grow under an air atmosphere
Anaerobic: Lacking oxygen
Anaerobic respiration: Respiration under anaerobic conditions involving a terminal electron acceptor other than oxygen
Anion: A negatively charged ionic species
AOA: Airport operating authority
APB: Acid producing bacteria
ARD: Acid rock drainage (also called acid mine drainage, AMD)
Arsenic demo: A smaller demonstration-scale BCR-based EEB System at Teck Cominco
ASGM: Artisanal and small-scale gold mining
Autotrope: A microbe that uses inorganic materials as a source of nutrients
BBR: BREW Bioreactor, the basic aerated gravel-bed EEB
BCDI: Benzotriazole-derived corrosion inhibitor in ADF and AAF ADPACs
BCR: Biochemical reactor, the basic anaerobic EEB

Berm: A dyke surrounding a lagoon or wetland cell
Bioaugmentation: The practice of adding imported, actively growing, specialized microbial strains into a microbial community in an effort to enhance the ability of the microbial community to respond to process fluctuations or to degrade certain compounds, resulting in improved treatment
Biocontrol: A method to control and enhance the operation of an EEB using biostimulation or bioaugmentation
Biofilm: Microbial biota attached to a surface in a gelatinous matrix
Bioinformetrics: A subdiscipline of metagenomics involving the use of computational procedures to extract meaningful data from the very large data sets produced by metagenomics
Biomass: The total mass of biological components
Biosolids: Municipal or industrial by-product sludges
Biosorption: The surface uptake of a contaminant by an organism
Biostimulation: The deliberate enhancement of the population of a particular bacterium already present in a microbial population
Biotic: Involving biological processes
BLM: US Bureau of Land Management
BNIA: Buffalo Niagara International Airport (also defined under its IATA code: BUF)
BOD: Biochemical oxygen demand concentration (mg/L)
BOF: Basic oxygen furnace
BREW: Bioreactor Engineered Wetland – the *BREW Project* was the R&D program that initiated the EEB concept
BREW Bioreactor: A BBR, the basic kind of aerated gravel bed EEB
BTEX: Benzene, toluene, ethyl benzene, xylenes
BUF: The IATA airport code for BNIA
C: Concentration (mg/L) or carbon
C*: Background concentration (mg/L)
CAPEX: The capital cost of a project
Cation: A positively charged ionic species
CAWT: Centre for Alternative Wastewater Treatment, Fleming College, Lindsay ON
cBOD: Carbonaceous BOD
CCA: Copper chrome arsenate (wood preservative)
CDB: Cellulose degrading bacteria
CDF: Central de-icing facility
CEC: Cation exchange capacity, a measure of the ability of a material to bind positively charged ions
CEF: IATA Code for Westover Air Reserve Base (Maine, USA)
Cell: A single basin of a CW, EEB or an associated 1° or 3° basin in an EEB System
Chalcophilic: "Sulfur-loving" metal(loids), for example, Cd, Cu, Pb, Se, Zn
Chelation: The solubilization of a metal by a chemical that "cages" it

CoC: Contaminant (or chemical) of concern
COD: Chemical oxygen demand concentration (mg/L)
CP: Closure plan
Cr(III): Trivalent chromium, also Cr^{3+}
Cr(VI): Hexavalent chromium, also Cr^{6+}
CRA: Conestoga-Rovers & Associates
CSR: Completely stirred reactor, a kind of kinetic modelling approach, sometimes also called a CSTR or complete mix reactor
CW: Constructed Treatment Wetland
d: Day
DNB: Denitrification bacteria
DNBR: Denitrifying bioreactor
DEG: Diethylene glycol
DO: Dissolved oxygen (mg/L)
EB: Engineered Bioreactor, an older name for an EEB
EBPR: Enhanced biological phosphorus removal process
ECE: Engineers cost estimate
EEB: Eco-Engineered Bioreactor (formerly called an EW or an EB)
Ecotechnology: Any environmental/biological technology, process or methodology used in ecological engineering
Effluent: Water exiting a WW treatment unit
EG: Ethylene glycol
Eh: Redox potential (millivolts). See also ORP
EIA: Edmonton International Airport (IATA Code: YEG)
Engineered Bioreactor: EB, an older name for an EEB, particularly a BBR
Engineered substrate: An artificial material such as a by-product or waste material (e.g., steel slag) used as the medium in a primary or tertiary cell of an EEB System
Engineered Wetland: EW, an older name for a BBR
Envirogenomics: A subdiscipline of metagenomics involving environmental samples
EPA: United States Environmental Protection Agency
EPS: Extracellular polymeric substance
ERAA: Edmonton Regional Airports Authority, an AOA
ET: Evapotranspiration
ETDC: Environmental Technologies Development Corporation, see Appendix F.2
Ethoxylated surfactants: Alkyl phenol ethoxylate surfactants in ADF and AAF ADPACs
ETP: Environmental Treatment Plant
F: Mass flux rate ($g/m^2.d$)
Facultative: Able to operate aerobically or anaerobically
FC: Fecal coliform bacteria
Feedstock: Wastewater influent to an EEB System or cell
Fe(II): Ferrous iron, also Fe^{2+}

Fe(III): Ferric iron, also Fe^{3+}
Ferrate: Hypervalent iron, Fe(VI) or Fe^{6+} as a cation, or FeO^{2-} as an oxyanion
FTT: Ferrate Treatment Technologies Inc.
FWS: Free water surface, a kind of CW in which there is open water flow on the surface
g: Grams
GAP analysis: A comparison of actual performance with desired performance
GCSW: Glycol-contaminated stormwater runoff
Genome: The complete set of genes or genetic material present in a cell or organism
Genomics: The branch of molecular biology concerned with the structure, function, evolution and mapping of genomes
GMP: Airport Glycol Management and Mitigation Plan
GPD: US gallons per day
h: Water depth (m)
HCO_3^-: Bicarbonate
Heavy metals: Heavier metals that are above atomic number 21, excluding the alkali and alkaline earth elements
Heterotroph: A microbe that uses organic materials as its source of energy
HLR: Hydraulic loading rate (cm/d), also referred to as q
HRT: Hydraulic (or nominal) residence time (d or y), also known as τ
HSSF: Horizontal SSF
Hydraulics: The science and engineering of water movement
Hydrology: The science dealing with the properties, distribution and circulation of water on ground surfaces, soils and the atmosphere, and its interaction with biota
Hydrolysis: Decomposition that degrades a compound into others by taking up elements of water
IATA: International Air Transport Association
ILN: IATA code for Wilmington Airpark, OH, USA
IMC: International Minerals Corporation
Influent: WW into a WWT unit
Inorganic: Any chemical that does not contain organic carbon
IRB: Iron-reducing bacteria
ISO: International Organization for Standardization
ISP: IATA code for Long Island McArthur Airport
Jacques Whitford: Jacques Whitford Environment Limited, a Stantec legacy firm
k: Reaction rate constant for a particular contaminant (day^{-1} or m/y)
K: Thousand(s) or hydraulic conductivity (mS)
K_N: Half-saturation constant for ammonia (mg/L)
K_{O2}: Half-saturation constant for dissolved oxygen (mg/L)
Kinetics: Pertaining to the rates at which changes occur in chemical, physical or biological processes

KISS: Keep It Simple Stupid (an injunction not to complicate things!)
K_{sp}: The solubility product constant, the equilibrium constant for a solid substance dissolving in an aqueous solution. It represents the level at which a solute dissolves in solution. The more soluble a substance is, the higher the K_{sp} value it has
L: Length
L: Liter
Labile: Relatively easy-to-treat
Leachate: A WW resulting from the percolation of infiltrating precipitation or groundwater through a waste accumulation and/or the exfiltration of porewater out of it
LHA: IATA Code for London Heathrow Airport
KIA: Kelowna International Airport
LNAPL: Light, nonaqueous phase hydrocarbons
LSRCA: Lake Simcoe and Region Conservation Authority
m: Meter
M: Molar or mass of limestone
mA: milliAmperes
MBI: Metagenome Bio Inc.
MD: Mine drainage
MDL: Minimum (or method) detection limit
mg: Micrograms
Me: General symbol for a metal or metalloid
Metabolism: The chemical oxidation of organic compounds by an organism resulting in the release of energy for growth
Metalloid: A semimetal, for example, As, Se, B
Metal(loid): A collective term for (heavy) metals and metalloids
Metagenomics: The study of a collection of genetic material (genomes) from a mixed community of organisms. Usually refers to the study of microbial communities
Meteoric water: Stormwater runoff and snowmelt
Microorganism: An animal or plant that can only be viewed by a microscope
Mineralization: The complete conversion of an organic material into inorganic compounds
MIW: Mining influenced water
mg: Milligram
MM: Million
MPWs: Mineral processing waters, a kind of alkaline MIW
MSW: Municipal solid waste (mixed domestic and commercial garbage and other wastes)
N: Nitrogen
NAWE: North American Wetlands Engineering, later Jacques Whitford NAWE and Stantec NAWE

n-BOD: Nitrogenous BOD concentration (mg/L)
NFTA: Niagara Frontier Transportation Authority, an AOA and BNIA's owner
Naturally Wallace: The successor firm of Stantec NAWE Principal, Scott Wallace, after Stantec NAWE closed
Nature Works: Nature Works Remediation, Corp.
Nelson: Nelson Environmental Ltd. (supplier of SAGR Bioreactor-based EEB Systems as well as aeration systems for BBRs)
Nexom: A successor company to Nelson
NH_3: Ammonia concentration (mg/L)
NH_3-N: Ammonia nitrogen concentration (mg/L)
Nitrification: The microbial oxidation of ammonia
NMD: Neutral mine drainage
NO_2: Nitrite concentration (mg/L)
NO_2-N: Nitrite nitrogen concentration (mg/L)
NO_3: Nitrate concentration (mg/L)
NO_3-N: Nitrate nitrogen concentration (mg/L)
NOx: Nitrite + nitrate concentration (mg/L)
NPDES: US National Pollutant Discharge Elimination System
O & M: Operating and maintenance
OCWA: Ontario Clean Water Agency
OGS: Oil–grit separator
OLC: Open limestone channel
OLD: Oxic limestone drain
OGL: Ontario Graphite Limited
OPEX: Operating and maintenance costs
Org-P: Organic phosphorus concentration (mg/L)
o-PO_4: Ortho-phosphorus concentration (mg/L)
Organic: Pertaining to chemical compounds that contain reduced carbon bonded with hydrogen, oxygen and a variety of other elements
Org-N: Organic nitrogen concentration (mg/L)
ORP: Oxidation reduction potential, also represented by Eh
OWS: Oil–water separator
Oxidation: A chemical reaction in which one or more electrons are lost
PAH: Polyaromatic hydrocarbon
PAO: Phosphorus-accumulating organism
Passive Treatment: Involving a natural WWT system requiring little attention
PCP: Pentachlorophenol
PCR: Polymerase chain reaction
Permeability: Capability of a porous medium to conduct fluid
PFR: Plug flow reactor
PG: Propylene glycol
Phytomining: Phytoextraction and recovery of a metal by plants to gain economic benefits

Phytotoxicity: Toxicity to plants
PLC: Programmable logic controller, see SCADA
POTW: Publicly Owned Treatment Works, a US designation for a municipal WWTP
PRB: Permeable reactive barrier
Primary treatment: The initial step in WWT, usually involving physical and/or chemical removal processes
Propagules: Bioengineered mixtures of seeds, shoots and vegetative cuttings
PROPASYS: PROPAgation SYStem, advanced ecotechnology to custom-design vegetation complexes
q: Hydraulic loading (cm/d or m/y), also represented as HLR
Q: W/WW flow rate (L or m^3 per day)
RAPS: Reducing and alkalinity producing system (see SAPS)
RBC: Rotating biological contactor
RBW: Reed Bed Wetland
RDX: Research Department Formula X (explosive stronger than TNT)
Recalcitrant: Hard-to-treat
Receiving waters: The water body into which a WW or treated effluent is surface discharged
Redox: Oxidation/reduction, see Eh and ORP
Redox potential: The potential of a soil or water to oxidize or reduce chemical substances
RI: Refractive index
RGM: Rosebel Gold Mine N.V. (Gold mining firm in Suriname originally owned by Cambior)
RO: Reverse osmosis, a membrane process
RTD: Residence time distribution
Runoff: Stormwater and snowmelt resulting from precipitation (meteoric water)
SAD: Strong acid dissociated (type of cyanide complex with certain metals)
SAE: Society of Automotive Engineers
SAG: Semi-autogenous grinding mill
SAGR: Submerged aerated growth reactor, Nelson's SAGR Bioreactor
SAPS: Successive alkalinity producing System (SAPS Bioreactor), a kind of ABR
SBR: Sequencing batch reactor
SCADA: Supervisory control and data acquisition, a remote monitoring and control system for operating process facilities. Involves a number of PLCs
Sedimentation Pond: An EEB System impoundment (pond, vault, tank) in which suspended solids and precipitates can settle out by gravity with or without chemical or biological help
Semi-passive treatment: Involving a normally passive treatment system that is manipulated, controlled or addressed in some more active way

SeRB: Selenium-reducing bacteria
SESI: SES (Soil Enrichment Systems) Inc.
SES: School of Environmental Science, University of Guelph
SFNW: Swamp Forest Natural Wetland
SJAA: St. John's International Airport Authority, an AOA
Sludge: The accumulated solids separated from WW during a treatment process
SOB: Sulfur-oxidizing bacteria
SRB: Sulfate-reducing bacteria
SS: Suspended solids
SSF: Sub-surface flow
Stantec: Stantec Consulting Limited, an Edmonton, AB, Canada HQ'd engineering consulting firm
Siderophile: "Iron-loving" metal(loid), for example, As, Co, Cr, Ni
Substrate: Usually granular, porous media in bioreactors and wetland cells
Systems analysis: The initial phases of an EEB System project, usually consisting of a feasibility study, a conceptual design and a treatability test
SW: Stormwater
T: Temperature
TCE: Trichloroethylene
TDS: Total dissolved solids (mg/L)
Teck Demo: The demonstration-scale BCR-based EEB System at Teck Cominco
ThOC: Theoretical oxygen content (mg/L)
TIN: Total inorganic nitrogen concentration (mg/L)
TKN: Total Kjeldahl nitrogen (mg/L), a measure of Org-N + NH_3-N
TMA: Tailings management area
TN: Total nitrogen concentration (mg/L)
TNT: Trinitrotoluene (explosive)
TON: Total organic nitrogen concentration (mg/L)
Toxicity: Adverse effects on the emergence, growth or reproduction of organisms
TOC: Total organic content (mg/L)
TP: Total phosphorus concentration (mg/L)
Transformation: The degradation or removal of aerobically susceptible species in an EEB
Translocation: The movement of a species from roots to shoots and aboveground tissue in a plant
TSS: Total suspended solids concentration (mg/L)
Train: A set of cells and other components of CW or EEB Systems forming a flow path
TRB: US Transport Research Board
TT: Treatability test

TWSP: Treated Water Storage Pond
Type I glycol: The main kind of ADF used at airports
Type II glycol: A kind of AAF sometimes used at smaller airports
Type III glycol: Another kind of AAF sometimes used for the anti-icing of smaller aircraft at some airports
Type IV glycol: The main kind of AAF used at most airports
UoG: University of Guelph
UoT: University of Toronto
UoW: University of Waterloo
USG: United States Gallons
UV: Ultraviolet (disinfection method)
VFA: Volatile fatty acids (mg/L)
V: Volts
VOC: Volatile organic compound (mg/L)
VSSF: Vertical SSF
WAD: Weak acid dissociated (type of cyanide complex with certain metals)
WAS: Waste activated sludge
WL: Wetland
WW: Wastewater
WWT: Wastewater treatment
WWTP: A conventional, active (mechanical) wastewater treatment plant
X: Limestone purity (fractional)
YEG: IATA code for Edmonton International Airport
YYT: IATA code for St. John's International Airport
YYZ: IATA code for Toronto Pearson International Airport
ZVI: Zero valent iron
3H: Tritium
1°: Primary
2°: Secondary
3°: Tertiary

Equation Symbols

Exp: To the power of e
i: Subscript for inlet (influent)
ln: Natural logarithm
N: Number of elements in a Tank-In-Series Reactor Sequence
o: Subscript for outlet (effluent)
t: Time (year or day)
y: Year
[x]: Concentration of a species x (μg/L or mg/L)

Latin

Ex situ: Away from place
In situ: In place and on-site

Greek

ε: Porosity
δ: Thickness (m)
τ: Hydraulic residence time (day or year)
ρ: Density (kg/m^3)
μ: Micro, or viscosity (kg/m.day)

References*

ACRP, 2008. *Formulations for Aircraft and Airfield Deicing and Anti-Icing: Aquatic Toxicity and Biochemical Oxygen Demand.* University of South Carolina, Document 3 of the Airport Cooperative Research Program, Phase 1 Interim Report for ACRP Project 02-01, November 2008.

ACRP, 2009. *Deicing Planning Guidelines and Practices for Stormwater Management Systems.* ACRP Report 14 Prepared by CH2M Hill, Gresham Smith & Partners, and Barnes & Thornburg LLP for the Transport Research Board and Sponsored by the FAA, Washington, DC, 2009.

ACRP, 2012. *Guidebook for Selecting Methods to Monitor Airport and Aircraft Deicing Methods.* ACRP Report 72 Prepared by Gresham Smith & Partners and TERRA HTDR, Inc. for the Transport Research Board and Sponsored by the FAA, Washington, DC, 2012.

ACRP, 2013. *Guidance for Treatment of Airport Stormwater Containing Deicers.* ACRP Report 99 of the Airport Cooperative Research Program, Research for the Transport Research Board Sponsored by the US Federal Aviation Administration, Washington, DC, www.TRB.org

Adams D.J., 2010. Electro-Biochemical Reactor Patent. WO 2010/002503 A2, 7 January 2010.

Adams D.J. and Peoples M., 2010. New electrobiochemical reactor for removal of selenium, arsenic and nitrate. *Proceedings of International Mine Water Association Conference*, Sydney, pages 9–102.

Adams D.J., Peoples M. and Opara A., 2012. New electro-biochemical reactor for treatment of wastewaters. In Drelich J. (ed.), *Water in Mineral Processing. Proceedings of the February 19–22, 2012 Meeting.* The Society of Mining, Metallurgy and Exploration Seattle: WA, pages 143–153.

Adams J. and Miller, J.D., 2010. Materials for Removing Contaminants from Fluids Using Supports with Biologically-Derived Functionalized Groups and Methods of Forming and Using the Same. US Patent 2010/0176053 A1, 7 January 2010. Also Adams J., Miller J., Newton N., Nanduri M. and Peoples M., 2010. Electrobiochemical Reactor, International Filing WO 2010/002503.

Anderson C., Meech J., Viega M. and Krisnayanti D., 2014. Can phytoremediation support the gold mining industries in developing countries? Case study for Indonesia. *First International Symposium on Sustainable Mining Operations, Shechtman International Symposium*, Cancun, 29 June–4 July 2014.

Anderson C., Moreno F. and Meech J., 2005. A field demonstration of gold phytoextraction technology. *Minerals Engineering*, 18, pages 385–392.

Arakaki A., Nakazawa H., Nemoto M., Moori T. and Matsunaga T., 2008. Formation of magnetite by bacteria and its application. *Journal of the Royal Society Interface*, 5(26), pages 957–1118, doi: 10.1098/rsif.2008.0170.

* As is noted in the text, along with design, the terminology used to refer to Eco-Engineered Bioreactors has been evolving, and allowance for this in some of the references needs to be taken. For example, many of the earlier references refer to EEBs as EWs or engineered bioreactors.

Arias C.A. and Brix H., 2004. Phosphorus removal in constructed wetlands: Can suitable alternative media be identified? In Liénard A. and Burnett H. (eds.), *Proceedings of the 9th International Conference on Wetland Systems for Water Pollution Control*, pages 26–30, September 2004; IWA Publishing: Avignon, pages 655–661.

Austin D., Maciolek D., Davis B. and Wallace S., 2006. Damköhler number design method to avoid plugging of tidal flow constructed wetlands by heterotrophic biofilms. *Proceedings of the 10th International Conference on Wetland Systems for Water Pollution Control*, Lisbon, 23–29 September 2006.

Barrie Johnson D., Kanao T. and Hedrich S., 2012. Redox transformations of iron at extremely low pH: Fundamental and applied aspects. *Frontiers in Microbiology*, 3, Article 96, March 2012, www.frontiersin.org

Bass Becking L., Kaplan I. and Moore D., 1939. Limits of the natural environment in terms of pH and oxidation–reduction potentials. *Journal of Geology*, 68(3), pages 243–284.

Béchard G., Yamazaki H., Gould W.D. and Bédard P., 1994. Use of cellulosic substrates for the microbial treatment of acid mine drainage. *Journal of Environmental Quality*, 23, pages 111–116.

Becking L., Kaplan I. and Moore D., 1960. Limits of natural environment in terms of pH and oxidation–reduction potentials. *Journal of Geology*, 68(3), pages 243–284.

Behrends L., Sikora F., Coonrod H., Bailey E. and Bulls M., 1996. Reciprocating subsurface-flow constructed wetlands for removing ammonia, nitrate, and chemical oxygen demand: Potential for treating domestic, industrial, and agricultural wastewaters. *Proceedings of WEFTEC '96; The 69th Annual Conference and Exposition of the Water Environment Federation; Water Environment Federation*, Alexandria, VA.

Benner S., Gould W. and Blowes D., 2000. Microbial populations associated with the generation and treatment of acid mine drainage. *Chemical Geology*, 169, pages 435–448.

Berdjeb L., Pollet T., Domaizon I. and Jacquet, S., 2011. Effect of grazers and viruses on bacterial community structure and production in two contrasting trophic lakes. *Microbiology*, 11, page 88.

Blowes D., Bain J., Smyth D. and Ptacek C., 2003. Environmental aspects of mine wastes. In Jamboor J., Blowes D. and Ritchie A. (eds.), *Treatment of Mine Drainage Using Permeable Reactive Materials*, Chapter 17, Vol. 31, Mineralogical Association of Canada, Ottawa.

Boadi N., Twumasi S. and Ephtraim J., 2009. Impact of cyanide utilization in mining on the environmental. *International Journal of Environmental Research*, 3(1), pages 101–108.

Bowmer K., 1987. Nutrient removal from effluents by an artificial wetlands: Influence of rhizosphere aeration and preferential flow studies using bromide and dye tracers. *Water Research*, 21(5), pages 591–599, May.

Brix H., 1999. How 'green' are aquaculture, constructed wetlands and conventional wastewater treatment systems? *Water Science and Technology*, 40(3), pages 45–50.

Brooks A., Rosenwald M., Geohring L., Lion L. and Steenhuis T., 2000. Phosphorus removal by wollastonite: A constructed wetland substrate. *Ecological Engineering*, 15, pages 121–132.

Brown M., Barley B. and Wood H., 2002. *Minewater Treatment, Technology, Application and Policy*. IWA Publishing, London.
Brown R. and Herendeen R., 1996. Embodied energy analysis and emergy analysis: A comparative view. *Ecological Economics*, 19, pages 219–235.
CEPA 1994. *Glycol Guidelines*. Order in Council under the Canadian Environmental Protection Act, Department of the Environment, PC 1994-106, 20 January 1994, Canada.
Christianson L.E., 2011. *Design and performance of denitrification bioreactors for agricultural drainage*, PhD thesis, Iowa State University, Ames, IO.
Clifton J., Rossiter W. and Brown P., 1985. Degraded aqueous glycol solutions: pH values and the effects of common ions on suppressing pH decreases. *Solar Energy Materials*, 12, pages 77–86.
Constantine T., 2008. *An Overview of Ammonia and Nitrogen Removal in Wastewater Treatment*. Presentation of 19 February 2008 by CH2M Hill.
Cordy P., Veiga M., Salih I., Al-Saadi S., Console S., Garcia O., Mesa L., Velasquez-Lopez P. and Roeser M., 2011. Mercury contamination from artisanal gold mining in Antioquia, Columbia: The world's highest per capita mercury pollution. *Science of the Total Environment*, 410–411(2011), pages 154–160.
Corsi S., Gels S., Loyo-Rosales J., Rice C., Sheesley R., Failey G. and Cancilla D., 2006. Characterization of aircraft deicer and anti-icer components and toxicity. *Environmental Science & Technology*, 40(10), pages 3195–3202.
Cotton A., Wilkinson G., Murillo C. and Bochmann M., 1999. *Advanced Inorganic Chemistry*. Sixth Edition, John Wiley & Sons, New York, ISBN 0-471-19957-5, pages 920–925.
Davies T. and Hart B., 1990. Use of aeration to promote nitrification in reed beds. *Proceedings of the International Conference on the Use of Constructed Wetlands in Water Pollution Control*, Cambridge, September 1990, as presented in *Constructed Wetlands in Water Pollution Control*, Cooper P. and Findlater B. (eds.), pages 383–389, Pergamon Press, Oxford, 1991, also in *Advanced Water Pollution Control*, 11, pages 77–84, 1990.
De Beer D., Stoodley P., Roe F. and Lewandowski Z., 1994. Effects of biofilm structure on oxygen distribution and mass transport. *Biotechnology & Bioengineering*, 43, pages 1131–1138.
Dow 2004. *SAE AMS 1424 Propylene Glycol-Based Type I Fluids, UCAR Concentrate and "55/45" PG Aircraft Deicing Fluids*. Dow Chemical Product Information Bulletin No. 183-00024-0704 AMS, July 2004.
Dow 2009. SAE AMS 1424 Ethylene *Glycol-Based Type I Fluids, UCAR Concentrate, XL 54 and "50/50" EG Aircraft Deicing Fluids*. Dow Chemical Product Information Bulletin No. 183-00021-0709 AMS, July 2009.
Duncan W., Mattes A., Gould D. and Goodazi A., 2004. Multistage biological treatment system for removal of heavy metal contaminants. *CIM Conference Proceedings, Hamilton ON, 2004, in Wastewater Processing and Recycling in WCFral and Metallurgical Industries*, Hamilton, ON, pages 469–484.
Dwyer D. and Tiedje J., 1983. Degradation of ethylene glycol and polyethylene glycol by methanogenic consortia. *Applied and Environmental Microbiology*, July, pages 185–190.
El-Naggar M.Y. and Finkel S.E., 2013. Live wires. *The Scientist*, May, pages 38–43.
EPA, 1994. Technical Report, Treatment of Cyanide Heap Leaches and Tailings. United States Environmental Protection Agency, Office of Solid Waste, EPA 530-R-94-037, September 1994.

EPA, 2000a. Preliminary Data Summary on Airport Deicing Operations, US Environmental Protection Agency (EPA).

EPA, 2000b. Manual, Constructed Wetlands, Treatment of Municipal Wastewaters. EPA/625/R-99/010, US Environmental Protection Agency (EPA), September 2000.

EPA, 2002. Anaerobic Lagoons Fact Sheet. United States Environmental Protection Agency, Solid Waste and Emergency Response, EPA 832-F-02-009, September 2002.

EPA, 2003. Evapotranspiration Landfill Covers Fact Sheet. United States Environmental Protection Agency, Solid Waste and Emergency Response, EPA 542-F-03-015, September 2003.

Field R., O'Shea M. and Brown P., 1993. The determination and disinfection of pathogens in storm-generated flows. *Water Science and Technology*, 28(3–5), 311–315.

Findall R. and Basran S., 2001. Design of a constructed wetland for the treatment of glycol-contaminated stormwater. *Journal of Water Management Modeling*, 15 February 2001 issue, pages 201–211, doi: 10.14796/JWMM.R207-14.

Fitch M. and Schoenbacher J., 2009. Performance of full-scale horizontal flow wetland for zinc. *2009 National Meeting of the American Society of Mining & Reclamation*, Billings, MT, 30 May–5 June 2009.

Fortin D., 2000. Seasonal cycling of Fe and S in a constructed wetland: The role of sulphate-reducing bacteria. *Geomicrobiology Journal*, 17, pages 221–235.

Franks A., 2012. What's current with electric microbes? *Journal of Bacteriology & Parasitology*, 3(9).

Fraser L., Garris H., Baldwin S., Van Hamme J. and Gardner W., 2015. Using genomics in mine reclamation. In Fourie A., Sawatasky L., and Van Zyl D. (eds.), *Mine Closure 2015*, Infomine Inc., Vancouver, BC, pages 1–10, 978-0-9917903-9-3.

Glymph, 2015. Slime, slime, slime. *WEEB Presentation by Tony Glymph, Environmental Toxicologist of the Wisconsin DNR*, 15 December 2015, available at www.doc.foc.com

Gould W., Francis M., Blowes D. and Krouse H., 1997. Biomineralization: Microbiological formation of sulfide minerals. In McIntosh J. and Groat L. (eds.), *Biological – Mineralogical Interactions*, Volume 25, Mineralogical Association of Canada Short Course, Wiley & Sons, NY, pages 169–186.

Gould W. and Kapoor A., 2003. The microbiology of acid mine drainage. In Jambor J.L., Blowes D.W. and Ritchie A.I.M. (eds.), *Environmental Aspects of Mine Wastes*, Volume 31, Mineralogical Association of Canada Short Course, Wiley & Sons, NY, pages 203–226.

Grady C., Daigger G., Love N. and Filipe C., 2011. *Biological Wastewater Treatment*. Third Edition, IWA Publishing, London.

Gruyer N., 2012. Traitment biologique des effluents de serre par des marais filtrants artificiels et des bioreactors passif, PhD thesis, Department de Phytologie, Faculte des Sciences de L'Agriculture et de L'Alimentation, Laval University, Quebec.

Gusek J. 2008. Passive treatment 101: An overview of technologies. *2008 US EPA/National Groundwater Association's Remediation of Abandoned Mine Land Conference*, Denver Co, USA, 2–3 October, 2008.

Gusek J. and Wildeman T., 2002. A new millennium of passive treatment of acid rock drainage: Advances in design and construction since 1988. *National Meeting of the American Society for Mining and Reclamation*, Lexington, KY, 9–12 June 2002.

Gusek J., Wildeman T., Miller A. and Fricke J., 1998. The challenges of designing, permitting, and building a 1,200 gpm passive bioreactor for metal mine drainage, West Fork Mine, Missouri. *15th National Meeting of the American Society for Surface Mining and Reclamation*, St. Louis MO, May 17–22, 1998.

Hammer D., 1991. *Constructed Wetlands for Wastewater Treatment, Municipal, Industrial, Agricultural.* Lewis Publishers, Chelsea, MI.

Hammer D., 1997. *Creating Freshwater Wetlands.* Second Edition, Lewis Publishers, Boca Raton, FL.

Hartwell I., Jordahl D., Evans J. and May E., 1995. Toxicity of aircraft deicer and anti-icer solutions to aquatic organisms. *Environmental Toxicology and Chemistry*, 14(8), page 1375.

Hays, 1935. Sewage Treatment Process. US Patent 1,991,896, filed 15 October 1931.

Hedlin R., Nairn R. and Kleinmann R., 1994. Passive Treatment of Coal Mine Drainage. US Department of the Interior, Bureau of Mine Information, Circular 9389.

Higgins J., 1997. BREW, an economically viable alternative for wastewater treatment. *MOE Environment and Energy Conference for Ontario*, Toronto, November 1997.

Higgins J., 1998. Engineered wetlands for the treatment of landfill leachates. *EPIC Symposium*, Toronto, May 1998.

Higgins J., 2000a. An advanced engineered wetland process for treating recalcitrant contaminants. *R 2000, 5th World Conference on Integrated Resource Management*, Toronto, 6 June 2000.

Higgins J., 2000b. The treatment of landfill leachates with engineered wetlands. *7th International Conference on Wetland Systems for Water Pollution Control*, Lake Buena Vista, FL, 11–16 November 2000.

Higgins J., 2002a. Ecological engineering, A Consultant's Perspective. *American Ecological Engineering Society Annual Meeting*, Burlington, VT, 30 April 2002.

Higgins J., 2002b. The use of a very large constructed sub-surface flow wetland to treat glycol-contaminated stormwater from aircraft deicing operations. *Water Quality Research Journal of Canada*, 37(4), pages 785–792.

Higgins J., 2003. The use of engineered wetlands to treat recalcitrant wastewaters. In Mander U. and Jenssen P. (eds.), *Constructed Wetlands for Wastewater Treatment in Cold Climates*, Chapter 8, pages 137–139, Advances in Ecological Sciences, WIT Press, Southhampton.

Higgins J., 2006. Treating reclaimed water for re-use and discharge in a South American Gold Mine. *Mining Environmental Management Magazine*, October, pages 21–23, issue, Mining Communications Ltd., London.

Higgins J., 2008. Treatment wetlands for wastewater treatment in small communities in Northern Canada. *43rd Central Canadian Symposium on Water Quality Research*, Burlington, 13 February 2008.

Higgins J., 2009. Utilisation de marais en mileux agricole et piscicole – études de cas. *9ième Journée "Pisciculture/Phosphore" Poly/IRBV*, Jardin Botanique de Montréal, 15 décembre 2009.

Higgins J., 2014. Engineered bioreactor systems. In Jorgensen, S. (ed.), *Book forming part of Encyclopedia of Environmental Management*, Taylor & Francis, Group LLC, New York, doi: 10.1081/E-EEM-12005 1659, published 3 December 2014, http://www.tandfonline.com.

Higgins J. and Anderson C., 2014. Treating runoff and leachates from artisanal gold mining operations. *First International Symposium on Sustainable Mining Operations, Shechtman International Symposium*, Cancun, 29 June–4 July 2014.

Higgins J., Aube B., Wildeman T., Gusek J., Bayer H. and Zinck J., 2006a. Treating mine drainage: A review of available options and their advantages. *Short Course at 7th International Conference on Mine Drainage (ICARD)/2006 Society of Mining Engineers (SME) Annual Meeting*, St. Louis, MO, 25 and 26 March 2006.

Higgins J., Dechaine L., Minkel K., Liner M. and Wallace S., 2007c. Sub-surface flow wetlands to treat glycol-contaminated stormwater from aircraft deicing operations, BATEA. *Proceedings of WEFTEC 80th Annual Technical Exhibition and Conference*, San Diego, CA, 13–17 October 2007.

Higgins J., Hard B. and Mattes A., 2003c. Bioremediation of acid rock drainage using sulphate reducing bacteria. *Proceedings of the Sudbury 2003, Mining & the Environment Conference*, Sudbury, May 2003.

Higgins J., Hurd S. and Weil C., 1999. The use of engineered wetlands to treat recalcitrant wastewaters. 4th International Conference on Ecological Engineering, Aas, Norway, June 1999. Also in *Advances in Ecological Science Health*, A35(8), pages 1309–11334, 2000.

Higgins J., Kinsley C., Bachard A. and Wallace S., 2006c. Very high ammonia nitrogen removals in aerated VSSF engineered wetlands. *31st Annual Meeting & Conference of the Canadian Land Reclamation Association (CLRA) & 9th Meeting of the International Affiliation of Land Reclamationists (IALA)*, Ottawa, August 2006.

Higgins J., Langan J. and Hildebrand M., 2010b. The upgrading of lagoon-based wastewater treatment systems in northern areas using high performance aerated lagoon-aerated engineered wetland system. *IWA 12th International Conference on Wetland Systems for Water Pollution Control*, Venice, 4–9 October 2010.

Higgins J. and Liner M., 2007. Treatment wetlands for wastewater treatment in the Arctic. *Northern Territory Water and Waste Association 2007 Conference*, Iqaluit, 3 November 2007.

Higgins J., Liner M., Baxter D. and Mattes A., 2009a. Use of engineered wetland technology for long term minewater treatment at closed mines. *CIM Conference, 09*, Toronto, 12 May 2009.

Higgins J., Liner M., Baxter D. and Mattes A., 2009b. The Kearney graphite mine closure plan: A proposal for an innovative, long term treatment system to manage reclaimed tailings area meteoric water. *CLRA/OMA Mine Reclamation Symposium*, Timmins, 9 June 2009.

Higgins J., Liner M. and Kroeker M., 2009c. Cold weather municipal wastewater treatment using aerated lagoon, aerated engineered wetland systems. *3rd Canadian National Conference and Policy Forum on Wastewater Management, CWWA*, Niagara Falls, 12 September 2009.

Higgins J., Liner M., Verkuijl S. and Crolla A., 2006b. Engineered wetland pilot-scale treatability testing of ammonia- and cyanide-contaminated South American gold mine reclaim water. *31st Annual Meeting & Conference of the Canadian Land Reclamation Association (CLRA) & 9th Meeting of the International Affiliation of Land Reclamationists (IALA)*, Ottawa, August 2006.

Higgins J., Liu G., Dechaine L., Minkel K. and Kinsley C., 2007d. Treatability testing in the design of large sub-surface flow wetlands to treat glycol-contaminated stormwater from aircraft deicing operations. *4th Central Canadian Symposium on Water Quality Research*, CCIW, Burlington, 13 February 2007.

Higgins J, Liu G., Dechaine L., Minkel K., Liner M. and Kinsley C., 2007a. Treatability testing in the design of large sub-surface flow wetlands to treat glycol-contaminated stormwater from aircraft deicing operations. *Proceedings of the 4th Central Canadian Symposium on Water Quality Research*, CCIW, Burlington, 13 February 2007.

Higgins J. and MacLean M., 1999. Constructed wetland treatment systems for airport stormwater which do not attract waterfowl. *Presentation to Society of Automotive Engineers (SAE) G-12 Committee, Glycol Deicing*, Washington, DC, 9 November 1999.

Higgins J., MacLean M. and Worrall P., 2001. The use of a very large constructed sub-surface flow wetland to treat glycol-contaminated stormwater from aircraft deicing operations. *Proceedings of the International Ecological Engineering Association Conference*, Mason University, Christchurch, November 2001.

Higgins J. and Mattes A., 2003. The use of engineered wetlands to treat mine drainage. *German-American Environmental Conference, The Rehabilitation of Industrial Wasteland and Post Mining Landscapes*, Gorlitz, 12 April 2003.

Higgins J. and Mattes A., 2009. The removal of very low levels of dissolved metals using advanced anaerobic bioreactors as parts of engineered wetland systems. *34th Annual Meeting and Conference of the Canadian Land Reclamation Association*, Quebec City, 25 August 2009.

Higgins J. and Mattes A., 2014. Keynote address: Kinetics and genomics from a pilot-scale study on the use of a SAPS bioreactor and a biochemical reactor to remove dissolved metals from ARD at various temperatures. *Ontario Genomics Institute Science and Mining Workshop, Vale Living with Lakes Centre*, Sudbury, 6 May 2014.

Higgins J., Mattes A. and Kawaja J., 2004. Engineered wetlands to treat high aluminum acid rock drainage. *Ontario MEND Workshop: Sludge Management and Treatment of Weak Acid or Neutral pH Drainage*, Sudbury, May 2004.

Higgins J., Mattes A., Wootton B. and Chin C., 2013. Use of an engineered wetland system for the long term treatment of ARD seepage from a closed tailings management area. *6th Annual CLRA/OMA Reclamation Symposium*, Cobalt, 18 June 2013.

Higgins J., Sencza W., Mattes A. and Kinsley C., 2007b. Biological removal of low concentrations of dissolved metals in an engineered wetland system. *Proceedings of Mining & the Environment IV, International Mining & Environment Conference*, Sudbury, 23 October 2007.

Higgins J., Smith S., Liner M. and Dumond L., 2009d. Developing engineered stormwater technology to enhance water quality improvement by stormwater management facilities. *New Directions '09 Stormwater Management Conference*, Vaughan, 1 December 2009.

Higgins J., Wallace S., Minkel K., Wagner R. and Liner M., 2010a. The design and operation of a very large vertical sub-surface flow engineered wetland to treat spent deicing fluids and glycol-contaminated stormwater at Buffalo Niagara International Airport. *IWA 12th International Conference on Wetland Systems for Water Pollution Control*, Venice, 4–9 October 2010.

Higgins J., Wallace S., Walters M., Smith S., Dummond L. and Mihail A., 2010c. Developing engineered stormwater wetland technology to better manage stormwater runoff quality. *IWA 12th International Conference on Wetland Systems for Water Pollution Control*, Venice, 4–9 October 2010.

Higgins J.P., Myatt T. and Mattes A., 2012. The use of ecotechnologies in the closure plan for a graphite mine. *Canadian Institute of Mining & Metallurgy, Edmonton 2012 Conference*, Edmonton, 7 May 2012.

Hill P., Wever M. and Galbraith-O'Leary B., 2015. Lagoon ammonia treatment – Evaluation of alternatives. *WEOA 2015 Conference*, Toronto, 21 April 2015.

Hurd S., Weil C. and Higgins J., 1999. The use of engineered wetlands to treat farmstead runoff. *4th International Conference on Ecological Engineering*, Oslo, June 1999.

Hussain S., Blowes D., Ptacek C., Jamieson J., Wootton B., Balch G. and Higgins J., 2014. Mechanisms of nutrient and pathogen removal in a pilot-scale constructed wetland/BOF slag. *Wastewater Treatment System, Environmental Engineering Science*, submitted for publication, August 1, 2014.

Hylander L., Plath D., Miranda C., Lucke S., Ohlander J. and Rivera A., 2007. Comparison of different gold recovery methods with regard to pollution control and efficiency. *Clean*, 35, pages 52–61.

INOTEC, 2011. Electrifying New Way to Clean Water. http://www.physorg.com/news/2011-01-electrifying-dirty.html. See also the INOTEC web-site at www.inotec.us

IRAP, 1999. *Development of Bioreactor Engineered Wetland (BREW) for Wastewater Treatment, Project 27886U*. Final Report to the National Research Council, Industrial Research Assistance Program, Ottawa, Fall, 1999.

ITRC, 2013. Biochemical Reactors for Mining Influenced Waters, BCR-1. *Guidance Document Prepared for the Interstate Technology and Regulatory Council*, November, 2013.

Ivanow V., 2016. *Environmental Microbiology for Engineers*. Second Edition, CRT Press, Taylor Francis Group, Boca Raton, FL.

Jage C., Zipper C. and Noble R., 2001. Factors affecting alkalinity generation by successive alkalinity producing systems: Regression analysis. *Journal of Environmental Quality*, 30, pages 1015–1022.

Jamieson T., Stratton G., Gordon R. and Madani A., 2003. The use of aeration to enhance ammonia nitrogen removal in constructed wetlands. *Canadian Biosystems Engineering*, 45, pages 1.9–1.14.

Johnson J., Varney N. and Switzenbaum M., 2001. Comparative Toxicity of Formulated Deicers and Pure Ethylene and Propylene Glycol. Report for United States Department of the Interior, Geological Survey, Water Resources Center, University of Massachusetts/Amherst, August 2001.

Jørgensen S., 2012, *Introduction to Systems Ecology*. CRT Press, Taylor & Francis Group, Boca Raton, FL.

Kadlec R., 2001. *Feasibility of Wetland Treatment BP-Amoco Refinery Remediation*. Phytokinetics Inc.

Kadlec R., 2003. Effects of pollutant speciation in treatment wetlands design. *Ecological Engineering*, 20, pages 1–16.

Kadlec R. and Knight R., 1996. *Treatment Wetlands*. Lewis Publishers, Boca Raton, FL.

Kadlec R. and Wallace S., 2008. *Treatment Wetlands*. Second Edition, CRC Press, Boca Raton, FL.

Kepler D. and McCleary E., 1995. Successive alkalinity-producing systems (SAPSs). *West Virginia Surface Mining Task Force Symposium*, Morgantown, 4 and 5 April 1995.

Kickuth R., 1989. US Patent 4,855, 040.

Kinsley C., Crolla A. and Higgins J., 2002. Ammonia reduction in aerated sub-surface flow constructed wetlands. *Proceedings of the IWA 8th International Conference on Wetland Systems for Water Pollution Control*, Arusha, 16–19 September 2002.

Kinsley C., Higgins J., Crolla A. and Pepper T., 2000. Nitrogen reduction of landfill leachate using an engineered subsurface flow wetland. *Proceedings of the Wetland Technology for Water Treatment Session of the Special Symposium of the International Society of Ecologists*, Quebec City, 6–12 August 2000.

Knowles G., 1965. Determination of kinetic coefficients for nitrifying bacteria in mixed cultures with the aid of an electronic computer. *Journal of General Microbiology*, 38(2), pages 263–278.

Knowles P., Dotro G., Nivala J. and Garcia J., 2011. Clogging in sub-surface-flow treatment wetlands: Occurrence and contributing factors. *Ecological Engineering*, 37(2), pages 99–112.

Krisnayanti B., Anderson C., Utomo W., Feng X., Handayanto E. and Mudarisna N., 2012. Khususiah: Assessment of environmental mercury discharge at a four-year-old artisanal gold mining area on Lombok Island, Indonesia. *Journal of Environmental Monitoring*, 14(10), pages 2598–2607.

Krohn J.P., 2007. *Performance analysis of a successive alkalinity-producing system treating acid mine drainage at Simmons Run in Coshcocton County, Ohio*. MSc thesis, Russ College of Engineering Technology, of Ohio University, March 2007.

Lanfergraber G., Haberl G., Laber R. and Pressl A., 2003. Evaluation of substrate clogging processes in vertical flow constructed wetlands. *Water Science & Technology*, 48(5), pages 23–34.

Lawson C., Strachan B., Hanson N., Hahn A., Hall E., Rabinowitz B., Mavinic D, Ramey W. and Hallam S.J., 2015. Rare taxa have potential to make metabolic contributions in enhanced biological phosphorus removal ecosystems. *Environmental Microbiology*, 17(12), pages 4979–4993.

Lee Y., Cho M., Kim J. and Yoon J., 2004. Chemistry of ferrate (Fe[VI]) in aqueous solution and its application as a green chemical. *Journal of Industrial and Engineering Chemistry*, 1, pages 161–171.

Lemon E., Bis G., Rozema L. and Smith I., 1996. SWAMP pilot scale wetlands – Design and performance. *Proceedings of Symposium on Constructed Wetlands in Cold Climates*, Niagara-on-the-Lake, 5 June 1996.

Liner M., 2011. Edmonton airport upgrades its deicing fluid treatment system. *Environmental Science & Engineering Magazine*, September, pages 59–62.

Liner M., 2013. Aerated gravel beds. *2013 ACI Deicing and Stormwater Management Conference, New and Emerging Technologies*, Crystal City, VA, 31 July 2013.

Liner M. and Kroeker M., 2010. Submerged attached growth reactors. *Environmental Protection Magazine*, 22 March.

Lopez O., Sanguinetti D., Bratty M. and Kratochvil D., 2009. Green technologies for sulphate and metal removal in mining and metallurgical effluents. *Enviromine 2009*, Santiago.

Lovley D.R., 2012. Electromicrobiology. *Annual Review of Microbiology*, 66, pages 391–409.

MacDonald D., Cuthbert I. and Outridge P., 1992. *Canadian Environmental Quality Guidelines for Three Glycols Used in Aircraft Deicing: Ethylene Glycol; Diethylene Glycol; and Propylene Glycol*. EcoHealth Branch, Environment Canada, Ottawa, 1992.

Mattes A., 2013. *The Trail biochemical reactor and the removal of arsenic with a focus on the specific role of iron*. PhD thesis, University of Guelph, August 2013.

Mattes A., Gould W.D. and Duncan B., 2002. Multi-stage biological treatment system for removal of heavy metal contaminants. *Proceedings of Environmental Services Association of Alberta, Remediation Technologies.* Banff, 16–18 October 2002.

Mattes A., Gould W. and Higgins J., 2003. Biological metal removal in an engineered wetland system, *US EPA International Applied Phytotechnologies Conference,* Chicago, IL, 3–5 March 2003.

Mattes A., Duncan W. and Gould W., 2004. Biological removal of arsenic in a multistage engineered wetlands treating a suite of heavy metals. *Proceedings of the BC WCF Reclamation Conference,* Cranbrooke, June 2004.

Mattes A., Higgins J. and Gould W., 2007. Passive biologically-based anaerobic treatment systems for the removal of metals – An overview of current research with examples. *6th International Copper/Cobre Conference Cu 2007, CIM Conference of Metallurgists,* Toronto, 29 August 2007.

Mattes A., Higgins J. and Sencza W., 2010. Biological removal of low concentration metals in an engineered wetland system. *IWA 12th International Conference on Wetland Systems for Water Pollution Control,* Venice, 4–9 October 2010.

Matthys A., Parkin G. and Wallace S., 2000. A comparison of constructed wetlands used to treat domestic wastes: Conventional, drawdown and aerated systems. *Proceedings of 7th International Conference on Wetland Systems for Pollution Control,* Orlando, FL, 11–16 November, pages 629–636.

MEND, 1999. *Review of Passive Systems for Treatment of Acid Mine Drainage.* MEND Report 3.14.1, Revised 1999.

Metcalfe and Eddy, 2004. Chapter 7. In *Wastewater Engineering, Treatment and Reuse.* Fourth Edition, revised by Tchobanolglous G., Burton F.L., and Stensel H.D. (eds.), Tata McGraw-Hill Publishing Company, New Delhi.

Mitsch W. and Jørgensen S., 2003. Ecological engineering: A field whose time has come. *Ecological Engineering,* 20, pages 363–377.

Molle P., Lienard A., Boutin C., Merlin G. and Iwena A., 2004. *How to treat sewage with constructed wetlands: An overview of the French systems.* In Lienard A. and Burnetts H. (eds.), *Proceedings of the 9th International Conference for Water Pollution Control,* September 2004, Avignon, France, IWA Publishing, Colchester.

Moshiri G., Ed., 1993. *Constructed Wetlands for Water Quality Improvement.* Lewis Publishers, Boca Raton, FL.

MT 2005. Literature Review, Evaluation of Treatment Options to Reduce Water-Bourne Selenium at Coal Mines in West-Central Alberta. Report prepared by Microbial Technologies, Inc. for Alberta Environment, Water Research Users Group, Edmonton, AB.

Mückshel B., Simon O., Klebensberger J. Graf N., Rosche B., Altenbuchner J., Pfannstiel J., Huber A. and Hauer B., 2012. Ethylene glycol metabolism by pseudomonas putida. *Applied and Environmental Microbiology,* 78(24), pages 8531–8539, December.

Mudder M. and Botz M., 2004. Cyanide and society: A critical review. *European Journal of Mineral Processing and Environmental Protection,* 4(1), 1303–0868, pages 62–74.

Murphy C., Wallace S., Knight R., Cooper D. and Sellers T., 2014. Treatment performance of an aerated constructed wetland treating glycol from de-icing operations at a UK Airport. *Ecological Engineering,* 80, pages 117–124, doi: 10.1016/j.ecoleng.2014.05.032.

Neculita C., Zagury G. and Bussiere, B., 2007. Passive treatment of acid mine drainage in bioreactors using sulphate-reducing bacteria: Critical review and research needs. *Journal of Environmental Quality (on-line)*, 36, pages 1–16.

Nelson, 2013. Corporate Brochure, Nelson Environmental Inc., Lagoon Based Wastewater Treatment Systems. OPTAER & SAGR, www.environmental.com

Nguyen T., Roddick F. and Fan L., 2012. Biofouling of water treatment membranes: A review of underlying causes, monitoring techniques and control measures. *Membranes*, 2, pages 804–840.

Nivala J., 2005. *Treatment of Jones County landfill leachate in a pilot-scale horizontal SSF engineered wetland*. MSc Thesis, University of Iowa, 2005.

Nivala J., Diederik P., Dotro G., Garcia J. and Wallace S., 2012. Clogging in sub-surface-flow treatment wetlands: Measurement, modeling and management. *Water Research*, 46(6), pages 1625–1640.

Nivala J., Knowles P. and Rousseau L., 2009. Reversing clogging in sub-surface-flow constructed wetland by hydrogen peroxide treatment: Two case studies. *Water Science and Technology*, 59(10), pages 2037–2046.

Odum H., 1996. *Environmental Accounting: Emergy and Environmental Decision Making*. Wiley Publishers, New York, 1996.

Opara A., Adams D.J. and Peoples M.J., 2013a. Pilot system successfully removes metals, inorganics. *Industrial Water World*, 13.

Opara A., Adams D, Peoples M. and Martin A., 2013b. Electro-biochemical reactor (EEBR) technology for selenium removal from British Columbia's Coal-Mining Wastewaters. *Minerals & Metallurgical Processing*, 31(4), pages 209–214, 2014.

Opara A., Peoples M., Adams D. and Maehl W.C., 2013c. *The Landusky Mine Biotreatment System: Comparison of Conventional Bioreactor Performance with a New Electro-Biochemical Reactor (EEBR) Technology*. Society of Mining Metallurgy, and Exploration, Salt Lake City, UT, February 2014.

Oremand R.S., Hollibaugh J., Maest A., Presser T., Miller L. and Culbertson C., 1989. Selenate reduction to elemental selenium by anaerobic bacteria in sediments and culture: Biogeochemical significance of a novel, sulphate-independent respiration. *Applied and Environmental Microbiology*, 55(9), pages 2333–2343.

Palmer S., Breton M., Nunno T., Sullivan, D. and Supreant N., 1988. *Metal/Cyanide Containing Wastes: Treatment Technologies*. Noyes Data Corporation, Park Ridge, NJ.

Parker D., 1975. *Process Manual for Process Control*. US EPA, Washington, DC.

Pick U. and Weiss M., 1991. Polyphosphate hydrolysis within acidic vaculoes in response to amine-induced alkaline stress in the halotolerant alga, *Dunaliella salina*. *Plant Physiology*, 97, pages 1234–1240.

Postgate J., 1984. *The Sulfate-Reducing Bacteria*. Second Edition, Cambridge University Press, Cambridge.

Qureshi N, Annous B., Ezeji C., Karcher P. and Maddox I., 2005. Biofilm reactors for industrial bioconversion processes: Employing potential of enhanced reaction rates. *Microbial Cell Factories*, 4, 24, doi: 10.1186/1475-2859-4-24.

Redmon D., Boyle W. and Ewing L., 1983. Oxygen transfer efficiency measurements in mixed liquor using off-gas techniques. *Journal WCPF*, 55(11), pages 1338–1346, November.

Redmond E., 2012. *Nitrogen removal from wastewater by an aerated subsurface flow wetland. 2012*, MSc thesis, University of Iowa, May 2012.

Reed S., Crites R. and Middlebrooks J., 1995. *Natural Systems for Waste Management and Treatment*. Second Edition, McGraw-Hill, New York.

Rich L.G., 1999. High performance aerated lagoon systems. *American Academy of Environmental Engineers*, pages 20–27.

Richter K., 2003. *Constructed eetlands for the treatment of airport de-icer.* PhD thesis, University of Sheffield, November 2003.

Richter K., Guymer I., Worrall P. and Jones C., 2004. Treatment performance of heathrow constructed wetlands. Liénard A. and Burnett H. (eds.), *Proceedings of the 9th International Conference on Wetland Systems for Water Pollution Control*, 26–30 September 2004, IWA Publishing, Avignon, pages 125–132.

Rossiter W., Godette M, Brown P. and Galek K., 1985. An investigation of the degradation of aqueous ethylene glycol and propylene glycol solutions using ion chromatography. *Solar Energy Materials*, 11, pages 455–467.

Sasowsky I., Foss A. and Miller C., 2000. Lithic controls on the removal of iron and remediation of acidic mine drainage. *Water Research*, 34(10), pages 2742–2746.

Schueler T., 1992. *Design of Stormwater Wetland Systems. Guidelines for Creating Diverse and Effective Stormwater Wetlands in the Mid-Atlantic Region.* Metropolitan Washington Council of Governments, Washington DC.

Skousen J., 1998. Overview of Passive Systems for Treating Acid Mine Drainage. Center for Agricultural & Natural Resources Development, https://www.osmre.gov/resources/library/ghm/hbtechavoid.pdf, from *Acid Mine Drainage Control and Treatment*, in Reclamation of Drastically Disturbed Lands, American Society for Agronomy, and the American Society for Surface Mining and Reclamation.

Sonstegard J., Harwood J., Pickett T. and Kennedy W., 2010. ABMet: Setting the standard for selenium removal for selenium removal. *Proceedings of IWC 10*, Pittsburg, PA, pages 189–203.

Stell S., 1998. Oxygen Demand of Aircraft and Pavement De-Icing Compounds. U.S Air Force, AFCEE/CCR-D, infohouse.p2ric.org/ref/22/21298.pdf

Stell S., 2010. Environmental Aspects of Aircraft and Airfield Deicing – An Air Force Perspective. U.S Air Force, AFCEE/CCR-D, infohouse.p2ric.org/ref/22/21298.pdf

Stoodley, P., Sauer K., Davis D. and Costerton J., 2002. Biofilms as complex differentiated communities. *Annual Review of Microbiology*, 56, pages 187–209.

Strong-Gunderson J., Weelis S., Carroll S., Waltz M. and Palumbo A., 1995. Degradation of high concentrations of glycols, anti-freeze and deicing fluids. *International Symposium on In Situ and On-Site Bioreclamation (3rd)*, San Diego, CA, pages 24–27, April 1995. Sponsored by Department of Energy, Washington, DC.

Switzenbaum M., Veltman S., Mercias D. and Wagoner B., 2001. Best management practices for airport deicing stormwater. *Chemosphere*, 43(8), pages 1051–1062, Publication No. 173.

Thoren A., Legrande C. and Hermann J., 2003. Transport and transformation of deicing urea from airport runways in a constructed wetland system. *Water Science and Technology*, 48(5), pages 283–290.

Transport Canada 1999. *The Environmental Impact of Urea on Airport Runways.* Report TP 10069E, Transport Canada Airports Group, January 1990.

Utomo W., Anderson C., Kusomo B., Krisnayanti B., Feng X. and Adams J., 2010. Environmental Risk Assessment and Sustainable Development of Artisanal and Small-Scale Gold Mining Areas in Indonesia. ASGM Research Proposal: Sekotong Mining Area.

Valentini A., Pompanon F. and Taberlet P., 2009. DNA barcoding for ecologists. *Trends in Ecology and Evolution*, 24, pages 110–117.

Velasquez-Lopez P., Veiga M., Klein B., Shandro J. and Hall K., 2011. Cyanidation of mercury-rich tailings in artisanal and small-scale gold mining: Identifying strategies to manage environmental risks in Southern Ecuador. *Journal of Cleaner Production*, 19, pages 1125–1133.

Versol W., 2007. *Artisanal gold mining in Suriname*. PhD thesis, Eindhoven University of Technology, accessed from http://alexandria.tue.nl/extra2/afstversl/tm/Versol2007.pdf

Viega M., 1997. *Introducing New Technologies for the Abatement of Global Mercury Pollution in Latin America*. UNIDO/UBC/CETEM/CNPq, Rio de Janeiro, page 94.

Veiga M. and Hinton J., 2002. Abandoned artisanal gold mines in the Brazilian Amazon: A legacy of mercury pollution. *Natural Resources Forum*, 26, pages 13–24.

Veiga M., Maxson P. and Hylander L., 2006. Origin and consumption of mercury in small-scale gold mining. *Journal of Cleaner Production*, 14, pages 436–447.

Vinci B. and Schmidt T., 2002. *Passive Periodic Flushing Technology for Mine Drainage Systems*. www.asmr.us.

Vymazal J. 2001. Types of constructed wetlands for wastewater treatment: Their Potential for Nutrient Removal. In Vymazal J. (ed.), *Transformation of Nutrients in Natural & Constructed Wetlands*, Backhuys Publishers, Leiden, pages 1–94.

Walker, R, 2010. *Final Report, Passive Selenium Bioreactor – Pilot Scale Testing*. Report for Bureau of Reclamation Science and Technology Program by Golder Associates, Project 4414, 10 March 2010.

Wallace S., 2001a. Patent: System for Removing Pollutants from Water. Minnesota, United States 6,200,469 B1.

Wallace S., 2001b. Advanced designs for constructed wetlands. *BioCycle*, June, pages 40–44.

Wallace S., 2002. Use of constructed wetlands for nitrogen removal. *NOWRA, 2002 Annual Conference and Exposition Proceedings*, Edgewater, MD, pages 277–298.

Wallace S., 2004. Engineered wetlands lead the way. *Land & Water*, 48(5), pages 1–9.

Wallace S. and Austen D., 2008. Emerging models for nitrogen removal in treatment wetlands. *Journal of Environmental Health*, 71(4), pages 10–16, November.

Wallace S., Higgins J., Crolla A. Kinsley C., Bachand A. and Verkuijl S, 2006. Very high rate ammonia nitrogen removal in aerated engineered wetlands. *10th International Conference on Wetland Systems for Water Pollution Control, International Water Association*, Lisbon, September 2006.

Wallace S., Higgins J., Liner M. and Diebold J., 2007. Degradation of aircraft deicing runoff in aerated engineered wetlands. *Proceedings of Multi Functions of Wetland Systems: An International Conference*, University of Padova and International Water Association, Padova.26–29 June 2007.

Wallace S. and Kadlec R., 2005. BTEX degradation in a cold-climate wetland system. *Water Science and Technology*, 51(9), pages 165–171.

Wallace S. and Knight R., 2006. Feasibility, Design Criteria, and O&M Requirements for Small Scale Constructed Wetland Wastewater Treatment Systems. Water Environment Research Foundation Project 02-CTS-S, Alexandria, VA.

Watzlaf G., Schroeder K. and Kairies C., 2000. Long term performance of alkalinity-producing passive systems for the treatment of mine drainage. *Proceedings of the 2000 National Meeting of the American Society for Surface Mining and Reclamation*, Tampa, FL, 11–15 June 2000, pages 269–274.

WEF, 2010. *Biofilm Reactors, Water Environment Federation Manual of Practice No. 35.* Biofilm Task Force, WEF, Alexandria VA.

WEF 2011, *Biofilm Reactors, WEF Manual of Practice No. 35,* Biofilm Reactors Task Force of the Water Environment Federation, ISBN 978-0-07-173707-4, WEF Press, Alexandria VA/McGraw-Hill Companies, NY.

WEF 2013, *Wastewater Treatment Process Modeling, Second Edition, WEF Manual of Practice No. 31,* Water Environment Federation, ISBN 9780071798426, WEF Press, Alexandria VA/McGraw-Hill Companies, NY, August 2013.

Wildeman T., Brodie, G. and Gusek, J., 1993. *Wetland Design for Mining Operations.* BiTech Publishers, Richmond, ISBN 0 921095 27 9.

Wildeman T. Gusek G. and Higgins J., 2006. Passive treatment of mining influenced waters. *Short Course Presented in Association with 7th International Conference on Acid Rock Drainage (ICARD) and the International Mine Water Association Symposium,* St. Louis, MO, March 2006, www.imwa.info/imwa-proceedings.2006.html

Wildeman T., Neculita, C., Zagury G. and Bussiere B., 2007. Passive treatment of acid mine drainage in bioreactors using sulphate reducing bacteria. *Journal of Environmental Quality,* 36, pages 1–16.

Wildeman T. and Schmiermund N., 2004. Mining influenced waters: Their chemistry and methods of treatment. *2004 National Meeting of the American Society of Mining & Reclamation and the 25th West Virginia Surface Mining Task Force,* 18–24 April 2004.

Wile I., Miller G. and Black S. 1985. Design and use of artificial wetlands. In Kayner E., Pelczarski S. and Benfardo J. (eds.), *Ecological Considerations in Wetlands Treatment of Municipal Wastewater,* Van Nostrand Reinhold, New York, pages 22–37.

Williams, R.D., Gabelman J., Shaw S., Jepson W., Gammons C. and Kill Eagle J., 2009. Zortman-Landusky: Challenges in a Decade of Closure. www.asmr.us

Wilkin R., Wallschlager D. and Ford R., 2003. Speciation of arsenic in sulfidic waters. *Geochemical Transactions,* 4(1), pages 1–7.

Wooley J., Godzik A. and Friedberg I., 2010. A primer on metagenomics. *PLoS Computational Biology,* 6(2): e1000667, doi: 10.1371/journal.pcbi.1000667.

Wootton B., Walters M., Higgins J., Hussain I. and Blowes D., 2010. The integration of an advanced phosphorus removal technology into an engineered wetland system. *IWA 12th International Conference on Wetland Systems for Water Pollution Control,* Venice, 4–9 October 2010.

Yates C., Wootton B., Jørgensen S. and Murphy S., 2013. *Wastewater Treatment: Wetlands Use in Arctic Regions.* Book forming part of Encyclopedia of Environmental Management, Published on-line 3 May 2013, Taylor & Francis, Group LLC, New York, www.tandfonline.com

Younger P., 2000. The adoption and adaption of passive treatment technologies for mine waters in the United Kingdom. *Mine Water & the Environment,* 19, pages 84–87, 2000.

Zagury G.J. and Necultia C., 2007. Passive treatment of acid mine drainage in bioreactors: Short review, application and research needs. *Proceedings of Ottawa Geo 2007 Conference,* Ottawa, ON, pages 1439–1446.

Zitomer D., 2001. Waste Aircraft Deicing Fluid: Management and Conversion to Methane. Report for State of Wisconsin Division of Energy, June 2001.

Index

Note: Page numbers followed by *"fn"* indicate footnotes.

U

V

W

Z

Printed and bound by CPI Group (UK) Ltd, Croydon, CR0 4YY

24/10/2024

01778301-0013